Borderscapes

BORDERLINES

A BOOK SERIES CONCERNED WITH REVISIONING GLOBAL POLITICS
David Campbell and Michael J. Shapiro, Series Editors

For more books in this series, see page 312.

Borderscapes

Hidden Geographies and Politics at Territory's Edge

PREM KUMAR RAJARAM AND CARL GRUNDY-WARR, EDITORS

BORDERLINES, VOLUME 29

University of Minnesota Press

Minneapolis

London

Earlier versions of chapter 7 have been published as "Signifying Boundaries: Detours around the Portuguese–Spanish (Algarve/Alentejo-Andalucía) Borderlands," *Geopolitics* 7, no. 1 (2002): 139–64, and as "The Poetry of Boundaries: Reflections from the Portuguese–Spanish Borderlands," in *B/ordering Space*, edited by Houtum et al. (Ashgate Publishing, 2005). Reprinted with permission.

"The Distance of a Shout," from *Handwriting*, by Michael Ondaatje, copyright 1987 by Michael Ondaatje. Reprinted by permission of Ellen Levine Literary Agency/Trident Media Group.

"The Border," from *Outposts*, by David Chorlton and published by Taxus Press at Stride, Exeter, England, copyright 1994 by David Chorlton. Reprinted by permission of David Chorlton.

"No Room at the Inn," by Yasmin Alibhai-Brown, was previously published in *New Internationalist Magazine* 350 (October 2002): 16. www.newint.org.

Published by the University of Minnesota Press
111 Third Avenue South, Suite 290.
Minneapolis, MN 55401-2520
http://www.upress.umn.edu

Library of Congress Cataloging-in-Publication Data

Borderscapes : hidden geographies and politics at territory's edge / Prem Kumar Rajaram and Carl Grundy-Warr, editors.
    p.   cm. — (Borderlines ; v. 29)
Chiefly papers from a conference on security and migration held in Chiang Rai, Thailand, in December 2004."
Includes index.
ISBN: 978-0-8166-4925-9 (hc : alk. paper)
ISBN-10: 0-8166-4925-1 (hc : alk. paper)
ISBN: 978-0-8166-4926-6 (pb : alk. paper)
ISBN-10: 0-8166-4926-X (pb : alk. paper)
1. Illegal aliens—Congresses. 2. Emigration and immigration—Congresses.
3. Border patrols—Congresses. 4. Boundaries—Congresses. I. Rajaram, Prem Kumar. II. Grundy-Warr, Carl.
JV6033.B59 2008
325'.1—dc22
                                                                2007018940

Printed in the United States of America on acid-free paper

The University of Minnesota is an equal-opportunity educator and employer.

12 11 10 09 08 07   10 9 8 7 6 5 4 3 2 1

Contents

Acknowledgments

This book started with a conference on security and migration held in Chiang Rai, Thailand, in December 2004. The conference was part of the Asia–Europe Workshop series organized jointly by the Department of Geography, National University of Singapore; Sciences-Po in Paris; the Centre for European Studies, Chulalongkorn University; and the Faculty of Law at Nijmegen. We should like to acknowledge the funding offered by the Asia Research Foundation and the Asia Research Institute of the National University of Singapore.

Most of the chapters in this book were presented during that conference. We benefited not only from the contributors published here, but also from presentations by Alessandro Dal Lago and Salvatore Pallida, as well as from the input of a number of graduate students from the National University of Singapore and Chulalongkorn University. The conference, and hence this book, would not have been possible without the assistance of a number of people. Charit Tingsabadh of Chulalongkorn University, as well as the administrative expertise offered by Kanika Bunrod of the Centre for European Studies, were vital in ensuring the smooth running of the conference. Bertrand Fort and Caroline Say of the Asia Europe Foundation helped at every stage of the conference planning, and thereafter as the book evolved. At the Asia Research Institute at the National

University of Singapore, director Anthony Reid facilitated a late application for extra funding, as did his administrative team, particularly Connie Teo and Valerie Yeo. Victor Savage, head of the Department of Geography at the National University of Singapore, supported and contributed to the conference.

We owe special thanks to May Tan-Mullins and her incredible on-site organizational skills at Chiang Rai, as well as to our local team of Chia Peng Theng, Tham Chen Fye, and Sam Kalyani.

In working with the University of Minnesota Press, we received the fine support of the series editors, Michael Shapiro and David Campbell, as well as of Carrie Mullen, Jason Weidemann, Katie Houlihan, Nancy Sauro, and Paula Friedman, who helped at various stages of this book.

Introduction

PREM KUMAR RAJARAM
AND CARL GRUNDY-WARR

We want to emphasize that the study of borders and migration centers on questions of justice and its limits. The border is not a neutral line of separation; borders between nation-states demarcate belonging and nonbelonging and authorize a distinction between norm and exception. The authority accorded by the territorial border vindicates a curtailed conception of justice, one that is particularly telling in its circular claim to being an exhaustive representation of human need. Justice operates to outline the limits of a spatial utopia, an attempt at a purposive unity enveloped by the nation-state. The territorial border thus functions to distinguish an arena within which justice may operate to entrench a space of utopic unity. This results in a grounded base for thinking and responding to chaotic heterotopia in the world (Foucault 1989a; Decha, chapter 10 in this volume).

We also want to emphasize that the clearing of utopic space and the limiting of justice rest on a desire to conceal or not hear voices and experiences of heterotopia and dystopia—not out there, but within. The border is not empty or readily pliable; it is a paradoxical zone of resistance, agency, and rogue embodiment. The enforced order of the utopic space of territorial justice is premised on a series of abstractions, including *the citizen* and *the community*. In these abstractions, that which remains inchoate, which cannot be abstracted, is placed "outside," consigned to the border, at the edges of the norm. This operative placing beyond the margins should not be taken at its word. The placing beyond of that deemed not to fit inaugurates a bicameral system of the political where the "inside," the norm, has a symbiotic relation with the "outside," the exception. The border is a zone in between states where the territorial resolutions of being and the laws that prop them up collapse. It is a zone where the multiplicity and chaos of the universal and the discomfits and possibilities of the body intrude. We use the term *borderscapes* to indicate the complexity and vitality of, and at, the border.

This book is a cross-disciplinary conversation between scholars of Asian and European borders in light of a set of security discourses and associated bureaucratic and punitive practices that have been prevalent in different forms and at different scales over the last five years. The focus of these discourses and practices has been the territorial border; the border is conceived as a tool of exclusion, one that can be strengthened and fostered to protect a community and a society against a phantasmic threat of otherness that tends to become flesh in the demonized and abject figure of a migrant or refugee. The book studies legal, bureaucratic, and punitive discourses and practices that center on creative practices of and at the border. Such border practices invade and permeate everyday sites; the border becomes not the imaginary line of separation but something camouflaged in a language and performance of culture, class, gender, and race (Soguk, chapter 12 in this volume).

Such camouflage reproduces the border in the multiple localities and spatialities of state and society. Borders, as Nevzat Soguk mentions here, can fold inward, enveloping and containing individuals and groups in societies within particular regiments of governmentality. Or they can fold outward, restricting entry and expelling irregular migrants. The term *borderscapes* is an entry point, allowing for a study of the border as mobile, perspectival, and relational. From this entry point, we study practices, performances, and discourses that seek to capture, contain, and instrumentally use the border to affix a dominant spatiality, temporality, and political agency. We also study those hidden geographies of association and digressions, concealed by instrumental captures of the border that demarcate a coherent inside from a chaotic outside, that question the anchoring of society, community, and politics to the phantasmic figure of the nation-state. Indeed, we point to the phantasm of the nation-state, to the performance of coherence centered on representations of the border as something static and stable, clearly outlining belonging and nonbelonging, the very substance and vindication of the nation-state (Sidaway, chapter 7 in this volume). Such phantasms are set against experience and ways of being that cross over and intrude into spaces unintended. The authors explore borders in Thailand–Burma, Australia, and Portugal–Spain, among other places, pointing to the vibrancy at sites taken to be "of zero width" (O'Toole 1997): empty lines demarcating the end of one territory and the beginning of another. We do not claim, though, that nation-states are entirely discursive constructs, to be blown away by a deconstructive wind. The phantasmic nature of their bordering strategies rests on ongoing processes of appropriation, of "accumulation by dispossession" (Harvey 2003). Such appropriation—of means of production, distribution, and financing, and of the meaning of spaces—enacts a form of foreclosure. Accumulation by dispossession involves, aside from disenfranchisement, also the reinscription of spaces and their material infrastructure. They become located within particular registers of meaning and action; they are understood in particular ways and acted on in particular ways. This dual register, giving meaning and outlining action, transforms accumulated spaces into known and utilizable places that have particular meanings and functions attributed to them in ways that preempt and foreclose the consideration of other meanings and functions of spaces. Moreover, such reinscription ties material infrastructure to a wider global or regional network of actions and processes; the inscription of the meaning of capital and other material infrastructure connects these to wider networks and processes of global capital movement and accumulation. This anchoring of domestic infrastructure to global networks and processes radically transforms their nature and strengthens the ongoing processes of inscription, dispossession, and accumulation.

Finally, in this book we also explore the constitution of "a new political" (Soguk, chapter 12 in this volume), centering on community and agency that work within and against, and are given life by, fluid border practices that fold inward and outward. We seek to recover a politics of becoming (Connolly 1999) that sees politics as process, community as disconnected from the rigid territorial spatialities of the nation-state and as making overtures to the global. Such a politics is insurrectionary; it resounds onto the staticity of the nation-state and the regulative practices that enforce stasis, pointing to a politics as ongoing process, not definitively contained by a particular spatiality but forming new, irregular, and fluid spatialities and communities as it operates.

JUSTICE, BELONGING, TERRITORY

The territorial arrangement of the world, what Ana María Alonso calls the spatialization of being, is based on ontological and epistemological resolutions (Alonso 1994). One leads from the other. The ontological resolution is one where the meaning of being a human being becomes tied up with the meaning of the nation-state. This location of the meaning of what it is to be human vindicates an understanding of political and moral belonging centering on and giving form to a "state" that is more an aspiration to "unity, coherence, structure, and intentionality" than a reflection of these as preexisting forms (Sayer 1994). The state-centered ontological resolution is then the fundamental basis by which a claim that the state represents and acts to guarantee order, security, and justice is made and vindicated. Simultaneously, this ideological claim conceals relations of domination that cohere the state, as well as the fragmented and tenuous nature of this domination. Identity and the nature of its relation to the state are not given but are contested and rely on forms of policing, often less than fully efficient, to appear coherent. The resolution is thus processual and contested; it requires ideological and political domination for its vindication.

The border here is a transformative and creative instrument; it marks the transition from a state of anarchy to one of order, thus enabling a narrative of justice and recognition centering on the clarification of what form of life or living constitutes belonging and what constitutes nonbelonging. Justice, in the territorial model, is centered on generalizable rules and laws. The administration of these rules and laws is the same as the administration of justice. Norms of justice are taken to have an independent standing; they do not depend on the extent of actors' commitments to specific values (Fraser 2001, 22).

In the normal model, norms of justice stem from the value accorded to territory (not to nation). Here the question of the relation among territory, state, and nation is important. In the normal model, *nation* is understood in a way that allows it to be used by the state as a means of legitimating itself. That is, ethnic or nation-based claims (for example, of indigenous populations) are worded in ways that reassert the priority of the sovereign law of the territorial state. The imposition of the state over nations (such as indigenous nations) has its historical origins both in violent conquest and in epistemological displacement. The modern bureaucracy of the state collects and collates facts about its given population. This bureaucratization of knowledge about population distances that population from its cultural or historical bases and defines it in terms of categories and concepts allowed by the territorial state. This creates a veneer of objectivity; it allows for the articulation of knowledge about and policies for an objectively distinct population. The creation of knowledge about population allows for a mode of governing that separates the act of ruling from individual actors, making organizations independent of particular settings or individuals. The state's dominant imposition over territorial space is vindicated through the exercise of the law and through the bureaucratic creation of subjectivities that allow for a self-justifying mode of governing (Foucault 1977; Foucault 1990; Andersen and Denis 2003; Escobar 1995; Hyndman 1999; Appadurai 1993).

The law, however, should not be taken as the final or sole arbiter of belonging or nonbelonging. Homi Bhabha writes of the ambivalence underpinning the rational edifice of modern politics (Bhabha 1994). The narratives of justice emanating from this edifice are not solely rooted in abstract law. Bhabha suggests that the narratives of justice rest on a violent suppression of (often racialized or gendered) otherness. This otherness shadows the norm and results in the radical relation of the norm with what it would exclude. It also results in a demand that belonging be determined not solely in terms of legal recognition, such as through the bestowal of citizenship, but also through forms of social performance. Belonging must be performed; groups must prove their cultural or social belonging through effective identity performance. The way people dress, speak, and socialize all have effects on recognition at particular points in society.

Recognition is thus not something exclusively undertaken by an abstract managerial state. Neither is it something restricted to the bestowal of citizenship. Recognition involves also social performance;

it indicates that the mark of belonging does not begin and end with the bestowal of citizenship, and that groups make appeals for recognition before different societal actors or institutions (Carruthers 2003). This differentiation and the nuance inherent in the generic terms *belonging* and *recognition* are also reflected in the concept of *citizenship*, which is often graduated and is to be distinguished in terms of the limits on its use and in terms of the rights it confers (Ong 2001). The mark of "citizenship" may not allow effective or equal opportunity for all; racial distinctions remain important in restricting access to housing, government loans, education, and government jobs in Israel and Malaysia, for example. Other less official forms of discrimination exist toward migrants (citizens or otherwise) in a number of countries; the assumption of apocrypha in discrimination claims is perhaps an indication of the force of the public discourse of recognition and the law.

Conferring belonging involves both *state* and *society*. If the analysis of recognition cannot be said to begin and end with the bestowal of rights, then the role of *society* in leavening and influencing the territorial narrative of justice emanating from a territorial resolution of ontology is important. What is the relation, in this ontological resolution, between *the state* and *society*? We suggest that a particular representation of society is used to legitimize the state and makes it knowable. Society here becomes a zone or realm within which the state operates; society is constructed as a zone more or less independent of the state but amenable to intervention by the state. The state intervenes in society to solve society's problems: in this conception, the state envelops society. *Society* is a discursive construct, understood as a zone of intervention where power is ordered and where that ordering is made apparent (Feldman 2003, 62). The construction of *society* thus is a means of making known, and legitimating, territorialized power and its embodiment in the state.

Society should not in itself be taken as a final or autonomous unit of analysis. The discursive and instrumental construction of society as a zone amenable to intervention is probably a secondary level of analysis. The construction of society in this way rests on a more fundamental construction of bodies and agencies and their location within society. The capacity to act on bodies, to distinguish legitimate embodiments and agencies from illegitimate ones, shapes the contours of society. Bodies are fundamental sites of discipline and punishment. The extent and force of such discipline and punishment vary across different bodies. Some bodies, as Perera, Haddad, and Toyota all show in this volume, are more exemplary sites than others for the coming together of the forces of domination (and creation) that reveal the structure of territorial power. What are particularly apparent are the processes of inclusion and exclusion upon which this power rests. These more exemplary bodies are often racialized or gendered (recognized or misrecognized primarily in terms of race or gender) or are migrants from particular areas. At the core of the territorial ordering of humanity are thus not only the abstract operation of law but also a hierarchy of bodies. This is the ambivalence that Bhabha speaks of, the emotive and violent conceptions of impurity and relations of fear and desire underpinning the rational edifice of the modern state.

The conferral of belonging and the mechanism of recognition arising from territorial conceptions of justice are thus not to be analyzed solely in terms of the abstract law. Such analysis repeats the ruse of sovereign control and rationality integral to the maintenance of sovereign power. The structure of order given by the territorial border is complex, precisely because that border is not a zone or space given to ready instrumentalization. The border is a landscape of competing meanings with a range of actors: the order, justice, and knowledge that emanate from it can only be viewed in abstract judicial terms if the element of semantic competition is removed. If it is not, then that order and that justice may be seen to be underpinned by ambivalent engagements with dispossessed and disenfranchised otherness. And the resultant knowledge may be understood as the bureaucratic or legalistic co-option and cajoling of identities, institutions, and forces into concepts readily digestible.

Territorial knowledge is centered thus on an explication of meaning that, in a circular manner, justifies its taxonomic role in society. Knowledge operates by making perceptible that which has reason to be seen (and seen in a particular light), while making imperceptible that which has no reason to be seen (Rancière 1999). Society becomes subject to a taxonomic knowledge where the land, its meaning, and people within its political borders are affixed to a particular trajectory. Knowledge is thus performed in ways that repeat and vindicate the ontological resolution upon which it is based.

The question of ontology is fundamental to our investigation. The efficacy and persuasiveness of statist conceptions of society and society's narratives of order, justice, community, belonging, temporal continuity, and "an aspiration to a rigorous methodological access to truth and totality" (Dillon 1999, 159) rest on the deferral of the question of human beings and their location. The rigor of the conceptions rests on a postponement of the question of what and where it is to be effectively and properly human. This rigor frames recognition (and misrecognition) in a formal public discourse (of the law) and orients us to a solution framed in public discourse. This occludes the thinking of a fundamental intertwining of self and Otherness, citizen and noncitizen, existing either prior to or as a result of the conferring of political subjectivity that distinguishes and separates. Indeed, stating the question of belonging and recognition tends to reinforce distinct and separate identities; it reinforces the sense that there is a group who already belongs and another seeking belonging from an external frame; it reinforces the territorial restrictions of community.

The framing of questions about belonging and recognition (and so about justice) becomes thus a second-order relation between two (or more) already distinct groups. The argument is *not* that there is no distinction in the level of inclusion or exclusion among groups at any given moment or space; the problem is rather that the tying of groups to a particular landscape is enabled through a fundamental relation with groups excluded. The identity of the included can only be thought with reference to the excluded; the framing of questions of belonging and of appeals for recognition thus acts as though there were no fundamental relation between those included and those excluded. We presume that it is important to do work at the border between citizen and noncitizen, those who belong and those who do not belong, rather than taking the distinct categories as given. The landscape of justice is vastly changed if we presume a fundamental relation. Not least, the purposive spatial utopia of the territorial state loses its trajectory; its goal of self-perpetuation in the same form can no longer be vindicated, and the maintenance of state sovereignty *for its own sake* becomes a problematic issue.

In the purposive and narcissistic sovereign state, the question of belonging is thought in terms of the extent to which an individual may be reasonably expected to adhere to the state's norms.

The community of the nation-state emphasizes, as its bottom line, a commonality that cannot be put under direct questioning. This community is teleological or narcissistically self-referential; community as commonality is both a point of departure and an end point (Ahmed and Fortier 2003, 253). The conception of justice that sets for itself a border or limit that is reenacted and strengthened by each enactment of its laws refuses or makes unrecognizable the presumption of another justice. The operation and performance of belonging that is set in motion by territorially secured justice reinforces its basis or rationales. Mirroring the self-referential teleology of community, territorially secured justice both inaugurates and reinforces a state-centric notion of what it is to be human. In this reinforcement, it asserts that the structure of justice and authority given by the territorial state is a creative rendering of the human being as an accountable and politicized subjectivity: a useful, necessary, and productive transformation from states of nature and bare lives.

Territorially secured justice is thus not merely one possible structure of justice, but makes a claim to being the authoritative and exhaustive form; it sets the basis and becomes the ground that renders the human being thinkable and accountable. Yet this proclamation of an authoritative and incontestable justice relies, ultimately, on shaky and contingent foundations. This is so in at least two senses.

First, the proclamation relies on unruly lines that demarcate security from anarchy, norm from exception (Walker 2003). In the territorial reading, the border is a transformative line, distinguishing politicized subjectivity from chaos outside (and maintaining this distinction). This instrumental use of the border rests on the capacity of the territorial imagination to fill the space of the border (Donnan and Wilson 2003, 13). The border, however, is not a zone that is readily given to such space filling. The border, or borderscape, is replete with actors and agents responding to, resisting, and relaying the set of discourses and practices that envision the border as an empty and pliable instrument of separation. That is, those elements who are rendered unaccountable, who have no reason to be visible, do not simply disappear; it takes an act of some will, as well as a certain amount of institutional force, to effect their invisibility and erase their relation to the norm.

Second, the claim to an exhaustive rendition of justice relies on a varying capacity to contain difference. Territorial claims to authority present themselves as coherent means of thinking and responding to plurality in the world (Walker 2003). The priority accorded to the particular over the universal allows for the co-option of difference into nation-building strategies. This co-option of difference relies on the efficacy of the naming and incarcerating strategies that lead to both a semantic and a physical exclusion of difference. This efficacy again rests on a capacity to disregard the ambivalence and paradox at borders; it relies on a disregard or containment of the fluid interplay of global and local, and of the fundamental relation between excluded and included.

In the preceding section, we have tried to show that the thinking of the question of belonging, recognition, and community within state and society rests on a number of abstractions; three are particularly important. The first is that of the autonomous and public individual. This abstraction enables and encourages the thinking of belonging in a public, or legal, sense; it corresponds to an ontological resolution where the problematic question of what (and where) it is to be human is deferred indefinitely. The way of thinking involved in this deferral encourages the second abstraction, that of a self-referential and narcissistic community. The deferral of the question of humanity at the threshold of the state leads to the thinking of belonging in terms of the extent to which the commonality of the community, which is both a starting point and a goal, may be maintained: the end (and beginning) of justice is its maintenance within the borders of a given community. The third abstraction is thus the system of justice itself. Justice is wedded to the state, through the abstractions of humanity and community; justice becomes a constricted set of rules or norms designed to preserve these.

These abstractions are complex and fearsome. In an attempt to rethink the politics of belonging, many of us focus in this book on a rethinking of the notion of community through the figure of the migrant and the border. This involves also a questioning of the restriction of justice to the state; does justice have also an emancipatory and even cosmopolitan calling? What is fundamentally at stake is the possibility of thinking community, humanity, and justice in ways that are not preemptively curtailed by the exigencies of territoriality. As a starting point, authors in this book find the political philosophies of Giorgio Agamben and Jacques Rancière important.

THE MIGRANT, TERRITORY, AND RECOGNITION

We, this volume's contributors, examine systems of recognition and the structures of intelligibility they underpin. We examine the differential operation of sovereign power toward different migrants in different social and political contexts. We examine specifically the fragmented nature of this sovereign power, how structures of recognition do not center on and emanate unproblematically and coherently from a managerial state. We suggest that such structures are broken up and fragmented at different points as they are relayed to different migrants. A common purpose of the diverse studies is the uncovering of modes of recognition and the hidden geographies that are concealed by these. Some authors analyze the operation of these structures and systems; others focus more on the uncovering of hidden geographies. Thus, while some of us concentrate on the construction of the border (that is, on territorial spatialization and its systems of recognition), others pay greater attention to the interstitial and in-between space of the border demarcating sovereign space. This conceptual division of labor mirrors a theoretical division. Many of us focusing on the critique of territoriality find Agamben important. Many of us paying greater attention to recovering hidden geographies find Rancière's study of perceptibility significant.

Agamben's critique of sovereignty derives from Carl Schmitt's critique of the state, the political, and liberalism, as well as from Foucault's biopolitics. Agamben understands the border as a concept distinguishing norm from exception. This exception is cast out from the norm, but in the act of casting out, the exception is brought into the system of the nation-state. The exception exists in a relation to the norm. In this relation, the exception is rendered recognizable and brought into a *system* of sovereignty. Sovereign power is not localized within a given territory; it extends outward to those it excludes. The extension outward of sovereign power, Agamben says, is not alien to sovereign law but is part of a system of law that legislates for its own removal.

Agamben theorizes the excluded as the surplus or by-effect of the production of governable identities (and here is his Foucauldian biopolitics). The production of the political subject has the by-effect of bringing into being also a "remainder," a by-effect of the production that cannot be incorporated into the structure of sovereign power. These by-effects, "no-longer human" (Schütz 2000, 121),

are consigned to zones of exception where the sovereign law no longer applies.

The distinction between natural and politicized lives cannot be
thought; birth marks an entry into a world where the nation, the
state, territory, and citizenship preemptively set the basis by which
one is recognized as human. The by-effects or remainders, that which
cannot be rendered sufficiently human (as territorial sovereign power
recognizes the concept), are consigned to zones of exception, where
the sovereign law does not apply, and are "bare lives."

The possibility of bare life is, Agamben says, the fundamental
condition of modern political existence. Following Hannah Arendt,
Agamben notes that the refugee, bereft of citizenship and thus thrust
into the barest of lives, ostensibly has the greatest possible claim on
human rights (Agamben 1998). Yet the denial or restriction of a full
complement of these rights to the refugee, and the provision of this
full complement for the citizen, emphasizes the conjoining of what it
is to be human with the marker of citizenship. The human qua human
has no particular claim on rights; they are not vested in his or her
body. The human qua citizen, on the other hand, has a clear claim to
a full range of human rights. The common aspect of existence is its
contingency: that which makes us human may be withdrawn.

Agamben suggests that "the camp" (by which he means both
refugee camps and concentration camps) is the archetypal zone of
exception. The juridico-political structure of the camp is important
for Agamben. The camp is a zone where the normal law does not
apply. Precisely because the structure of intelligibility given by sovereign law is not evident, it becomes difficult in the camp to distinguish fact from law, right from wrong. This is not the end of
Agamben's analysis of the camp, but the beginning. Agamben notes
that the sovereign law legislates for the camp. That is, the sovereign law legislates for its own removal. This understanding leads
Agamben to suggest that the zone of exception is not external to
the norm but is bound up with it. The trajectory of sovereign power
cannot be thought without also thinking the nature of this relation.
Sovereignty is not to be taken at its word: it extrudes from its declared territorial location to encompass that which it has ostensibly
cast out. What is singular about this inclusion is that it is an extreme
(and fundamental) form of relation, where bare life is included solely
through its exclusion.

This declaration of a zone of exception, this legislation by the law for its own removal, is thus a means of clarifying the ostensible borders of the norm. It is a declaration of the restricted scope of sovereignty and is in part due to the limited conception of humanity upon which sovereign power rests. Agamben says that the sovereign interiorizes that which it can and exteriorizes that which it cannot. Politics thus is an ongoing power struggle, an ongoing exercise in clarifying that form of life which is interiorizable and that which is not. But this relation between interiorized and exteriorized means precisely that what is ostensibly cast out in a zone of exception is actually incorporated into the norm. Casting out is not a careless expulsion, but a careful placing outside of the declared boundaries of the norm; it is a holding in semantic and physical stasis of that which can clarify, and continue to clarify, the boundaries of the norm. The rule or norm, Agamben argues, "lives off the exception alone" (1998, 27). This is the extreme form of relation where something is included solely through its exclusion that vivifies and makes coherent the norm. This extreme form of relation operates, first, to create a zone of exception to which bare life may be consigned, and then to collapse this zone and the distinction between norm and exception.

Agamben's Schmittian conception of sovereignty has led to a critique that he overly concretizes the place of sovereign power. The critique is that Agamben's notion of biopower (in contrast to Foucault's) is equated with sovereign power (Rancière 2001, 4). Sovereign power, in Rancière's critique of Agamben, is that which has the power of decision over life, creating and distinguishing bare lives from politicized lives, consigning the former to states of exception. There is a polarity here that Didier Bigo identifies in his chapter, where bare life is, in extremis, that condition of abjection from which no thought of resistance is possible. Power and resistance are separated by the decisionist sovereign who identifies the space of the law and its limits. There is thus no difference between sovereign power and biopower: sovereign power is the decisive exercise of control over subjects, including the confinement of subjects to a position of bare abjection. It is because of this decisionism that Agamben has been accused of a certain pessimism about the possibility of resistance (Vacarme 2004).

Agamben's notion of power, however, does not necessarily preclude resistance. His intention is not to demonstrate that there are clear zones of power and zones of abjection but rather to clarify the focus of resistance. At its most clear, this involves the identification of the nature of the relation between rule and exception. Such identification allows for the questioning of the bordering practices of the norm. For Agamben, the refugee is the limit concept that "radically calls into question the fundamental categories of the nation-state . . . and that thereby makes it possible to clear the way for a long-overdue renewal of categories in the service of a politics in which bare life is no longer separated and excepted, either in the state order or in the figure of human rights" (1998, 134). His is probably a messianic project: the aim is not to continue to remain within the conditions of subjectivity and resistance given by territorial power but rather to investigate the limits of these and thereby think a notion of the political that strives to go beyond territoriality and toward the global.

Rancière focuses on a study of the border between, in his terms, "man" and "citizen" (between, that is, bare and politicized life). Rancière reads these categories as ambiguous, noting that the maintenance of the border depends on a particular reading of human rights. Rancière emphasizes what might be called the play of rights. Rights are not vested in a clearly predefined group. For Rancière, rights have two existences, one as written rights, the other as rights that are used. As written or inscribed rights, declarations of human rights are a part of the community. They are not abstract ideals; the identification of conditions of rightlessness might demonstrate the impotency of the claim that rights are a part of the community, but the fact is that some political communities are informed not only by conditions of inequality but by a fundamental statement about intrinsic equality. The second existence of human rights is in its usage. Perhaps for Agamben (and almost certainly for Arendt) the exercise or use of human rights is undertaken within a predetermined sphere (that of citizenship); in opposition, Rancière suggests that the taking up and exercising of human rights is done by political subjects who are not definitively defined, individually or collectively. The taking up of rights calls into question and brings into dispute the meaning of predicates such as *freedom* or *equality:* "political predicates are open predicates: they open up a dispute about what they exactly entail and whom they concern in which cases" (Rancière 2004, 303). In other words, political predicates, such as human rights, are taken up by subjects who are not prerestricted to the exercise of these rights within a particular predetermined and limited sphere (such as citizenship). The taking up of these rights thus involves putting the limits or restrictions of the rights under question.

Rancière's understanding of rights stems from his understanding of politics. For Rancière, oppressive politics takes up a form of counting where the community is understood as a sum of its parts, where each part has a predetermined function. This is politics as police. Politics as process involves the counting of "a part who have no part," those who are imperceptible (Rancière 2004, 305). This form of counting separates the community from itself; it separates the community from its parts, places, and functions. This form of counting is based on the idea that political subjects are not predetermined. Territorial resolutions of ontology are never definitive, but set out a question about who is to be included and who is not (Rancière 2004, 302). For Rancière, the constitution of the political subject always has a surplus; political names are always "litiginous"; they always carry the possibility of extension because they are premised on forms of counting that "inscribe the count of the uncounted as supplement" (2004, 305). Because politics is about making visible through forms of counting, it is necessarily relative; it can only be understood in terms of what is not counted. This does not mean a condition of abstract ontological solidarity or intertwinement, but a practical sense of the openness of human subjectivity, society, and politics. Such an openness leads to a sense of the inherent disputability of rights and is perhaps strongly evident in the way that colonial struggles were fought in courts of law from first principles of equality and the like: colonial law was ambiguously both the means of oppression and the means of emancipation (Comaroff 2001).

In this volume, Nevzat Soguk finds in Rancière's (and Etienne Balibar's) work the possibility of an insurrectional politics that stretches the borders of belonging through the assumption of fundamental equality in communities. There is, Soguk says, a fundamental ontological equality in Rancière's theory of politics that displaces the priority of territorial inscriptions of identity. It is through a Rancièrian politics that the borders that entrench sites of exclusion and inequality can be transformed and can hark to the "cosmopolitical." This shifting of borders precisely embraces migrants into the community. Rancière argues, "The point is, precisely, where do you draw the line separating one life from the other? Politics is about that border. It is the activity that brings it back into question" (Rancière 2004, 303).

Rancière's reading of the law as instituting, paradoxically, conditions of inequality and equality, conditions located not in a predefined group or individual but in a subject who is always in the process of becoming, leads to a critique of the contingency of contemporary social order. Rancière provides a lens that disrupts the sense that society and politics may be located in definitive zones. Justice thus cannot be understood as the managed administration of law emanating from these distinct zones. Rancière demonstrates that society is in process, that political order is always contingent, and that the border between norm and exception, belonging and nonbelonging, is in a state of flux and dispute.

LANDSCAPES OF DIS-"PLACE"MENT

In our attempts to (re)conceptualize the multiple agencies and temporalities at, and of, the border, the concept of *landscapes* is fruitful. But before we discuss landscapes, it is helpful to briefly consider all space as constructed; in this sense, space and spatial relations should be considered in terms of processes of change, and of landscapes as always *in the process of becoming* rather than temporally fixed spaces. We believe that it is essential to examine different forms and exercises of power to understand the role and significance of the border in relation to mobile human life. We also think of space in Lefebvrian terms: socially constructed, with various "spatial practices" (perceived space) producing and (re)producing the "conceived space" (Lefebvre 1991, 33). Of course what is perceived and conceived cannot be thought of without references to forms of power, as noted below. "Representational space" ("lived space") is "space as directly *lived* through its associated images and symbols, and hence the space of 'inhabitants' and 'users'" (Lefebvre 1991, 39, original emphasis). As argued here and in one of the contributions to this volume (by Dean, chapter 8), these three moments of space are highly relevant to conceptualizations of *the border,* which is neither natural nor neutral but is always a social, cultural, and political construct (Paasi 2005). Lefebvre's notions of space are relevant to our thinking about landscapes and borderscapes, particularly as these conceptions hold out possibilities for counterhegemonic spatial and nonspatial practices together with alternative ways of visualizing space and society.

Human landscapes are contested creations, and as such are subject to an ongoing attempt to fix identities and meanings to a particular place—for example, through expressions of political and social belonging (Massey 1992; Kong and Law 2002). Lily Kong and Lisa Law, citing Doreen Massey, note that the processes of place-making that lead to a proprietary sense of the meaning of a landscape are located within particular sets of social relations interacting at a given moment and space. The particularity of any given place is due to the specificity of the interaction at that place (Massey 1992, 11; Kong and Law 2002, 1504). This means that landscapes do not have a fixed meaning but are always "dynamic, provisional, and contingent" (Kong and Law 2002, 1504). Kong and Law suggest that the meaning of landscapes is subject to processes of contestation that proffer a particular understanding of the meaning of the landscape as normal while thereby insinuating the deviancy of other meanings.

Cultural geography understands landscapes as repositories of contesting interpretations of the meaning of a piece of land and of its appropriate use. Landscapes denote different and contesting technologies of the self (Bunnell 2002). They assert particular moral geographies that denote a hierarchy of land use, and in this way act as an instrument of governmentality, attributing a sense of correct and incorrect behavior. For instance, "national" landscapes are constructed, both physically and semantically, to lend a particular sense of belonging (and thus outline the possibility of distinguishing deviant readings or uses of the landscape) (Cerwonka 2004). Janet Sturgeon (2005) has examined "border landscape" dynamics over time, examining how Akha people across the modern borderlands of China, Burma, and Thailand have adapted their land uses, local environments, livelihoods, intervillage connections, and cross-political space linkages, patron–client relations, and so on, to processes of enclosure, state-induced agriculture and forestry changes, and changes in property rights to land and resources. She views the complex changes "over time in response to local needs, state plans, and border possibilities" as examples of "landscape plasticity" in which numerous actors and agencies are involved (Sturgeon 2005, 9). However, we can also see that state-centered issues of belonging and ways of seeing have done much to transform the political land-scape of citizenship (Toyota, chapter 4 in this volume), which relates to issues of political geography and power in the landscape.

There are many ways in which practices of power influence, shape, manipulate, and construct the landscape. Politics operates "*in* and *through* places" and is often specifically of or about places (Jones, Jones, and Woods 2004, 115). As Foucault (1984, 252) noted, "space is fundamental in any form of communal life; space is fundamen-tal in any exercise of power," and landscapes are "produced and re-produced" as sites and outcomes of "social, political, and economic struggle" (Keith and Pile 1993, 24). Thus, we are not simply con-cerned with landscapes per se, but with landscapes of power. In other words, landscapes are more than just an assemblage of sites and places of struggle; they are actually essential to the analysis of power and politics. To quote from Jones, Jones, and Woods (2004, 116): "landscapes are *powerful* because of the role they play in structuring everyday lives. . . . We refer to landscapes that work in this way as *landscapes of power*. A landscape of power operates as a political device because it reminds people of who is in charge, or what the dominant ideology or philosophy is, or it helps to engender a sense of place identity that can reinforce the position of a political leader." This is the link among *border, power,* and *landscape.* Winchester, Kong, and Dunn (2003) also discuss "landscapes of power and power of landscapes," combining particular ideas about ideology and hege-mony in the process. They use J. B. Thompson's (1981, 147) reference to ideology as "a system of signification which facilitates the 'pursuit of particular interests' and sustains 'relations of domination' within society" (Winchester, Kong, and Dunn 2003, 66). And they combine this idea with Gramsci's (1973) notion of hegemony as a process "by which domination and rule is achieved" involving "hegemonic con-trols" or "sets of ideas and values which the majority are persuaded to adopt as their own" and are eventually "portrayed as 'natural' and 'common sense'" (Winchester, Kong, and Dunn 2003, 66). This con-cept of ideological hegemony can relate to notions of the conceived and perceived space of human landscapes—and, clearly, any *official* notions of boundaries and border space may be thought of as dimen-sions of a particular form of hegemonic control and ideology. "The most successful ruling group is the one which attains power through ideological hegemony rather than coercion," and aspects of this are

"through the control and manipulation of landscapes and the practices of everyday life" (Winchester, Kong, and Dunn 2003, 67).

Landscapes of power can help to make particular ideologies and
political practices more tangible, natural, familiar, acceptable, meaningful, and so on (Duncan and Duncan 1988), precisely because they
are inscribed in aspects of the landscape. As such, practices deemed
to be out of place help to reinforce that which is conceived to be *in
place*. From the perspective of our book, most of the contributors are
interested in aspects of the control, surveillance, and management of
migrants and of displaced persons. The very existence of boundaries
and the landscapes of "the state-citizen-nation nexus" means that
undocumented mobility and border crossings become transformed
into actions that are perceived as transgressions and threats (Soguk
1999). Dominant landscapes of power are effectively based upon
rigid territorial sovereignties, and humanitarian responses to *dis*-
"place"ment crises are mostly concerned with the perceived symptoms (rather than causes) of being *dis*-"placed." Humanitarian ob-
sessions with "safe return" imply that conditions of *dis*-"place"ment
are themselves a *dis*order. In other words, the dominant ideological
hegemony (which, in relation to refugees, exiles, and many other
groups of displaced people *is* implicated in the world political map)
helps to create the abnormality out of multiple conditions of *dis*-
"place"ment regardless of circumstances and multifarious causes. In
this sense, within dominant landscapes of power, once *dis*-"placed,"
you are automatically "lost" in "sovereign space," whether this is
on one side of an international border (as an internally *dis*-"placed"
person) or on the other (as a refugee, exile, irregular migrant, or
member of whatever category is applied) (Grundy-Warr 2002).

We would like to argue that conditions of *dis*-"place"ment help
to create new possibilities, lived spaces, and ideas about borders.
We have suggested elsewhere that landscape-oriented approaches
are useful for teasing out the complex and multilayered implications
and meanings of human movement, specifically in relation to ideas
about place, being *dis*-"placed" (Prem Kumar and Grundy-Warr
2005). Flows, mobility, and networks are dimensions that help to
(re)mold and creatively (re)shape in dynamic processes relating to
landscape plasticity (Sturgeon 2005).

Further, we wish to stress human mobility and conditions of moving from one place to another (or of being forcibly or voluntarily

[dis]-"placed") as a norm or, at least where coercion is involved, as a common aspect of human existence (Appadurai 1996) rather than as an aberration of human life. And it is in this context, and in relation to our readings of Agamben and Rancière, that we consider *borderscapes* as both a derivative dimension of human landscapes and as ways of thinking through, about, and of alternatives to dominant landscapes of power.

BORDERSCAPES

For both Agamben and Rancière, in our reading, politics is about the questioning of the border that would restrict the meaning of what it is to be human within a territorial frame. Their approach differs. For Agamben, sovereign power is encompassing; resistance has necessarily to be thought outside its scope if it is not to remain within sovereign power's categories of ethics and justice. For Rancière, politics is not about the acquisition and maintenance of power, but rather about processes of counting: thinking the border between different forms of life is to consider those forms of life that have been miscounted. Rancière argues that society is in process, that political order is always contingent, and that the border between norm and exception, belonging and nonbelonging, is in a state of flux and dispute over processes of counting.

We use the concept "borderscapes" to emphasize the inherent contestability of the meaning of the border between belonging and nonbelonging. The above discussions about the relation of state, society, and ontology emphasize that the conception of the political border as located in a specific zone at the edge of the nation-state is problematic. Such conceptions are attempts to clear territorial space of dissension and difference; they are instrumental means of asserting the limits of territorial justice and belonging. Yet the instrumentalization of the border, which clarifies a distinct space of politics and a space outside politics (a zone of exception), rests on an occlusion of the role that society plays in ameliorating and influencing territorial place-making. The borderscape is thus not contained in a specific space. The borderscape is recognizable not in a physical location but tangentially in struggles to clarify inclusion from exclusion.

The term is coined by Suvendrini Perera in chapter 9 in this book. Perera writes of the Australian border as something in flux. In Rancière's terms, it is always under process. The border is an open predicate; its meaning is not distinct. The border expands or contracts and operates differentially before different groups of migrants, giving rise therefore to "multiple resistances, challenges, and counterclaims." The borderscape cannot sit still, because it is a zone of multiple actors and multiple bodies each calling on different histories, solidarities, and discourses of protection, care, or security. Perera writes:

> There are multiple actors in this geo-politico-cultural space, shaped by embedded colonial and neocolonial histories and continuing conflicts over sovereignty, ownership, and identity. The bodies of asylum seekers, living and dead, and the practices that attempt to organize, control, and terminate their movements bring new dynamics, new dangers and possibilities, into this zone. Allegiances and loyalties are remade, identities consolidated and challenged, as border spaces are reconfigured by discourses and technologies of securitization and the assertion of heterogeneous sovereignties.

Border spaces are zones of specific social interactions that give a particular meaning. These social interactions are not contained entirely within a physical space outlined by the political border, but incorporate and speak to temporalities, solidarities, and cosmopolitanisms that refuse the categorizations of inside/outside generated by the border. Arjun Appadurai's typology of different "scapes" characterizing a disjunctive, fluid, irregular, and perspectival globality has affinities with our conception of borderscapes. Appadurai points to the subjective and contested nature of spatiality: its fluid irregularity constituting similarly irregular communities, temporalities, and agencies (1994).

The term *borderscape* reminds one of the specter of other senses of the border, of experiences, economies, and politics that are concealed. The instrumental usage of the border as a tool of governmentality must always be incomplete. It is through the identification of spaces and social practices that are alien or inappropriate ("out of place," Tim Bunnell says), that the dominant meaning is made known (Bunnell 2002). Yet the meaning of the landscape must not be traced back to the state. The landscape is not organized through a central body; landscapes are to be understood as an effect of power, disembodied from intentionality. The meaning of landscapes derives from its social interactions. The ruse of territorial power is centered on the concealment or disparagement of these social interactions.

The borderscape is thus not a static space. The border cannot be seen in purely instrumental terms. The usage of the border as a zone demarcating norm from exception depends, Rancière might argue, on a particular political accounting, one that reads border society as a sum of its parts, where each part has a particular function, aggregating into a static and usable border. This type of counting presents the border as one of many reference points from whence territorially centered narratives (and policies) of justice and belonging ensue. However, if the border between bare life and citizen, between belonging and nonbelonging, is imbedded in nothing so strong as contingency, then the instrumentalization of the border cannot be sustained. Rancière suggests that social space is not static, but episodic (Shapiro 2003). Society is made up, in Massey's (1992) terms, of a series of interactions that are temporally and spatially contingent, and that hence give a particular meaning to a place. Because society—in this sense the society of the border, the borderscape—is always in process, the meaning of the border can only be understood in terms of ongoing encounters.

The borderscape is thus a zone of varied and differentiated encounters. It is neither enveloped by the state nor semantically exhaustible. The borderscape is a zone of competing and even contradictory emplacements and of temporalities that hark to forms of spatial organization that refuse the territorial imperative. If the borderscape is understood as a zone of contingent meanings, then it may (and does) hark to conceptions of belonging that stretch across (and into) the territorial divisions that stand in the stead of our considerations of solidarity and justice. The borderscapes that we investigate in this book point to hidden geographies and insurrectionary political forms that question the territorial restriction on justice and belonging. We investigate these in two ways. We take note of the creation of a zone of exception where normal law does not apply, but we do not forget society. We focus on the condition of abjection into which migrants are placed, but we do not conflate the ruse of sovereignty with actuality: zones of abjection are not without resistance. Underlying our endeavors is the potential of a community premised not on the closed predicates offered by a prior organization of society but on the possibility of justice premised on a fundamental capacity to perceive and respond to otherness without, or outside of, the restrictions of territorial authority.

THE BOOK

This book is divided into four interrelated parts.

The first part contains those essays that focus on critiques of territorial sovereignty and the particular narratives and structures of justice, security, and belonging that ensue from such sovereignty. The focus is on the bureaucratic, legalistic, and discursive ways in which the state legitimates itself through practices of inclusion and exclusion. Authors point to the way that these contribute to the instrumentalization of the border and its deployment as a means of distinguishing norm from exception, belonging from nonbelonging.

The second part looks at moral representations of the border. Borderscapes are not only instrumental devices employed to distinguish insider from outsider; they also denote particular moral frameworks. Borders are invested with affective value, as barriers to threats or pollution.

The third and fourth parts of the book contain those essays that attempt to map the hidden geographies at the borderscape. In these sections, the authors focus on the problematic of perceptibility, of the possibility of making the imperceptible resound with meaning onto the territorial stage. Articles here attempt a remapping of borderlands, focusing on the plethora of actors, networks, and spatial organizations at the border; and they attempt to think the beginnings of an insurrectionary politics, one based on recovering the experience of those rendered imperceptible, while simultaneously trying to put into question those structures of recognition and politics of accounting that give these persons no reason to be seen.

The first part of the book begins with Didier Bigo's study of the governmentality underpinning the detention of foreigners in Europe and underlying declarations of states of exception. Bigo gives this specific governmentality a name: the *banopticon*. The detention of foreigners, Bigo argues, is not fully understood if it is explained in terms of a logic of surveillance. Rather than strict fidelity to Foucault's notion of the *Panopticon,* which can (though this is not necessarily done by Foucault himself) flatten the field of surveillance, Bigo argues that what is important about the contemporary detention of foreigners in Europe is the difference between surveillance for all and control, or detention, of the few. The logic of surveillance in contemporary Europe is a differentiating and graduated logic; it operates in different ways on differently identified groups of people. It is also a logic based on or derived from an affective community and from its notions of belonging and nonbelonging. Bigo argues that the banopticon operates because of the tacit acceptance, by communities, of control because that control only applies to foreigners. The banopticon is thus a device or technology for transforming and identifying otherness. Bigo argues that the banopticon shows the border as societal construct: it is not a static line of separation but a transformative process involving bureaucratic and everyday definitions of alienness. Bigo uses the metaphor of the Möbius strip to explain this border. The distinction between inside and outside is intersubjective; it is not clearly outlined but depends on societal interactions and bureaucratic and security professionals' policies and discourses. The consequence is a far-ranging and important reading of the sociology of sovereignty where notions of alienness and threat as derived form a "social and political optic" rooted in public participation in, and acceptance of, distinctions between outsider and insider.

In the second chapter, Alice Nah continues the process of mapping landscapes of internal control by which migrants or foreigners are defined as types of threat. Nah traces in depth the security landscape of the Malaysian nation. For Nah, in Malaysia the metaphor of the zone of exclusion in which irregular migrants are contained operates at a variety of scales. Regulative processes that place migrants outside of the norm do not operate solely at the sovereign or national scale. Nah traces complex and violent punitive practices of controlling the movement and identity of people deemed irregular migrants. She also, like Bigo, suggests that the identification of *alienness* is populist: it is derived from complex societal ideas of belonging and nonbelonging. She maps, then, a shifting and graduated security landscape. Malaysia's borders are not static but expand and contract in relation to external and internal pressures; the border is seen here as episodic, changing its features and meaning in response to different sorts of encounter over time.

The third chapter is Elspeth Guild's study of how the image of the border separating *people like us* from outsiders, and harked to by sovereign power in its justification of punitive action against foreigners, is the subject of litigious dispute in the United Kingdom. Guild studies how the imposition of European human rights norms on British courts extends the meaning of "the UK people." The "UK people" is that population that the state security apparatus distinguishes as a population it will protect. This definition is an attempt at distinguishing the norm from the exception while asserting the role of the state in clarifying and protecting this norm. Guild argues that European human rights norms impose on the United Kingdom a transformation of the concept of its people. Guild investigates specifically legal challenges to the arbitrary detention of foreigners for security reasons, noting that attempts to place certain individuals outside of the declared borders of the law is a contentious act precisely because the borders demarcating legality and illegality, belonging and nonbelonging, are not fixed but open predicates.

Mika Toyota ends the first part of the book by writing about the regulation of citizenship in Thailand. She describes a series of regulative processes whereby hill tribes in northern Thailand are distinguished as particular sorts of groups whose claims to belonging to the Thai nation are at best tenuous. This regulation masks how the Thai state relies on a stigmatic distinction between *upland* and *lowland* for its coherence and national cosmology. Toyota traces over time the governmentality of separation where the intertwinement of upland and lowland in Thailand, generating complex and changing episodic meanings of place, is concealed. What should also be remembered here is how the so-called hill tribes have criss-crossed mountain valley spaces for centuries so that, in many respects, the new borders, both within nation-states and separating them, are a very recent phenomenon for analyzing people who have networks, ties, and movements across political space and indeed define themselves as much by their migrations as by their situations within national spaces. Thus, notions of citizenship are denials of complex histories, geographies, and alternative narratives of space and place (see also van Schendel 2005).

Borders are also invested with a certain aesthetic or moral value. They can be seen as barriers to pollution, protecting the community from dangerous elements. A security landscape not only mobilizes bureaucratic or police methods of control and punishment, it also discursively asserts a particular moral framework within which punishment and control operates. Emma Haddad begins the second part of this volume by analyzing the European depiction of the border as a place of danger that provides a particular sort of moral challenge not only to the state but also to the community as a whole. This conflation of a particular threat (to the operational logics of the state) into a generalized threat faced by the community as a whole is integral to the maintenance of an instrumental representation of the border vindicating particular forms of state control over the limits of community.

Alexander Horstmann continues this part of the book with his ethnography of violence and hidden illegal economies animating the borderscape of southern Thailand/northern Malaysia. This is a paradoxical borderscape of sex tourism, Islamic resurgence, police killings as part of the central government's war on drugs, political insurgency, and violence stemming from Mafia-like rackets and contests over the control of drug and prostitution rings. This is a landscape difficult to represent in any authoritative way. Violence here has many shades. Horstmann traces the attempts of the central Thai government to assert control over the borderscape, to assert an authoritative representation that works to quell the ambiguous religious, political, economic, and criminal interplay in the borderscape. Although the Thai state seeks to depict the area as a hotbed for criminality, drug running, and radical Islam, Horstmann shows the nuance of the borderscape. The area is depicted by Horstmann as a complex interplay of claims and contests that themselves generate complex counterclaims and further contests. The purifying call of resurgent Islam does not merely sit alongside gangsterism but exists in a relation to it: gangsterism vindicates purification; the form of Islam being taught in the borderscape is a response to gangsterism as well as to the encroaching of the Thai state onto an area tenuously connected to Bangkok. There is no overall aggregation of meaning: the actors in the borderscape cannot be attributed with particular functions that cohere into a manageable or authoritative representation of the border. Rather, there are ongoing processes of contest and negotiation among actors contesting authority in the borderscape. Duncan McCargo (2006) argues that the so-called liminal zone of the southern border provinces of Thailand "have been thrust to the very center of Thailand's national politics," but in fact, our notion of borderscapes means that there is no *real* center or periphery in politics. Rather, we prefer to center the multiple voices, lives, and possibilities that are part of complex borderscapes, in addition to the project of "centering the border," in analyses that seek to avoid the "territorial traps" (Agnew and Corbridge 1995) of much thinking concerning all categories of *dis*-"placed" people, borderlanders, so-called minorities, and so on.

The third part of the book concerns rethinking borderscapes. It opens with James Sidaway's reading of the semiotics of borders. Sidaway looks at how the Portuguese–Spanish border has been affected by an imaginary geography of EU-centered connectedness. For Sidaway, borders and what they denote are structures or fields that require reading to have meaning. Borders do not in themselves, in their physical location on maps, convey meaning. They are read from particular vantage points. Such readings are more than discursive; they have material effects. Sidaway argues that the *harmonization* associated with EU readings of borders contributes to the "*disintegration* of ways of life based on the Portuguese–Spanish border as a space of liminality and regulation."

This part of the book continues with Karin Dean tracing the complexities of the meaning of the Thai–Burma border. Dean distinguishes between the border as a perceived space and as a lived space, demonstrating how the perception of the border by state and military authorities depends on a flattening out of complex economies, forms of violence, and social interactions that are evident when the border is understood as a lived space. This is important, as it is precisely in the lived spaces of the border that there are alternate ways of viewing space and place; there are also numerous examples of how people quietly and imaginatively circumvent and subvert aspects of "conceived space" in their everyday lives.

Suvendrini Perera continues the investigation of hidden geographies in her study of the Pacific borderscape. As noted earlier, Perera sees the borderscape as a fluid field of a multitude of political negotiations, claims, and counterclaims. The reading of the border as a marker of separation and difference conceals these. She argues that what is needed is a different reading of Australia's Pacific border, an "alternative conceptual and spatial frame for analyzing what are usually read as disparate elements—negotiations of Indigenous sovereignty, regional governance initiatives, security, and border control operations—within it."

The final three chapters explore the possibility of thinking a new political from the borderscape. Decha Tangseefa writes an unsettling chapter about the complex processes involved in silencing displaced people along the Thai–Burma border and in incapacitating our ability to listen to their sufferings. He argues that a selective political counting distinguishes perceptible lives from imperceptible ones and vindicates the normal order of things. Following Rancière, Decha argues that the sovereign political community headed by the Thai state affixes its logic, and its cosmology, through a selective mode of counting where those identities that make sense to its sensibilities are recognized as political subjects; those that are not are rendered as imperceptible subjects. Karen displaced people, in between sovereign territories, are an example of people rendered imperceptible as political subjects. Decha undertakes a Rancièrian politics, one that seeks to examine the limits of the sovereign state's capacity to render imperceptible some forms of political subjects. Decha shows that the sovereign state relies on and is vindicated by a perception of control of displaced Karens as naked lives, as politically imperceptible. He argues that, by pointing to the limits of this control, it is possible to make that which has no part in the sovereign partition of the political—that which has, in other words, no reason to be perceptible—resound onto the settled norm. From this possibility arises the possibility of a politics that refuses the static territorial logic and rather is motivated by an imperative to uncover those subjectivities that find no reason to be heard from within the dominant perception.

Prem Kumar Rajaram has a similar intention: to make that which is rendered silent or imperceptible resound onto the territorial norm. Whereas Decha focuses on the fragmented sovereignty of the Thai state (and its attempt to conceal this fragmentation), Prem Kumar looks at the ruse of sovereignty perpetrated by the Australian state and the part that temporality plays in the maintenance of this ruse. Prem Kumar demonstrates that integral to the maintenance of a ruse of control over territoriality is a control over temporality that translates to a control over which sorts of people, representing what sorts of histories, are able to speak and articulate political claims. The central aspect of Prem Kumar's chapter is a study of the Australian government's "excision" of Melville Island in response to the landing of a boatload of asylum seekers. He writes about the possibility of transforming a politics of *being* into a politics of *becoming*. The former rests on a particular mapping or gridding of social space where political subjects are understood as belonging because they signify fidelity to a particular history of that gridding; this thus forms a politics of being, where the question of the location of human being, and the ethical and political limits imposed, is foreclosed. The transformation to a politics of becoming is one based on a critique of the border as limit concept, as that which outlines the limit of the normal political and moral community. The idea is to think community in terms of irregularity and processuality, where preordained limits to action and ethics become difficult to vindicate.

Nevzat Soguk ends the book with his study of how migrants practice an insurrectional democracy amid the confines of territoriality. In a bold and sweeping tour de force, Soguk finds amid territorial cosmologies and powers, often taken to be totalizing, the radical and ongoing practice of a new politic. This practice centers on migrants using their subalternity as ways and means of demonstrating the hierarchies that commodify and incarcerate their bodies. Soguk finds an emancipatory and exhilarating process of democratization where migrants refuse the hierarchies of domination exerted on them, and refuse the territorial cosmologies that would perceive them as out of place. Soguk sees the border as an operative process, as a field of dynamic and contingent relations and processes best understood as *borderizations*. The border is thus not a material and fixed wall or fence, but the many-headed sum of policies and discourses generated by a variety of state and nonstate actors. Borderizations are contingent: they generate particular obstacles *and opportunities* for different sets of migrants. Borderizations thus not only conceal but also reveal; they provide spaces and trajectories for migrants to work within. Political community is not clearly demarcated; its mapping of legitimate agencies and political subjectivities is a spectacle or ruse. Sovereign authority over what may occur in the polity is dependent on a concealing of the acts of borderizations that allow, and indeed invite, migrants to participate in the fluid and ill-defined landscape of the polis. From this basis comes the articulation of a radical democracy. Soguk suggests that migrants use the border, as a field of borderizations, as a field of opportunity and definition to work within and against to resound onto the polis.

WORKS CITED

Agamben, Giorgio. 1999. *Homo Sacer: Sovereign Power and Bare Life.* Stanford, Calif.: Stanford University Press.

Agnew, John, and Stuart Corbridge. 1995. *Mastering Space: Hegemony, Territory, and International Political Economy.* London: Routledge.

Ahmed, Sara, and Anne-Marie Fortier. 2003. "Reimagining Communities." *International Journal of Cultural Studies* 6, no. 3: 251–59.

Appadurai, Arjun. 1993. "Number in the Colonial Imagination." In *Orientalism and the Postcolonial Predicament*, ed. Carol Breckenridge and Peter van der Veer. Philadelphia: University of Pennsylvania Press.

———. 1994. "Disjuncture and Difference in the Global Economy." In *Colonial Discourse and Postcolonial Theory*, ed. Patrick Williams and Laura Chrisman. New York: Columbia University Press.

Bhabha, Homi 1994. "Remembering Fanon: Self, Psyche, and the Colonial Condition." In *Colonial Discourse and Postcolonial Theory*, ed. Patrick Williams and Laura Chrisman. New York: Columbia University Press.

———. 1996. "Sovereignty without Territoriality: Notes for a Postnational Geography." In *The Geography of Identity*, ed. P. Yeager. Ann Arbor: University of Michigan Press, 40–58.

Bunnell, Tim. 2002. "Kampung Rules: Landscape and the Contested Government of Urban(e) Malayness." *Urban Studies* 39, no. 9: 1685–1701.

Carruthers, Ashley. 2002. "The Accumulation of National Belonging in Transnational Fields: Ways of Being at Home in Vietnam." *Identities: Global Studies in Culture and Power* 9, no. 4: 423–44.

Cerwonka, Allaine. 2004. *Native to the Nation: Disciplining Landscapes and Bodies in Australia.* Minneapolis: University of Minnesota Press.

Comaroff, John L. 2001. "Colonialism, Culture, and the Law." *Law and Social Inquiry* 26, no. 2: 305–405.

Dillon, Michael. 1999. "Another Justice?" *Political Theory* 27, no. 2: 155–75.

Duncan, J. S., and N. Duncan. 1988. "(Re)reading the Landscape." *Environment and Planning D: Society and Space* 6, no. 2: 117–26.

Feldman, Allen. 2003. "Political Terror and the Technologies of Memory: Excuse, Sacrifice, Commodification, and Actuarial Moralities." *Radical History Review* 85 (Winter): 58–73.

Foucault, Michel. 1977. *Discipline and Punish: The Birth of the Prison*, trans. Alan Sheridan. New York: Vintage Books.

———. 1984a. "Of Other Spaces." *Architecture /Mouvement/ Continuité* (October). Trans. Jay Miskowiec. Reproduced at http://foucault.info/documents/heteroTopia/foucault.heteroTopia.en.html (accessed April 7, 2005).

———. 1984b. *The Foucault Reader*, ed. P. Rabinow. New York: Pantheon.

———. 1990. "Governmentality." In *The Foucault Effect: Studies in Governmentality*, ed. P. Burchell, C. Gordon, P. Miller. London: Harvester Wheatsheaf.

Fraser, Nancy. 1989. *Unruly Practices: Power, Discourse, and Gender in Contemporary Social Theory.* Minneapolis: University of Minnesota Press.

———. 2001. "Recognition without Ethics?" *Theory, Culture, and Society* 18, nos. 2–3: 21–42.

Gramsci, A. 1973. *Letters from Prison*. New York: Harper and Row.

Grundy-Warr, Carl. 2002. "Lost in Sovereign Space: Forced Migrants in the Territorial Trap." *Asian and Pacific Migration Journal* 11, no. 4: 228–71.

Harvey, David. 2003. *The New Imperialism*. Oxford: Oxford University Press.

Hyndman, Jennifer. 1999. *Managing Displacement*. Minneapolis: University of Minnesota Press.

Jones, Martin, Rhys Jones, and Michael Woods. 2004. *An Introduction to Political Geography: Space, Place, and Politics*. London and New York: Routledge.

Keith, M., and S. Pile, eds. 1993. *Place and the Politics of Identity*. London: Routledge.

Kong, Lily, and Lisa Law. 2002. "Introduction: Contested Landscapes, Asian Cities." *Urban Studies* 39, no. 9: 1503–12.

Lefebvre, Henri. 1991. *The Production of Space*. Oxford, UK: Blackwell Publishing.

Massey, Doreen. 1992. "A Place Called Home?" *New Formations* 17 (Summer): 3–15.

McCargo, Duncan. 2006. "Rethinking Thailand's Southern Violence." *Critical Asian Studies* 38, no. 1: 3–9.

Ong, Aihwa. 2000. "Graduated Sovereignty in Southeast Asia." *Theory, Culture, and Society* 17, no. 4: 55–75.

Passi, Anssi. 2005. "The Changing Discourses on Political Boundaries." In *B/ordering Space*, ed. H. van Houtum, O. Kramsch, and W. Zierhofer. London: Ashgate, 17–31.

Pateman, Carol. 1989. *The Disorder of Women: Democracy, Feminism, and Political Theory*. Cambridge, UK: Polity Press.

Prem Kumar Rajaram and Grundy-Warr, Carl. 2005. "Protection or Control? Global Migration Policies and Human Rights." In *International Migration and Human Rights: Proceedings of the Sixth Informal ASEM Seminar on Human Rights Series*, ed. Bertrand Fort. Singapore: Raoul Wallenberg Institute and Asia-Europe Foundation, 71–109.

Rancière, Jacques. 2001. "Ten Theses on Politics." *Theory and Event* 5, no. 3.

———. 2004. "Who Is the Subject of the Rights of Man?" *South Atlantic Quarterly* 103, no. 2–3: 297–310.

Schutz, Carl. 2000. "Thinking the Law with and against Luhmann, Legendre, Agamben." *Law and Critique* 11, no. 2.

Shapiro, Michael. 2003. "Radicalizing Democratic Theory: Social Space in

Connolly, Deleuze, and Rancière." http://homepages.gold.ac.uk/psrpsg/
shapiro.doc (accessed May 8, 2006).

Soguk, Nevzat. 1999. *States and Strangers: Refugees and Displacements
of Statecraft*. Minneapolis: University of Minnesota Press.

Sturgeon, Janet C. 2005. *Border Landscapes. The Politics of Akha Land
Use in China and Thailand*. Seattle: University of Washington Press.

Thompson, J. B. 1981. *Critical Hermeneutics*. Cambridge: Cambridge
University Press.

van Schendel, Willem. 2005. "Geographies of Knowing, Geographies of
Ignorance: Jumping Scale in Southeast Asia." In *Locating Southeast
Asia: Geographies of Knowledge and Politics of Space,* ed. Paul H. Kra-
toska, Remco Raben, and Henk Schulte Nordholt. Singapore: Singapore
University Press, 275–307.

Walker, R. B. J. 2003. "Polis, Cosmopolis, Politics." *Alternatives* 28, no. 2:
267–96.

Winchester, Hilary, Lily Kong, and Kevin Dunn. 2003. *Landscapes: Ways
of Imagining the World*. London: Pearson, Prentice Hall.

I

Knowledge, Power, Surveillance

1

Detention of Foreigners, States of Exception, and the Social Practices of Control of the Banopticon

DIDIER BIGO

Anderton said: "You've probably grasped the basic legalistic drawback to precrime methodology. We're taking in individuals who have broken no law." "But surely they will," Witwer affirmed with conviction. "Happily they don't—because we get them first, before they can commit an act of violence. So the commission of the crime is absolute metaphysics. We claim they're culpable. They, on the other hand, eternally claim they're innocent. And, in a sense, they are innocent." ... "In our society we have no major crimes," Anderson went on, "but we do have a detention camp full of would-be criminals."

— PHILIP K. DICK, *THE MINORITY REPORT*

The hypothesis underlying this chapter is that the detention of foreigners is related to a specific form of governmentality: the banopticon. The banopticon may be considered a *dispositif*: the detention of foreigners considered *"would-be* criminals" in camps is, for the present time, the locus that concentrates and articulates heterogeneous lines of power diffracted into society. Beyond the sovereign logic of state territory, beyond the analysis of the "state of exception," and even beyond the biopolitics of liberalism, the logic is one of "permanent exceptionalism" or of derogation by the government of the basic rule of law in the name of emergency. This is rooted in the routinization of the monitoring of groups on the move through technologies of surveillance, and is further linked to a will to monitor the future.

I try to differentiate here discourses and cases about the permanent state of emergency. I further try to show that a number of discourses that focus only on the declaration of a state of exception (either to justify or to criticize it) are prisoners of the sovereign state's claim that it is right or viable to draw a boundary between norm and exception, as well as of the (il)liberal belief that there exists a clear dichotomy between inside and outside. This boundary between inside and outside distinguishes and separates a sphere where the words of government articulate the truth of the real, which is applied through law (and order) from a sphere where there is nothing, except anarchy and perpetual virtual war (Walker 1993; Walker 2002).

However, the everyday obedience of the population to rules of conduct emanating from the inside/outside dichotomy can be misleading if the study of law and politics is undertaken without also analyzing the sociology of bureaucracies and its technologies. This sort of study puts the emphasis on political discourses instead of on practices of control and surveillance that frame the conditions of possibility of these discourses (and their acceptance). It is my argument that the everyday routines of the detention camp for foreigners at the borders of the European Union, where foreigners are sent back to their home country as soon as possible, illumine what I refer to as the banopticon as a form of governmentality better than do the shocking cases of Guantanamo or Belmarsh Prison in the United Kingdom. The generalization or normalization of "individual jails" for suspected terrorists through electronic tagging and control orders is another example of the everydayness of the banopticon form of governmentality.

The detention camp for foreigners is for the banopticon the equivalent of what the prison was for the panopticon of Michel Foucault (Bentham and Foucault 1977; Foucault 1977). It is a way to analyze not the specific locus (of detention) but the society that produces it. The detention camp for foreigners accounts for the de-judicialization of punition and the end of a certain idea of punition as reform of the mind and the body (Garland David 2001). Detention is not related to penal law, it is de-judicialized. The focus on the "neutralization" or repatriation of aliens is seen as normal as long as these techniques are not used against citizens. The detention of aliens is linked to administrative law, not to penal law. Moreover, detention is correlated with practices of institutions that differentiate and reframe the borders of "foreignerness." This is done by delinking notions of foreigner and citizen from territory, and linking them with a vision of the *abnormal* (Foucault, Ewald, et al. 1999). Detention appears at the crossroads of internal criminal law and international law. Detention creates zones of uncertainty with new legal parameters established by government. These parameters destabilize existing rights and the common judicial understanding of the rule of law. Detention camps are often located in specific places at border zones where governments refuse to consider them to be under their sovereignty. They are, rather, places where different forms of coercion against detainees may be exercised. This, as we shall see, is as true for detention camps in Europe as it is for camps in Guantanamo Bay.

The detention camp of would-be criminals thus appears where the line tracing the border is unclear, where *inside* and *outside* are not delimited objectively as in a cylinder, but intersubjectively as in a Möbius strip (Bigo 2001). Camps are located in the "hors là," in the "in between," in the "contested trace of where is the boundary"(Bigo 2005). To illustrate the banopticon, my first section here will discuss the notion of *ban* and its relation to Foucault's panopticon. Succeeding sections detail the relation of the detention of suspect foreigners in Guantanamo, where breaches of human rights and of the rule of law against suspected terrorists occur, with the more mundane situation of detention centers or camps for people on the move in Europe. I will demonstrate in the conclusion that the banopticon approach accounts for parallels between the detention camps for foreigners in Europe, considered normal, and the high-security prison camps for suspected terrorists. The concept of the banopticon is then a way to analyze the relations among the claims of derogation and exception by the different powers in place, avoiding a theological (and/or decisionist) vision of the political by reinserting the illiberal practices of the liberal regimes into a global understanding of a specific form of governmentality in Western societies. By outlining relations of continuity and difference among the different situations (of the inside and the outside, the normal and the exceptional, the present and the foreseen), I will finish with an analysis of the topology of the boundaries of our understanding of what is the political and what is the identity of a community.

THE BAN, THE PAN, AND THE EXCEPTION

The banopticon form of governmentality is different from Bentham's *Panopticon,* reread by Foucault. The latter supposes that everyone in a given society is equally submitted to surveillance and control, that there exists a physical proximity between watchers and the watched, as well as an awareness of being under scrutiny (Mathiesen 1997; Haggerty and Ericson 2000). The banopticon, on the contrary, deals with the notion of exception, and the difference between surveillance for all but control of only a few.

The Ban is rooted in the belief of the people (whether citizens or foreigners), that control will be only for the "others" and that, if it happens to oneself, it is still a legitimate control for protection of self and society (Huysmans 2005). The Ban deals with the transformation of the concept of foreigner and otherness. It centers on the possibility of retracing the boundaries of alienness and the consequent treatment of "abnormal" aliens with a different set of rules and rights, including detention independent of any charge (Bigo 2005).

This essay may be understood as a development of Foucault's hypotheses on governmentality and power relations, but a development that refuses Carl Schmitt's decisionist vision, which is more and more accepted, on both the left and the right, as an adequate description of the nature of the political in the present time. This essay is also a specific critique of the thesis of the panopticon and of the role of the police and of surveillance in modern societies. This critique arises from within a particular contemporary context of emerging Europeanization and globalization, of the growth of electronic surveillance, and of the birth of a new kind of prison with the emergence of detention camps for foreigners.

Foucault's Panopticon

In *1984,* George Orwell showed that totalitarianism does not involve only explicitly coercive measures. Orwell demonstrated that governments can play with the media, and with a lack of awareness of history and of memory, to turn public opinion into a mass following of leaders without discrimination or attention to what these leaders may say or do. Orwell was also one of the first twentieth-century authors to analyze the normalization of large parts of the population and the possibility of giving these a certain kind of "non-

political freedom," and among the first to analyze the relation between surveillance and rationality. He has shown that if the idea of
security, even in the name of protection against an enemy, becomes
a society's main value and has no limits, it can destroy society and
democracy. Orwell's work is valuable in the context of September 11
and March 11; however, Orwell focused on state agencies and on
centralized policing. Orwell, that is, appears as a prisoner of a centralized view of surveillance. His focus is exclusively on a sovereign
surveillance—and here he is at odds with Foucault, with regard to
accounting for the rise of surveillance. Orwell is blind to the *pasto-
rale* of surveillance in the name of protection and social security. He
does not see the productive side of power. Foucault, on the contrary,
shows the "positivities of surveillance," the link with freedom and
protection, the relation of complicity of the watchees to their own
control; however, in comparison with Orwell, he does not analyze
with the same accuracy the imperative of freedom and the normalization of population as a constraint (Foucault and Gros 2001).

Foucault's microanalysis of surveillance, in opposition to accounts of sovereign surveillance, is based on an argument that the
prison is not a separate place from society, an institution closed in
on itself, and therefore exceptional or apart from the norm. On the
contrary, he emphasizes that the prison concentrates disciplinary
mechanisms that exist elsewhere though in a more diffuse way. The
prison is on a continuum with hospitals, schools, factories, and
army barracks. Prisons inform us about society. If the concept of
the panopticon has theoretical purchase, this is not because it had
effectively been applied to plans for the construction of the modern
prison. If Foucault's panopticon is an important theoretical model,
this is because it rationally diagrammed the devices of surveillance,
connecting their diversity and their heterogeneity with the variation
of their application points and the multiplicity of the institutions
engaged.

Haggerty and Ericson argue that, "for both Orwell and Foucault,
surveillance is part of a regime where comparatively few powerful
individuals or groups watch the many, in a form of top-down scru-
tinity" (Haggerty and Ericson 2000, 617). This argument is true,
however, only because the authors have first Weberianized Foucault
and focus only on the panopticon in a Goffmanian vision. Haggerty
and Ericson differentiate their research from Foucault's by insisting

on Deleuze's notion of *agencement,* which they translate as "assemblage." Haggerty and Ericson explain that an assemblage consists of a "multiplicity of heterogeneous objects, whose unity comes solely from the fact that these items function together" (Haggerty and Ericson 2000, 608). Rather than a critique of Foucault, however, this definition of *agencement,* or assemblage, is similar to Foucault's notion of *dispositif* (Davidson 1997).

For Foucault, the dispositif is heterogeneous. He argues that "resistances and power interact at the molecular scale." This may not always be apparent concerning the dispositif of *surveiller et punir* ("surveillance and punishment" is a more exact translation than "discipline and punish"), but is more developed in Foucault's discussion of the dispositif of sexuality in *la volonté de savoir.* Foucault describes his notion of dispositif (I refuse here the translation of dispositif as "apparatus" proposed by Grosrichard to avoid an Althusserization of Foucault):

> What I'm trying to pick out with this term is, firstly, a thoroughly heterogeneous ensemble consisting of discourses, institutions, architectural forms, regulatory decisions, laws, administrative measures, scientific statements, philosophical and moral propositions—in short, the said as much as the unsaid. Such are the elements of that dispositif. The dispositif itself is the system of relations that can be established between these elements. Secondly, what I am trying to identify in this dispositif is precisely the nature of the connection that can exist between these heterogeneous elements. Thus, a particular discourse can figure at one time as the program of an institution, and at another it can function as a means of justifying or masking a practice which itself remains silent, or as a secondary re-interpretation of this practice, opening out for it a new field of rationality. In short, between these elements, discursive or nondiscursive, there is a sort of interplay of shifts of position and modifications of function which can also vary very widely. Thirdly, I understand by the term "dispositif" a sort of—shall we say—formation which has as its major function at a given historical moment that of responding to an urgent need. The dispositif thus has a dominant strategic function . . . there is a first moment which is the prevalent influence of a strategic objective. Next, the dispositif as such is constructed and enabled to continue in existence insofar as it is the site of a double process. On the one hand, there is a process of functional overdetermination . . . on the other hand, there is a perpetual process of strategic elaboration. (Grosrichard 1980)

Foucault insists that the concept of panopticon is not fully coherent: it is not only an architectural form but also a form of governmentality that includes architecture and legal texts, the penal system, prisons, the role of the bourgeoisie, and the role of social science (Foucault 1994). The panopticon is fragmented, as every dispositif is, and is not limited to the prison; and at every moment, in every place, it changes, but has its own consistency and builds specific bridges between fragments. For Foucault, power is evident in the working of the dispositif and this power, and the contest over it, is what defines the political. However, the dispositif does not operate through repression or through ideology (Veyne 1979). In place of repression or ideology, Foucault has formulated a conception of normalization and discipline.

STATE OF EXCEPTION: SAFEGUARD OR SHIPWRECK?

The dispositif of the ban permits us to understand the relation among territory, exception, and routines. It destabilizes the common understanding of the state of exception as unrelated to routines and norms.

The Ban and the Territory

What exactly is the connection among the concepts of state frontiers, terrorism, and immigration control? At first glance, as John Crowley emphasizes, "this may seem a rather silly question. After all, immigration is most appropriately defined in legal terms as entry into a country, by a non-resident, for the purpose of residence, which obviously involves crossing frontiers or borders" (Crowley 2005), and terrorism is considered "the attack of the claim by the state to have the monopoly of legitimate violence upon a population in a specific territory"(Badie and Birnbaum 1994).

However, looking at social practices, the border of the state does not in fact have a privileged position in either immigration policy or antiterrorist policy, and there are strong theoretical reasons not to expect it to. If we avoid the general and normative statements about what a state needs to be, and instead look at the social practices of control by the different agencies, if we look at the technologies of government, we may draw another picture.

The reduced significance of the state border is not simply a consequence of the freer movement of people. It is rather that a differential freedom of movement (of different categories of people) creates new logics of control that for practical and institutional reasons are located elsewhere, at transnational sites (Bigo and Guild 2003). The border in its conventional territorial configuration is thus eroded relatively as a site of control. The border of the state is still, at the symbolic level, a powerful boundary because the state tries to configure all the other boundaries concerning identity, solidarity, and equality along the lines of its territory. However, increasing transnationalization, as described above, contradicts this alignment of boundaries along the state frontier and the consequent delimitation of inside and outside.

The transnational movement of capital, people, and ideas destabilizes the territorial state and the simplicity of the (Schmittian) formula of designating the sovereign. The frontiers of states, which were territorially based codes of obedience in a binary form—one against the other, ones to be protected and others to be mistrusted—in the nineteenth century, are now under question. The reactivation of border controls after September 11 is not the sign of efficiency renewed, but a sign of a ritual against a fear of the unknown, with fewer and fewer persons believing in the ritual and seeing in it a simulacrum.

Thus the Schmittian vision of politics is not the future of the twenty-first century, but an old grammar of a time of fixed frontiers and identities, an old vision and an old solution for an old time. In danger, the government of the United States has played the only traditional card it had: the mobilization of people through a state of emergency, the use of a clear distinction between "us" and "them," and the designation of *friends* and *enemies* without possibility to remain outside of the "with or against" game. But this policy is, in the long run, a failure. The functions of borders change over time, and it is important to say, against Schmitt and Samuel Huntington, that the delimitation between inside and outside is not natural, and that we have always had third parties and uncertainties about where to begin the inside and where to locate the outside. State frontiers can no longer be employed as before: to delineate who is with us and who is against us. The ghost of the enemy within is haunting this geopolitical vision in search of simple explanations and clear distinctions between "us" and "them," "friends" and "foes," "good" and "evil," "inside" and "outside." The territory of a state is not an organic political body (Bigo 2002).

The Ban and the Exception

If the control of territory is not a solution, if border controls are hopeless, is it possible for the state to justify the existence of that certain place, the detention camp for foreigners, set up under special laws to cope with the irruption of novelty in the form of violence or in the form of the displacement of people? Politicians put their hope in this "sovereign moment" more than in tighter controls of borders, desiring to create a special situation with special powers allowing them to act "at full speed" in the name of the emergency. They want to designate who and what are the dangers. They want to *ban* some people.

Jean Luc Nancy writes, "the notion of Ban comes from the old Germanic terminology which indicates both exclusion from the community and the command and insignia of the sovereign" (Nancy 1983). Giorgio Agamben in *Homo Sacer* develops, in part, Nancy's idea concerning the Ban to understand the nature of sovereignty when in relation to exceptionalism (Agamben 1998). He does not discuss in any detail the mechanisms of the boundaries of exclusion, but insists on the results of the depoliticization of life and the tendency of sovereign power to reduce life to bare life. Agamben is more interested in the framing of the exception from above than in its framing from below. Agamben also more or less abandons a relational approach for a decisionalist one. For him, the exception comes from above, not from the structuring of the relation to the victim, as Nancy argues. Agamben uses Schmitt's phrase, "the Sovereign is he who decides on the state of exception" (Agamben 1998, 5), to explain his point of view. For him, the Ban is the potentiality "of the law to maintain itself in its own privation, to apply in no longer applying" (Agamben 1998, 28). The Ban is the moment of exception as decided by the sovereign. The exception affects the life of everybody and is driven by the tendency to apply sovereign power to bare life, to eliminate the political from the life of the people and to concentrate the political in the moment of the specific and absolute exception of the naming of the enemy and its virtual or real elimination. The will of the sovereign power is to dissociate the relations of power and to localize it in the hands of the sovereign.

Agamben critiques the Foucauldian approach in its central point: power and resistance are undividable. For Foucault, it is not possible to localize power on one distinct site and resistance on another site.

Deleuze has insisted on this point against the Marxist followers of Foucault, who tried to return to a more classical vision (Deleuze 1986). But, for Agamben, Foucault's approach is precisely wrong in this argument of the undividability of power and resistance and cannot deal with the existence of concentration camps as a principle fact of modernity. For him, and contrary to Foucault, the polarization between power and bare life is possible and in fact drives all the contemporary practices of power, including those of liberal states and democracies. Carl Schmitt, Agamben thinks, has seen what Foucault has not seen: the sovereign moment (Agamben 1995; Agamben 2000; Agamben 2002).

Agamben also argues that Arendt has also better understood the contemporary tragedy than Foucault; only the later Foucault, acting for human rights, understood his own mistake. I shall come back to this critique of Foucault and the rehabilitation of Schmitt through an Arendtian critique that reverses the ethical point of view, but I want first to make clear that Agamben is reducing Nancy's notion of Ban by a double move: first, by not analyzing how the boundaries of the exception in society (before any state or sovereign state) are traced; second, by exaggerating the capacity of the actors speaking in the name of the sovereign and by essentializing sovereignty through a conception that plays against (yet with) the rule of law. Agamben, "captured" by the theological view of Schmitt, forgets "society" and forgets the web of power and resistance. This is particularly obvious in the pages of *Homo Sacer* concerned with the paradox of sovereignty. Agamben ignores the resistance of the weak and their capacities to continue to be humane and to subvert the illusory dream of total control.

Primo Levi and others have answered Agamben concerning the camps, and James Scott has magnified the resistance of the weak and the capacity of hidden transcripts (Scott 1990). "Bare life" is never obtained, not even in the concentration camp. It is the political dream of some bearers of power but is not the description of social practices. The decisionism of the sovereign is also an illusion. The sovereign is not a political body constrained (or unconstrained) by the law. Sovereignty is always "on the move," as Rob Walker has argued (Walker 2005). In brief, then, the Ban is not to be confused with the state of exception. The relation of exception and the relation of Ban are not synonymous. The abandonment of life by the law is always contested and resisted. The state of exception is the relation of the state or the "government" to the Ban; it is the relation of the excluded to the sovereign when the claim of the bearer of sovereignty to be the sole power has been successful (a claim that never in fact succeeds completely).

Is this to say that Agamben is wrong from the very beginning in differentiating between sites of power and of resistance, in coming back to an essentialization of power, a confusion of power and violence, and the idea that the state power express itself through the law (and the exception)? Is Agamben refusing the idea of power as a microphysics, conserving only from Foucault that power is exerted on bodies? It seems that Agamben falls into the critique of Deleuze against a certain left that tries to continue with some postulates concerning power (property, locality, subordination, essentialization, modality, legality) (Deleuze and Hand 1988). And this is why Agamben, entangled between Foucault and Schmitt, finishes by asserting a vision of power coherent with Schmitt at the methodological level, even if from the left. And it is against this vision that, it seems to me, we have to come back to Foucault, by refusing to have a statist view of governmentality, a legal-theological, essentialist view.

Thus the distinction between the state of exception and the Ban is important. The state of exception is only the tip of the iceberg, the visible moment of the Ban, the moment where arbitrariness is not a routine. But in the Ban the norm is the routine of the exception. Therefore the state of exception exists where the state apparatus is strong and well-organized and refuses to tolerate any breach in its claim to having a monopoly of legitimate violence. In this case only, the Agambenian approach is quite a useful one—but in competition with the liberal approach, which also has merits at this level. The two approaches, moreover, cannot deal with *relations* of power, where governmentality links power to resistance (Foucault 1975).

For Agamben, as well as for Schmitt, the relation of exception is a relation of the suspension of the law, giving to the law its real meaning. The exception is logically bordering the law, determining what the law is and is not, even if the strategies of the lawyers are to create the illusion that the law includes the exceptions and circumvents the exception, by agreeing with the idea of a hole of arbitrariness or decisionism controlled by the law. This vision of what is a state of exception is highly different from the traditional liberal vision of the constitutionalists in Germany, France, the United Kingdom, and the United States.

Jef Huysmans developed a better understanding of the debate of the 1930s in Germany, in the light of September 11 and the renewal of a Schmittian approach (Huysmans 2004). He has shown the importance of liberal thinking in the face of a Schmittian attack and the liberals' success in establishing checks and balances and the refusal of a "supreme sovereign," of a "führer."

Pasquale Pasquino has shown in detail the liberal vision concerning the *roman dictature* and the "safeguard of democracy" by the "exceptional moment." He has insisted on the difficulties of nonwritten constitutions to cope with the moment of exception inside the rule of law, but also of the greater flexibility of nonwritten constitutions and the specific role of judges where these exist (i.e., the United Kingdom). He has developed the different doctrines of monism and dualism in the moment of exception (Ferejohn and Pasquino 2004). For me, what is important is that the liberal vision that was developed in Germany and France is based on a *moment* of exception that is in fact a derogation to the law admitted previously by the law under certain circumstances (Guild 2003a; Guild 2003b). This vision is especially connected with military action, as outlined in legislation regulating the involvement of the French military inside France. In the liberal model, civilians accept that the rules of the military, which normally only apply "outside," may supercede the rule of law "inside." Nevertheless, they try to put the military under control concerning ends, if not means. The question of the exception is then more the involvement of the military inside as a means to combat the enemy than it is a regime change; it is more a civil-military relation than an endangerment of the rule of law.

In focusing on the legal elements and not on the organic or institutional elements of a state of exceptionalism, liberal critics do not pay heed to the importance of technologies, taking refuge instead in constitutional doctrine to cope with the irruption of violence inside societies and the search for a way to constrain it. I shall insist, in what follows, on the role of these technologies, which are less visible but more effective than is the political debate about the relative size of the "loophole" allowing the derogation through exception of the law by the sovereign law itself in certain situations.

This downplay of the role of exception as incorporated into the law is often criticized by human rights lawyers. The role of exception in the law is the basis for a coup d'état. François Mitterrand used the term "permanent coup" *(le coup d'état permanent)* in his critique of the use of Article 16 by de Gaulle during the Algerian war, and of the Constitution of 1958 as revised in 1962. The idea of *special powers* in cases of emergency was a way to describe, in a different manner, the state of exception, but when one refuses the argument that the military does not do politics, doubt and suspicion about the justification of the military's role remains. It is important to understand the safeguards and the checks and balances regarding the state of exception, and to clarify who is in charge, to define concretely the boundaries of the exception.

The Schmittian vision is opposed to this liberal discourse of the roman dictature considered as a lie or as a mythology. Schmitt insists that the exception justified as a duty to save the republic exists alongside the exception as key moment of the political: the tracing of the border between friend and foe. The moment of decision and who decides about whom needs to be targeted—and excluded from within or combated as enemy—are the key questions that law can never anticipate. The political is not reducible to the lawmaking process and the rule of law. The political is veiled in the routines of the law and its masks; it is, however, unveiled when open violence irrupts inside society or between two countries.

Schmitt sees a confusion in liberal thinking between the political and the rule of law. Schmitt critiques the liberal dilemma in the face of new situations where liberals are obliged either to withdraw from their principles or to declare their incapacity to act when confronted with novel irruptions (of violence or terrorism, for example). The incapacity to respond to novelty comes, Schmitt argues, from an incapacity to see the future because the sovereign and the political are constrained by the rule of law. The political is cut from its religious formulation and reduced to rationality and law. Agamben is not clear about this sharp-cut opposition between the two formulations of the state of exception; he finds favor with the Schmittian idea of the hypocrisy of law and is fascinated by the Ban in relation to sovereignty—by, that is, the capacity to decide who has to be excluded. "He who has been banned is not, in fact, simply set outside the law and made indifferent to it, but rather abandoned by it, that is exposed and threatened on the threshold in which life and law, outside and inside, become indistinguishable. It is literally not possible to say whether the one who has been banned is outside or inside the juridical order" (Agamben 1998, 28). Agamben, however, seems to agree with Schmitt that the political needs the sovereign, in cases of emergency, to differentiate from the rule of laws, the routines, the governmentality of everyday, because only at this moment can the borders of the polity be created by casting some outside and including others. If Schmitt and Agamben differ, it is more in their ontology than in their methodology.

From these premises, Agamben derived his analysis of the concentration camp and the famous distinction between bare life and biopolitics. I do not intend to comment on his work in general here, but it is important to emphasize the following points. First, there is a danger in a Schmittian fascination, as this can lead to the justification of the use of force in the name of the sovereign against any use of violence, and the incapacity of opposing this use of force, given an ethical point of view not sustained by the framing of the liberal state and human rights values. Second, I want, nevertheless, to use Nancy's notion of the Ban in correlation with a more Foucauldian approach to the notion of power, alongside a different approach to the notion of the border between inside and outside. I do this by developing the idea of a different topology to think the contemporary border and how controls still apply in a world on the way to globalization.

In defining the Ban, I use the notion of the Möbius ribbon, which is different from that of the cylinder. In the Möbius ribbon, the delimitation of inside and outside are subjective, or, more explicitly, intersubjective. People situated in different places disagree about the border and have difficulties situating it clearly. The Möbius ribbon destabilizes the idea of an objective border between inside and outside, friend and foe, law and exception (the liberal view of the border of a cylinder). Within the strip, zones of indetermination appear; zones of conflagration (of violence and of meanings) emerge, and they are not no-man's land: on the contrary, they are populated by individuals excluded from both the inside and the outside, from both friendship and enmity, from both law and exception. They are banned in places with no names and no status—or with multiple statuses creating a differential of forces and resistances between the capacity to use coercion and the capacity to be protected. They are considered as terrorists or bogus asylum seekers or illegals or "abnormals" and are neither accepted nor considered as normal enemies. They are called "enemy combatants" but not prisoners of war, or criminals, in Guantanamo. They are called "undesirable" or "unwelcome" in detention centers or in transit zones and international airport zones. They are people on the move trapped during their travel and sent back or detained in places historically related with death, and they are stocked there before being sent back as soon as possible, or are punished to deter others. They are like lumps in the sauce of the fluidity of the freedom of movement of persons. They are invisible and without clear status, imprisoned but without charges against them, outside of the scope of criminal law but also not war enemies to be judged and killed. They are often in the zone of indetermination between nationals and foreigners as im-migrants or as il-legals and, for suspected terrorists, as in-humans. They are placed under administrative decisions to avoid their defense along the lines of penal law or international law. The growth in every domain of life of administrative decisions, of a body of administrative laws and tribunals where the concept of laws and tribunals is disrupted by the idea that the state and the powerful must win these cases because it is necessary to defend society against the banned people, gives birth to the notion of a permanent state of exception. The exception is not contrary to the norm and the routine; the exception becomes the norm, becomes the routine.

In what follows, I will give some examples of why the logic of exceptionalism is so important to understand the concept and nature of the political today.

GUANTANAMO BAY: EXCEPTION IN ACTION

The situation at "Camp X-Ray" in Guantanamo is typical of this logic of Ban. The so-called enemy combatants in Guantanamo are not inside the normal juridical order of the United States, but suspended from it. They are neither outside of the reach of the sovereign power, nor inside the legal process: first, because, in the rhetoric of the U.S. administration, the place itself is derogatory to U.S. law; second, by becoming part of the category of "terrorist" taken outside of the United States, not terrorists or criminals or prisoners of war.

This juridical limbo is a threshold where the border between inside and outside is uncertain. Violence as force and order comes from both inside and outside to target detainees, but the protections of laws (penal and international) are repelled, creating a zone full of people where only a thin line separates existence from nonexistence. The order of inside and outside is merged and intertwined, creating a zone of indetermination.

In Afghanistan, some prisoners are sorted out and deported to Guantanamo, and others are not and instead stay in Afghanistan. There are no precise criteria differentiating the two categories (as shown by considering those liberated from Guantanamo, who are not necessarily foreign or Taliban). Some are U.S. citizens but nevertheless have been placed inside the same situation as other "enemy combatants." In the same move, the categories of prisoner of war under international law and of criminals in penal law are reduced, to open a third category allowing indefinite detention. This administrative decision by order of the president is not outside the law; it is not free from the hold of power but is free from control of the judiciary as a way to counterbalance its decisionism.

This zone is a zone of "limbo"' and is not limbo: it is a field of experimentation where there is a permanent regulation of specific cases that depends on opportunities, not on principles. The local initiatives of the military are not constrained by a top-down obedience to the sovereign. The military are obeying the president, but also are interpreting its will and are actively organizing a zone of conflagration. The military re-create the sensorial privation of high security jails with artisanal earplugs, blindfolds, and orange spacesuits. In Guantanamo, the technologies of power used by the U.S. administration edict new regulations and invent them permanently. It is a laboratory, not the reproduction of the *lager*. The military does not intend to exterminate the "enemy combatants," but to police the future through the control of detainees' bodies and knowledge. Detainees as "enemy combatants" are transformed into oracles who will give to the intelligence services the key to the future.

For the liberal view of the state of exception, the problem in Guantanamo is the fact that the status of detainees was not declared beforehand and is being retroactively enforced. It is also the fact that, even if these prisoners are not citizens, they are human beings and have rights bearing on their persons. They are not "bare life";

they are protected by international law, and the United States administration cannot free itself by declaring that they are outside the scope of humanity.

For the Schmittian view of the state of exception, the U.S. administration has the power and the duty to reframe the meaning of what constitutes law to cope with the new situation. In the Schmittian view, Guantanamo is the proof that the exception reframes the meaning of law and changes concepts of prisoner of war, foreigner, and human being. For the critical view of the state of exception, which Agamben and others try to develop, this will of the administration to reduce the detainees to their bare lives is a negation of the meaning of life. The ethical point of view does not come from the respect of rules of law, but from the memory of the horror of the Holocaust and from an Arendtian point of view. It is important to trace the genealogy of Guantanamo Bay and to show the analogy with the *lager* and the concentration camp.

However, this critique forgets the logics of contemporary capitalism in a globalized world of transnational political violence. This critique tries to resolve the ethical point with reference to the past, giving the possibility for the Schmittian to insist on the novelty of the case. I think it is more important to show how the indetermination of borders is developing and how this creates zones of conflagration where the language and the meanings of words are objects of struggles for the creation of new categories and new forms of institutional knowledge, with their correlates, as forms of institutional power; I shall do this, concentrating on the Department of Homeland Security.

To emphasize the border as zone of indetermination, it is also important to see the effects of the power of such mechanisms, or dispositifs, beyond the targeted group. Guantanamo Bay holds at least three hundred persons, including one or perhaps two U.S. citizens. But Guantanamo is the most visible tip of an iceberg that has led the American government to interrogate systematically more than five thousand nationals from Middle East nations on the sole basis of nationality and to arrest several of them, to practice kidnapping over the world against suspects, to send them abroad in a cynical attempt to delocalize liability on the use of torture, and also to carry out frequent "surveys" at the center of the population being targeted, encouraging allied countries (e.g., Saudi Arabia) to do the same (and to send the results for their profiles). In addition, the U.S. government has been moved to implement a series of practices derogatory to the international conventions already accepted in a bid to centralize and control the information technology databases containing the personal data of these foreigners or long-term residents, in order to "anticipate" who is moving and for what reasons.

This has had a negative effect not only on the civil liberties and individual rights of foreigners in the United States, but also on those of foreigners in Europe, through pressure imposed by the American government. The idea of the contamination of the categories of foreigners by their reframing through the exception is central.

Nevertheless, in Guantanamo the uncertainty concerning inside and outside does not exist at the territorial level, because the United States has chosen an island physically outside of the nation itself. Here, the link with Agamben's *lager* makes sense. Guantanamo is not an open space. Physically, a clear differentiation exists between inside and outside. There are barbed wires or more subtle technologies of separation. The notion of the state of exception has a correlate that is the closed space, the barbed wire, the territorialization of surveillance. Here, coercion exists to delineate and to mark the frontier between inside and outside.

But the Ban is not always linked with this case of closed borders, of an internal zone of coercion and impunity for official perpetrators of violence. The Ban may have no physical boundaries delimiting an inside from an outside. And I will discuss these cases, which are even more complicated, through the examples of the detention centers for illegal migrants in Europe, Sangatte, the waiting zone in the French airports, and the gated communities.

DETENTION CENTERS AT EUROPE'S BORDERS

I want to demonstrate that the Ban is not always linked with closed borders, with an internal zone of coercion and of impunity for the official perpetrators of violence. The Ban may have no physical boundaries delimiting an inside from an outside.

Normalization is more important than the disciplining of the excluded. It is not only suspected terrorists who are detained. For them, detention is indefinite by administrative order. But for so-called illegal migrants, even if detention is made definite by law, social practices of control underpinning the detention are quite indefinite. The development of administrative detention without trial for foreigners is now largely developed in European democracies. Here, the danger is to focus only on antiterrorist policies and to disregard the continuum of insecurity created by processes of (in)securitization, of fear, and of unease by management professionals.

The Ban is more than a declaration of a state of exception. The Ban is more than a decision by the sovereign. The Ban is linked with a general governmentality that I have called the management of unease (Bigo 2002). The Ban is developed through routines, through technologies, by (in)security professionals and not only by professionals of politics (to use the Max Weber formula) in their conflicts with the courts, lawyers, and the rule of law. I will try to demonstrate this, first by developing different examples of detention centers, and afterwards by analyzing the way technologies of control attempt to monitor people on the move and strive to monitor the future.

As the map of the detention camps in Europe shows (Map 1.1), we are not speaking of exceptional cases like Guantanamo or Belmarsh. We are speaking of thousands of people put in detention without trial, some on a simple suspicion by the administration that they may be illegal. With approximately one hundred centers in the European Union (EU), detention is routine, involving a lot of personnel but not provoking much debate, as though the banality of the situation makes the population think of this practice as normal. Yet we are speaking not of suspected terrorists but of mere tourists or economic migrants considered illegal who try to enter the EU to stay indefinitely. Some are already condemned by tribunals (often administrative tribunals) and are awaiting expulsion; others are inside the Schengen Information System and have tried to reenter the EU after being rejected once, others are entering for the first time but do not have the papers required to stay and travel inside the Schengen area. A small proportion claim asylum—when possible in practice, as police officers are often not very cooperative and are reluctant to extend the detainees their rights. Many are sent back before trial, and are detained only for about two weeks; they are sent back as soon as possible on a flight of the airline that brought them in. Approximately 30 percent of detainees are trapped in a detention center for more than two weeks. Much depends on the relevant legislation, on whether they have the right to see a judge or not before being sent back, on whether or not they can appeal the

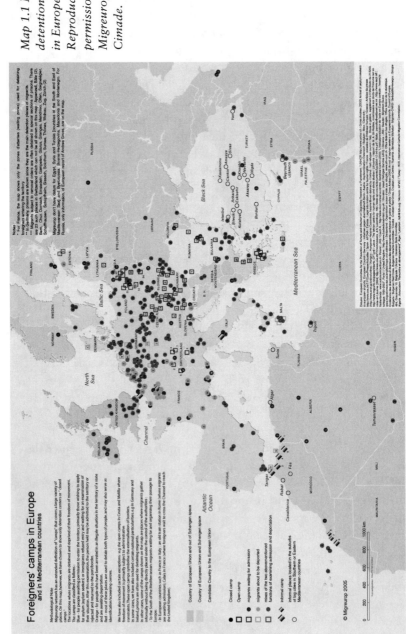

Map 1.1 Map of detention camps in Europe. Reproduced with permission from Migreurope and Cimade.

administrative decision, on whether or not they have access to lawyers and judges, and so on.

Thus the priority in these camps is not to detain people indefinitely but to send them back to their points of origin as soon as possible. The priority is to prevent people from settling, denying them
the possibility of staying and living inside a country not considered
their own. The aim of the centers is not to jail persons to correct
behavior or to defend society against them. The purpose is not disciplinary, even if in some countries the two ideas (discipline and
removal) are merged and work together upon the same populations
and in the same sites. The rationale is not one of punishment but
of keeping the detainees at a distance from a certain territory and,
sometimes, inside the given territory, from certain welfare benefits.
What is important to understand is the dialectical movement between the speed with which the detainees are sent back and the
desire to deter them from attempting reentry, which is sometimes
correlated with punishment and arbitrary conditions of detention.

To understand this dialectic, I will use the example of France,
not because I am French but because France was obliged, as a result
of specific resistance, to distinguish between two types of detention often confused in other countries. This distinction better shows
the two faces of the same coin. France experimented with detention
centers for foreigners in 1936 during the Spanish Civil War. These
centers were built to "protect" republicans fleeing Spain, but also to
"protect" the French population from a massive influx of foreigners.
The rhetoric is not new. Later on, during the Second World War,
some of the same camps were first to detain Jews before they were
deported to extermination camps, but the French camps were used
not like the Nazi camps for Jews in Poland, but rather as forced-
labor or transit camps. Iimmediately after the Second World War,
the population of these camps changed once again to include the
terrorists of that time (nationalists claiming independence from the
French empire). The administration saw no reason to close them; on
the contrary, the administration developed them. To avoid revolts,
forced labor was discontinued, and instead people asking for assistance and other populations were placed in these same centers to
optimize camp management. With the end of colonization and the
dismantling of the French empire, and with the end of the Algerian
war, what could be called the first generation of camps came to a

close, even though the same sites remained more or less active (some camps reopened four or five years later). The populations known as *pieds noirs* and the *harkis* (Algerians who fought for France during the war of independence) briefly populated the camps while being resettled into France after leaving Algeria.

In the 1970s, the idea of detaining in these sites foreigners who were considered illegal migrants took hold. One of the most (in)famous cases was the Arenc camp in Marseille. Hundreds of people were detained in a Marseille shipyard, for no legal reason and without trial, for the time it took the police to "filter" whom they would or would not admit—in other words, who was suited to the French labor market and who was not. The "undesirables" were expelled. In the mid-1970s, with the discourse calling for an end to legal migration, the number of detained people rose in dramatic proportion, and older camps were reopened and the major airports were put under surveillance. The Bonnet Law of 1980 tried to impose the idea of a right to detain foreigners as long as their entry status to France was not clarified. The prevailing discourse at the time claimed that these detentions were only administrative measures, but the judges of both administrative law and penal law wanted to review the process. Thus, paradoxically in France, two kinds of judges were involved in the detention of foreigners, thereby preventing the administration from operating these specific detention places alone. As one of our interviewees explained: it took many years for the administration to gain the freedom to do what it pleased and to do away with the varied judicial controls (but civil servants still complained that too many formalities were required and that lawyers were still prevented from sending "illegals" and "unwelcome" foreigners back quickly and efficiently).

Once in power, the French Socialist Party, which initially joined in the criticism of detention centers, established a distinction between detention centers for illegal aliens, from which people were deported once a trial proved that they did not have the right to stay in France, and "waiting zones" for people who had just arrived in the country and wanted to enter France. In the latter case, official discourse affirmed that the people could not be detained, because they had committed no crime, so they waited in lounges and hotels in the waiting zone until the administration handed down a decision. It was not long, however, before waiting zones became detention zones with closed doors and closed windows.

The rhetoric nevertheless claimed that these foreigners were still free to leave France, to choose to return voluntarily or to go to another country (except for the Schengen area) if they so desired. By thus distinguishing between the two cases—foreigners who wanted to enter the territory, and foreigners already inside the territory illegally and therefore liable to be expelled—the Socialists thought they had found a way of justifying the detention of foreigners without trial. They insisted that this practice was for a very short period of time and that the legality of the process and the rights of the persons detained had been ascertained. Their device also enabled them to criticize countries where the two functions were mixed (nearly all other European countries). The rhetoric was quite successful, as it made the detention of foreigners in waiting zones routine and at the same time justified the criminalization of illegal residence. Public opinion and sometimes members of parliament did not perceive the difference or the fact that some people became illegal aliens by a mere change and tightening of immigration law. They accepted the idea of an insecurity continuum between criminality and illegality, between illegality and foreignerness. The image of the illegal alien, especially from Middle Eastern Islamic countries and the Maghreb, was connected to a threat to the welfare state, a threat to law and order through petty crime and drug trafficking, and a threat to external security through transnational political violence (i.e., terrorism coming from Muslim countries and/or the Middle East). But this connection was only made possible by the transformation of control and surveillance technologies, their spread, and civil obedience to them. And the more sensitive cases are not the most visible Ban but the invisible Bans, the ones with no barbed wire, no walls or even guards.

The Ban can be applied as a technique of government without closed spaces. "Open" spaces are not antithetical to the Ban. The dispositif mechanism crosses over closed and open spaces.

International Zone as Waiting Zone

As I have explained, France has distinguished between detention centers for illegal migrants waiting to be expelled after a legal decision, and "waiting zones" in airports where people are detained if the border police consider they do not have the right to continue their journey.

The waiting zone defines the diagrammatic modes of contemporary imprisonment and the connection that this imprisonment has with the free circulation of people on the world scale. The waiting zone is at one and the same moment a unique, specific fragment of the logics articulating the manner of carrying out police work and, itself, the significance of a certain type of social relation that it condenses to the extreme.

The waiting zone is an exception within the legal order and appears as the point of closure of the legal territorial and sovereign system upon itself. The waiting zone is a fragment because one cannot read its economy except in relation to the mechanisms of visas, of the detection of fraud, of the computerization of data, of the proactive logic of policing at a distance. The waiting zone is exemplary because the maintenance in waiting, for entry into a given territory, is homologous with the congealment of time permitting the administration to manage a detainee's files and with the paradoxical liberty "to go anywhere except where one wants to go."

Following this line of thought, we could take up the Association nationale d'assistance aux frontières pour les étrangers (ANAFE) slogan "We are all in a waiting zone," even if we don't want to know what goes on at airport frontiers, these places where foreigners passing through are imprisoned. We in fact live under the same constraints as those in these jail hotels. Nomadization and "the empire of speed" make us want to circulate farther and faster, consequently putting us more and more frequently in waiting situations. The waiting zone is, to an increasingly globalized society, what prison was to modern society and the national state. It is at the same time a condensed locus and the most arbitrary place.

Sangatte: Frontier as Fish Net

The Red Cross camp at Sangatte was not even a waiting zone. But it looked like one. Here too there was no barbed wire, no clear-cut boundary between inside and outside, not even any border guards. Sangatte was an open space, a net more than a camp. It was a place where different migratory flows converged, where people arrived because the current of their path carried them there. People could come and leave. They had the freedom to go wherever they wanted (especially to leave France and cross the Channel). The traffickers who helped the asylum seekers get to the United Kingdom very often came from within the French police, explaining the best route by which to leave the country and overstating the advantages of being a refugee in the United Kingdom. The Red Cross was more aware of the difficulties people from Afghanistan or Kurdistan would face there, and they knew how high the accident rate was for people trying to leave (by entering the Channel Tunnel and the Eurostar, or by being stowaways on trucks and ferries, or by taking small boats).

Sangatte was an exception, a ban without specific borders delimiting a zone. It was a more or less indefinite geographical place (which is why the closing of the Red Cross camp was a nonevent: migrants now are simply less concentrated and more dispersed throughout the area between Calais and Dunkirk), but defined by the population arriving there (the most well-known illegal immigrants in Europe, well-known and well-documented by a multitude of administrations).

One last example of the Ban is an inversion of the logic of exclusion: the gated communities. This phenomenon can be seen as "imprisoning the outside," to borrow Blanchot's expression (Blanchot 1969, 18). Society becomes "abnormal," and those living in a gated community imagine that the people inside are the only real ones, the only true society.

The Ban is, then, more widespread than we care to see and acknowledge. It is easy to see when the government explicitly refers to an emergency situation. It is easy to see when a physically defined zone is needed for its operation: the concentration camp, the Guantanamo camp, or the materialization of the international zone. But the Ban is generalized, and we are active in promoting it.

TECHNOLOGIES OF CONTROL

The most characteristic feature of contemporary times is the destabilization of the categories regarding the perceptions of war. There is nowadays not any dual polarization simplifying choice and blocking third-party positions. There is not any special timescale where the world could be differentiated between friends and foes. The Schmittian momentum of the exception, allowing executive decisions while suspending the rule of law, is itself suppressed, distorted inside the aleatory timing of the struggle, and leads nowhere. It drags in time, waiting for a new randomized action from an enemy without definite shape, and has no interest in entering the time frame of continuous tension as a test of strength. On the contrary, it has great interest in playing with time, striking at will every two or three years without any possible anticipation of when and where.

Yet a permanent state of emergency cannot be maintained when confronting an enemy that cannot be localized, an enemy that could be inside the country as well as outside. Public attention has been focused on a particular enemy, an infiltrate but more or less recognizable: the radical Muslim. Here, there is the external enemy who has infiltrated inside the territory but is nevertheless identifiable as different from the *us* of the community. As noted by Gabriel Peries and Ayse Ceyhan, the enemy inside the territory is not the enemy within (Ceyhan and Peries 2001). There is a difference in quality between the two. Nevertheless, a slippage on either side is always possible. And this is that what seems to be happening with the discussion regarding those "converted to radical Islam." One is ever more suspicious of the black Americans converted to Islam or, worse, of the white boys embedded in the rules of their society but nevertheless potential traitors, who are even more difficult to detect and to catch. One is obliged to dress them up with particular signs that explain their "abnormality" (Foucault, Ewald, et al. 1999).

Yet there is a political paranoia, or, more precisely, a political demonology, that develops from places where the use of surveillance technology is supposed to find what the human intelligence cannot find: namely, signs and marks of a potential hostile intention (Rogin 1988). This has to do with something that goes beyond the racism and the transformation of the concept of foreigner that is taking shape at the center of these technological experiments.

One gets further from the notion of racial profiling that focuses on physical features indicating hostile intention and belonging to an ethnically identifiable population (Muslim, would-be Arabs), toward a technological concept of surveillance classified in a system of actions and in the anticipation of profiles. This gives less freedom to ordinary arbitrary forms of everyday racism, in which the personal enemy becomes the public enemy. In the new case, a look or an attitude does not constitute in such a stark manner the differentiation between the foreigner (or tolerated citizen of foreign origin) and the suspected terrorist. The power of the security control desk held by security control agents is also perceived as dangerous, and the controllers must themselves be controlled, yet, as it is impossible to give these agents practical codes of conduct limiting the category of the people they have to search, one aims at reconfiguring the searches around actions and their parameters. Of course, these mechanisms do not oppose each other symmetrically. The Immigration and Naturalization Service has tried to distinguish between citizens, foreigners subject to less severe control, and those coming from countries known as risks and subject to reinforced control; but, at the same time, the signal given has been to be suspicious of those looking too perfectly American, too white American. The legal category distinguishing citizens from foreigners can no longer be used as a reference point in this political demonology, even when foreigners are the most controlled and punished (all the more easily since they have fewer rights): citizens can also be dangerous. The recent detention in Guantanamo of the American citizen José Padilla, as well as the imprisonment of a U.S. citizen converted to Islam who acted as imam for the prisoners, U.S. Army captain James J. Yee, shows this quasi-infinite extension of the "fear of the contamination of minds" (Kershaw 2003).

How will it be possible to find again the boundaries, the distinctions between those who are hostile and those who are not, when everybody is inside the country? How can people be protected against those wanting to get in, and how are entrants' motivations to be clarified? How can somebody anticipate their actions? How can somebody control the fear of others, of all the others, including one's relatives?[1] These questions haunt the security services but remain unanswered, except for those questions relating to the most classical control procedures and to the indiscriminate use of information technologies linked to other identification technologies using digital imprints, photonumerical systems, and iris or genetic imprints. It is no longer really important to define the foreigner as the citizen from another nation, but rather as an "abnormal," whether citizen or not. Here, the alienness or abnormality is linked to specific statistical and technological processes building the profiles of what is normal and what differs from the norm. It is under this perspective that one must read the Total Information Awareness Project, renamed, to disguise the disproportionate ambition, the Terrorism Information Awareness Project (partly blocked by the U.S. Congress in one of its first acts of opposition to the administration on the question of combating terrorism).

For the (in)security professionals and the politicians, here is the image of the future enemy, of the new hostility that is being reconfigured through a capillarization of political conflict and of surveillance. It is no longer the foreigner as such who is being targeted, but all those, foreigners or not, who have an action profile that the behavioralists establishing the profiles have judged a sign of potential danger. The foreigner is no longer the noncitizen; the noncitizen is the one with the strange, bizarre or slightly deviant, abnormal behavior—or the opposite, the one with such normal behavior that it seems suspicious. The main advantage of this policy is that it hides itself behind technical neutrality. It appears reasonable and not subject to classic racism. It is inspired by the science of traceability and aims to anticipate, through an in-depth analysis, action sequences—an anticipation process in which the computer has no soul and, therefore, does not have the human defect of classifying persons according to skin color. But this technical redefinition of the foreigner does not solve at all the failure to determine the image of the enemy. It is all about a forward escape into technology as a last resort and is far from developing reasonable antiterrorist policies; it drags the politician into a world of fiction.

BOUNDARIES OF THE BAN: WHERE ARE THE LIMITS?

The Möbius strip models the limits of the ban. In the Möbius strip (or Möbius ribbon), the delimitation of inside and outside is, as I have argued, subjective—or, more explicitly, intersubjective. Zones of indetermination populated by people excluded from both inside and outside appear.

The political program of liberalism on a worldwide scale is dedicated to freedom of movement. But it is not applied in a homogeneous manner. The freedom of capital, goods, and services is a source of profitability; the freedom of human movement is not necessarily so. Those who are profitable and economically solvent are to be separated from those who are not, but this must be done without checking everybody. Such is the price of managing speed. However, sorting, anticipating, and simulating, while avoiding controlling everyone, paradoxically assumes that everyone is involved in the enterprise of sorting so as to supervise only some and to push these into an internal exile through relegation for deportation and repression in zones of indetermination in Guantanamo or in the detention camps for foreigners in Europe.

What and who is "exceptional" must be defined collectively. People will have to say who they believe is dangerous or threatening, who is a source of uncertainty, and point these people out to professional security agents, who in turn will be in charge of the Ban, of relegating and repressing the so-called undesirables. Here is a way of geographically, socially, culturally, and mentally keeping at a distance those that a social optic and a political act have designated undesirables. This is to hold the *hors là*[2] of each of these people that proximity designates nonetheless (or because of) as stranger, as suspect, as enemy or potential smuggler. The waiting zone is then the juridico-physical enrollment of a spatiotemporal compression (Harvey 1996) locally generating effects of unification and division isomorphic to those of globalization. Within the zones of waiting and transit, between exile and asylum, a long march is enacted for these banished people, where circulating is tolerated but where stopping to rest or settle is not allowed. The Ban is then the freedom to leave where one does not want to go. It is the time of endless circulation, of constant rotation. It is a way to bet on orbit and to manufacture a satellite population. The latter is held in weightlessness, at a distance and in "dis-time"; policing at distance has created a move to police beyond the geographical border, policing at "dis-time" is a move to police beyond the present, to try to police the future by anticipation, hence the move toward and rhetoric of prevention, proactivity, and preemptivity. However, this future is of course always a *futur antérieur* (future perfect) formed through profiling, traceability, and morphing based on assumptions embedded within stereotypes concerning race, poverty, inequality.

NOTES

1. See the 2003 play *Homeland Security* by Stuart Flack, directed by Sandy Shinner, Victory Gardens Theatre, Chicago.

2. *Hors là* best translates as "out there." It is a term used by Michel Serres in his book *Atlas* to signify presence and absence (Serres 1994).

WORKS CITED

Agamben, Giorgio. 1995. *Moyens sans fins: Notes sur la politique*. Paris: Editions Payot and Rivages.

———. 1998. *Homo Sacer: Sovereign Power and Bare Life*. Stanford, Calif.: Stanford University Press.

———. 2000. *Remnants of Auschwitz: The Witness and the Archive*. New York: Zone Books.

———. 2002. "L'état d'exception." *Le Monde*, September 12.

Badie, B., and P. Birnbaum. 1994. "Sociologie de l'etat revisitée." *Revue internationale des sciences sociales* 46, no. 2: 189–203.

Bentham, J., and M. Foucault. 1977. *Le panoptique*. Paris: P. Belfond.

Bigo, Didier. 2001. "The Möbius: Ribbon of Internal and External Security." *Identities, Borders, Orders*, ed. M. Albert, D. Jacobson, and Y. Lapid. Minneapolis: University of Minnesota Press, 91–116.

———. 2002. "Security and Immigration: Toward a Critique of the Governmentality of Unease." *Alternatives* 27 (supplement): 63–92.

———. 2005. "Global (In)Security: The Field of the Professionals of Unease Management and the Ban-opticon." *Traces : A Multilingual Series of Cultural Theory, no. 4: "Sovereign Police, Global Complicity,"* ed. Jon Solomon and Naoki Sakai. Hong Kong: University of Hong Kong Press.

———. 2005. *Policing (In)security Today*. New York: Palgrave.

Bigo, D., and E. Guild. 2003. "The Logic of the Schengen Visa: Remote Policing." *Cultures et Conflits* 49 (spring): 5–147.

Blanchot, M. 1969. *L'entretien infini*. Paris: NRF Gallimard.

Ceyhan, A., and G. Peries. 2001. "L'ennemi intérieur: Une construction politique et discursive, construire l'ennemi intérieur." *Cultures and Conflits* 43: 3–11.

Crowley, J. 2005. "Where Does the State Actually Start? The Contemporary Governance of Work and Migration." *Controlling Frontiers: Free Movement into and within Europe*, ed. D. Bigo and E. Guild. Aldershot, U.K.: Ashgate, 140–60.

Davidson, A. I. 1997. *Foucault and His Interlocutors*. Chicago and London: University of Chicago Press.

Deleuze, G. 1986. *Foucault*. Paris: Editions de Minuit.

Deleuze, G., and S. Hand. 1988. *Foucault*. Minneapolis: University of Minnesota Press.

Ferejohn, J., and P. Pasquino. 2004. "The Law of the Exception: A Typology of Emergency Powers." *International Journal of Constitutional Law* 2, no. 2: 210–39.

Foucault, M. 1975. *Surveiller et punir: Naissance de la prison*. Paris: Gallimard.

———. 1977. *Discipline and Punish: The Birth of the Prison*. New York: Pantheon Books.

———. 1994. *Dits et écrits, 1954–1988*. Paris: Editions Gallimard.

Foucault, M., F. Ewald, et al. 1999. *Les anormaux: Cours au Collège de France (1974–1975)*. Paris: Gallimard/Seuil.

Foucault, M., and F. Gros. 2001. *L'herméneutique du sujet: Cours au Col-lège de France (1981–1982)*. Paris: Gallimard/Seuil.

Garland, David. 2001. *The Culture of Control: Crime and Social Order in Contemporary Society*. Chicago: University of Chicago Press.

Grosrichard, A. 1980. "The Confession of the Flesh." *Power/Knowledge: Selected Interviews and Other Writings, 1972–1977*, ed. C. Gordon. New York: Harvester Press, 194–95.

Guild, E. 2003a. "Agamben before the Judges. Sovereignty, Exception, and Anti-Terrorism." *Cultures et Conflits* 51 (autumn): 127–56.

———. 2003b. "Exceptionalism and Transnationalism: U.K. Judicial Con-trol of the Detention of Foreign 'International Terrorists.'" *Alternatives* 28, no. 4: 491–515.

Haggerty, K. D., and R. V. Ericson. 2000. "The Surveillant Assemblage." *British Journal of Sociology* 51, no. 4: 605–22.

Harvey, D. 1996. *Justice, Nature, and the Geography of Difference*. Ox-ford: Blackwell.

Huysmans, J. 2004. "Minding Exceptions: Politics of Insecurity and Lib-eral Democracy." *Contemporary Political Theory* 3, no. 3: 321–41.

———. 2005. *Politics of Protection*. New York: Routledge.

Kershaw, S. 2003. "Army Chaplain in Detention Sought to Teach about Islam." *New York Times*, September 24, 2003.

Mathiesen, T. 1997. "The Viewer Society: Michel Foucault's *Panopticon* Revisited." *Theoretical Criminology* 1, no. 2: 215–34.

Nancy, J. I. 1983. *L'impératif catégorique*. Paris: Flammarion.

Rogin, M. P. 1988. *Ronald Reagan, the Movie, and Other Episodes in Po-litical Demonology*. Berkeley and Los Angeles: University of California Press.

Scott, J. C. 1990. *Domination and the Arts of Resistance: Hidden Tran-scripts*. New Haven, Conn.: Yale University Press.

Seress, M. 1994. *Atlas*. Paris: Editions Julliard.

Veyne, P. 1979. *Comment on ecrit l'histoire, suivi de Foucault révolutionne l'histoire*. Paris: Éditions du Seuil.

Walker, M. 2002. "America's Virtual Empire." *World Policy Journal* 19, no. 2: 13–20.

Walker, R. B. J. 1993. *Inside/Outside: International Relations as Political Theory*. Cambridge: Cambridge University Press.

———. 2005. "L'international, l'impérial, l'exceptionnel." *Cultures and Conflits* 58: 13–52.

2

Struggling with (Il)Legality: The Indeterminate Functioning of Malaysia's Borders for Asylum Seekers, Refugees, and Stateless Persons

ALICE M. NAH

For asylum seekers, refugees, and stateless persons in Malaysia, there are no clearly demarcated temporal or spatial limits to Malaysia's borders. Malaysia, along with their anxiety-filled relationship to it, does not begin and end until the time they permanently leave the country. Their unsettled relationship to the border will dominate and permeate every sphere of life while they remain in Malaysia; for many, it will shape the course of their lives afterward.

The first encounter that forced migrants have with the Malaysian border is when they make their entry into Malaysian territory. If they pass through successfully, a much longer and, for many, more traumatic phase emerges, that of negotiating the security landscape of the territory over which the Malaysian government exerts sovereignty. This internal landscape, by default if not by design, is a hostile one to asylum seekers, refugees, and stateless persons from a juridical-political standpoint: one that does not consider their circumstances different from other immigrants, one that too easily scripts and treats them as "illegals." As such, they are people who do not belong, individuals who should be punished, disciplined, and excised from Malaysia's territory.

This security landscape is specifically designed to ensure multiple exclusions for illegal immigrants. Part of Malaysia's mechanisms of "internal control" (Brochmann 1999) restricts these immigrants from attaining basic standards of living and from obtaining access to essential services. According to domestic law, undocumented immigrants are prohibited from getting jobs and renting accommodations; they are not allowed to send their children to schools and are reported to the police if they seek medical attention in government hospitals. The security landscape is also a punitive one: immigrants caught and charged in court for not possessing legal identity documents are sentenced to imprisonment and possible whipping. After periods in detention, many leave Malaysia with both physical and mental scars.

Nevertheless, Malaysia's borders are neither static nor stable; they have been and continue to be politically, socially, and economically constructed. Although there are some material markers (such as fences, gateways, and ports) to suggest the concreteness of these borders, they are primarily constructs of geographical and political imagination, sustained by discourses and practices of control. As constructs, Malaysia's borders have been subject to change—for example, on the basis of bilateral agreements, upon the decisions of ministers, and due to pressure from different stakeholders such as the United Nations High Commissioner for Refugees (UNHCR). The struggles over the definition of Malaysia's borders, and the resulting indeterminacy of its operation, are revealed through the disruptive figures of the asylum seeker, the refugee, and the stateless person.

Malaysia is a signatory of neither the 1951 Convention Relating to the Status of Refugees nor the 1967 Protocol Relating to the Status of Refugees, and its existing legal framework does not provide international protection for asylum seekers, refugees, or stateless persons. In recent years, the UNHCR has been these persons' strongest advocate in Malaysia, relying primarily on political negotiation, lobbying, and advocacy to ensure their protection and assistance. The UNHCR has been supported in its aims by growing networks of local and transnational actors, but remains the lead agency in promoting international protection. However, by virtue of its role as an intergovernmental organization within the territory of a sovereign state, as well as of its restricted mandate, its space for maneuver has been limited. The UNHCR's contemporary struggle to protect the persons it is concerned with has been a modest but important battle to *create exceptions* to existing state regimes concerning illegal immigrants: to convince Malaysia of the need to adapt its border practices so that certain classes of immigrants are exempt from ordinary rules. As this is an ongoing negotiation, it produces a condition of *(il)legality* for these groups of immigrants, an uncertain and unresolved socio-legal location in which they are possibly legal—through practices of exception—but remain illegal by default.

This chapter analyzes discourses and practices over borders, for it is through everyday performances that borders are constructed. There are three interrelated aspects of border construction vis-à-vis asylum seekers, refugees, and stateless persons that I explore here.

First, I examine the legal mechanisms and law enforcement practices that underpin Malaysia's management of borders in relation to undocumented immigrants. This is important in two specific ways. First, these mechanisms and practices act as the foundation upon which contests over borders occur in relation to asylum seekers, refugees, and stateless persons. Second, these activities reveal the biases and assumptions of the Malaysian state in relation to immigrants.

I argue that Malaysia's immigration policies and practices rest on several premises. The first is a distinct view of the world in which each and every individual possesses a place where he/she belongs. Not only do undocumented immigrants not belong in Malaysia (and therefore, in Malaysia's view, are legitimately subject to arrest, imprisonment, and whipping), but they have to be returned to specific places. The routes by which they are deported, ideally, are based on their country of origin; those from Indonesia are returned to Indonesia, Indian citizens to India, and so on.[1] A second premise is that all illegal immigrants are the same; they are criminals who should be subject to the full force of punishment without distinction or exception; trafficked victims, asylum seekers, refugees, and stateless persons by default are undifferentiated from other immigrants. A third premise is that immigrants are fully responsible for their own undocumented status, regardless of the political realities and the practices outside their control that resulted in their illegality; they are therefore fully answerable for their transgression.

Second, I argue that borders exist *in tension*; they not only have the capacity to produce but are, in themselves, produced (and reproduced constantly) through negotiation, conflict, and amity between different stakeholders. I examine the different aspects of such negotiated productivity through two cases: the arrest of hundreds of asylum seekers outside the UNHCR in August 2003, and the crackdown on illegal immigrants launched in March 2005. I point out several observations from these cases: first, that, aside from the Malaysian government, there are at least two stakeholders inextricably involved in the operation of the border upon asylum seekers, refugees, and stateless persons: the states of which the immigrants are citizens, and the UNHCR. Furthermore, there is realpolitik to border construction. As we can see in the case of the August 2003 arrests in Kuala Lumpur, this takes on a different dimension when the media is involved, for it forces border negotiation into the public sphere. Put in the spotlight, states have to be careful about how they articulate official positions for these have implications on bilateral ties with other states. The crackdown on illegal immigrants in 2005 also illustrates the pragmatism that drives Malaysia in its management of asylum seekers, refugees, and stateless persons. To address a severe shortage in labor, it announced that it would legalize these groups of forced migrants and allow them to work. In addition, struggles over the operation of borders can also lead to crisis and internal change within stakeholders. Subsequent to the August 2003 arrests, the UNHCR transformed itself, taking on a different engagement strategy with the Malaysian state. It reformed its practices and began to be more proactive, politely assertive, and publicly firm about the principles of international protection. This placed it in a better position to negotiate with the Malaysian government in responding to the 2005 crackdown.

My third argument is that borders include and exclude on the basis of *identity,* and therefore processes of identity creation, reification, and performance are central to the operation of borders. They occur at different levels. Individuals, as subjects of the border, are constantly being read in terms of their identity to determine their place in relation to Malaysia's borders. Two objects act as the repository of individuals' identities—the *documents* they hold, which confirm their politico-legal identity, and their *bodies,* which point to their citizenship, ethnicity, religion, gender, and age. The UNHCR has been more involved in constructing the former—that is, in reinforcing a politico-legal recognition of difference based on documents. The message has been that asylum seekers, refugees, and stateless persons may be undocumented immigrants, but they are nevertheless

of a different sort and therefore should be exempted from the normal operation of law. The significance of identity as read through
the body is more apparent at an *interpersonal* level—for example,
at the point of arrest. The decentralized nature of law enforcement
in Malaysia gives individual officers great flexibility in determining
who should be arrested. Whether or not asylum seekers, refugees,
and stateless persons are arrested depends on the meaning of their
identities and how their identities are performed vis-à-vis officers
empowered by the law.

For these reasons, borders are in constant flux. Asylum seekers,
refugees, and stateless persons are subjects that occupy an indeterminate position, that of being "(il)legal." The temporal and spatial
limits of borders shift continuously for them; they are often unsure
of where the reading of the border places them. They remain fearful,
for the consequences of reading can lead to great vulnerability and
suffering.

I elaborate on these arguments in the following sections, after a
brief introduction to Malaysia and its border practices in relation
to immigrants. I also review the groups of asylum seekers, refugees,
and stateless persons in its territories and examine the central role
of the UNHCR in ensuring their protection. I end this chapter with
a discussion on the concept of (il)legality and suggest how this uncertain space has been negotiated in Peninsular Malaysia.

MALAYSIA'S HUMAN MEGAPROJECTS:
IMMIGRATION CONTROL AND IMMIGRANT FLOWS

Malaysia lies at the heart of Southeast Asia. Formed in 1963 and
comprising Peninsular Malaya and two eastern states, Sarawak and
Sabah,[2] Malaysia has been, and continues to be, an important node
in regional and global human mobility. From the 1970s, Malaysia
embarked on a series of modernization projects, repositioning itself
in terms of global flows of capital (Bunnell 2002). These transformations needed foreign labor and Malaysia has drawn a ready supply
from surrounding poorer countries, primarily Indonesia, Philippines,
Bangladesh, India, and Sri Lanka. Prior to the 1997 economic crisis,
it had up to 1.43 million foreign workers (ILO 1998). This rose to an
estimated 2.6 million in 2004, an addition of more than a quarter of
its domestic labor force,[3] half of whom were suspected to be undocumented. It has also been a recipient of asylum seekers, refugees, and

stateless persons, most of whom arrive from the same countries that supply Malaysia with foreign workers (a point elaborated below).

Malaysia attempts to regulate the inflow of foreign workers through a complicated (and often inconsistent) combination of bilateral agreements with sending countries, complex bureaucratic procedures involving multiple agencies, and ad hoc policies. The selling and buying of labor generates tremendous profits in this "immigration industry" (Hugo 1991, 131), often at the expense of the migrants themselves. Many are cheated by unscrupulous agents, leaving the migrants in debt and undocumented in Malaysia (Jones 2000; Spaan 1994).[4] Some cannot afford the prohibitively high costs of entry through legal methods and therefore make their way into Malaysia through unofficial channels. To discourage incoming flows and to control existing numbers, Malaysia continuously arrests, detains, imprisons, whips, and deports undocumented immigrants. Between 1992 and 2000, an estimated 2.1 million undocumented immigrants were apprehended (Hugo 2002). In 2003 alone, 111,893 were arrested, of whom 62,939 were from Indonesia.[5] That year, the Prisons Department spent about 2.2 million ringgit (about US$590,000) per month on managing ten immigration depots in the Peninsula.[6] After the Immigration Act was amended in 2002, more than 18,000 undocumented immigrants were whipped, over a two-year time period, with another 16,900 sentenced for other punishment.[7]

ASYLUM SEEKERS, REFUGEES, AND STATELESS PERSONS IN MALAYSIA

Most asylum seekers, refugees, and stateless persons in Malaysia originate from neighboring countries. As Malaysia does not conduct refugee status determination, all asylum claims are lodged with the UNHCR. In the past five years, there has been an exceptional growth in new applications. The UNHCR's active caseload increased by 2,513 individuals in 2002 and a further 14,747 individuals in 2003.[8] In addition, the UNHCR began the documentation of Rohingyas, stateless persons expelled from their homelands in Myanmar. By the end of 2004, the UNHCR had an active caseload that comprised the protection of 35,227 individuals,[9] which rose to 42,075 individuals at the end of July 2005.[10] Of these, 49.3 percent were from Aceh, 46.1 percent from Myanmar, and the remaining 5 percent from forty other countries.[11] Of these active cases, 82 percent were men and an estimated 13 percent were minors aged eighteen and below. All of these persons of concern are considered an urban caseload; they do not reside in refugee camps.

While the numbers registered with the UNHCR substantially increased from 2002 onward, they do not represent the full number of asylum seekers, refugees, and stateless persons in Peninsular Malaysia. Many remain undocumented despite their intense desire to lodge an asylum claim, as the UNHCR registers only a limited number of people per day, setting differing quotas and registration procedures for groups of different origins.[12] This results in great competition at the gates of the UNHCR among asylum seekers; some sleep for several days at a time outside the UNHCR in order to be first in line when the gates open for registration. They are also prey to unscrupulous police officers, who, in the early hours of the morning, threaten them with arrest and exact bribes from them.

Some of those currently registered with the UNHCR did not mount an asylum claim for several years. When questioned as to why, different ethnic groups give the same answers. Rohingyas, Chins, and Acehnese who have stayed for long periods in Malaysia say that they either did not know about the UNHCR or, if they had heard of it, did not know how to lodge an asylum claim. Those who had heard about it were also aware that the process of refugee status determination was protracted and were not convinced that the UNHCR would intervene effectively if they were arrested. The material benefits of registration, in their perception, were therefore minimal. This was to change, however, after the August arrests in 2003.

CONTROLLING MALAYSIA'S BORDERS: IMMIGRATION LAW AND ENFORCEMENT PRACTICES

In this section, I examine the legal framework and law enforcement practices through which Malaysia's borders are constructed. As stated, Malaysia is a party neither to the 1951 Refugee Convention nor to the 1967 Protocol.[13] It has also acceded neither to the 1954 Convention on Stateless Persons nor to the 1990 International Convention on the Protection of the Rights of All Migrant Workers. There are no state institutions specifically responsible for asylum seekers, refugees, and stateless persons, and there are no domestic laws or regional mechanisms of protection enacted in relation to them.

Malaysia's immigration mechanisms are governed primarily by the Immigration Act of 1959/1963 (revised in 1975 and 2002). The Immigration Act outlines the scope and methods of immigration control, confers powers of enforcement to civil servants, and lists offenses and associated punishments for persons who enter, reside in, or depart from Malaysia. In Malaysia's framing of borders, there are a limited number of *authorized* points of entry, through which immigrants can pass only if they *already* possess valid travel documents that specify their individual identity.[14] The burden of proof that these requirements have been fulfilled lies on the immigrant, who is held responsible for his/her status while in Malaysia.[15] Entry and stay in Malaysia without a legal document is punishable by a fine of up to 10,000 ringgit and/or up to five years imprisonment; offenders are liable to whipping of up to six strokes of the cane.[16] Once in Malaysia, immigrants remain legal only as long as their travel documents are valid.[17] Overstaying is punishable by a fine of up to 10,000 ringgit and/or up to five years imprisonment.[18]

Power to enforce the Immigration Act is vested in immigration officers. When raids are conducted, immigration officers are often assisted by police officers as well as by civilian groups, such as the People's Volunteer Corps (Ikatan Relawan Rakyat, or RELA) and neighborhood watch groups (Rukun Tetangga, or RT). Immigration and police officers have the right to ask for identity documents from anyone suspected of committing an immigration offense. Failure to comply is sufficient grounds for immediate arrest; no warrant is necessary.[19] There is also a standing order in government hospitals that undocumented immigrants are to be reported to the police. Because of this, undocumented immigrants often approach hospitals only for absolute emergencies and in the late stages of their illnesses. Undocumented pregnant women who approach hospitals for delivery are not exempt from arrest; they, along with their newborn babies, have been taken into custody immediately after delivery.

Once detained, immigrants may be held for up to fourteen days before being brought before a magistrate,[20] who is empowered to extend detention for the purposes of further investigation. In practice, immigrants who fail to produce valid documents within this period are either transferred directly to immigration depots to await deportation or are charged in court and sentenced. The discretion to bring them to court lies with the immigration officer in charge of the case. Immigrants sometimes plead guilty under duress, even if innocent, for refusal to do so (requesting a trial) may result in their being detained in remand facilities for several months awaiting their court dates. They are not permitted to post bail. These periods of detention can exceed the total length of imprisonment they would have faced if they pleaded guilty. Those who plead guilty are typically sentenced to between two to six months of imprisonment and/ or one to two strokes of the cane. The severity of their punishment is at the discretion of the magistrate; repeat offenders have harsher sentences.

Unsurprisingly, these laws place asylum seekers, refugees, and stateless persons in a very awkward position. Most are either unable or afraid to obtain travel documents based on individual identity from their country of origin before embarking on their journey to Malaysia. Acehnese and Myanmarese, for example, flee from repressive military operations authorized by their own states, regimes that use identity documents as a form of control and discrimination. Thus, when charges of illegal entry are read in court, they have no permitted counterarguments. The relative few who are able to secure the necessary documents before travel are still not safe from the Immigration Act: the protracted processes of refugee status determination and resettlement usually results in their overstaying, an offense still punishable by imprisonment. Over the years, hundreds of asylum seekers, refugees, and stateless persons have been prosecuted for immigration offenses;[21] those who have lived for long periods in Malaysia have been sentenced and deported repeatedly.

After serving prison sentences, immigrants are brought to immigration depots, where they await deportation. Each immigrant is processed individually, their countries of origin determined prior to expulsion. There are two types of deportation: in a group, or on an individual basis. Group deportation occurs regularly, and there are two destination countries: Indonesia and Thailand. Citizens of Indonesia are brought to seaports along the west and south coasts of the Peninsula and deported by sea, typically on ferries and private sea vessels (but occasionally on specially prepared military warships). Citizens from Myanmar and Thailand are taken overland to the north of Malaysia and deported at the Thai–Malaysia border.[22] Immigrants from other states generally travel by air back to their homelands. By default, they have to provide their own tickets,

although, in certain scenarios (usually after several months of detention), tickets have been provided for them. Whenever possible, illegal immigrants are returned to where they "belong." Deportations
to Thailand are the exception: interviews with deportees suggest
that detainees from African, Middle Eastern, and some Asian countries are also deposited at the Malaysia–Thai border.

After returning to Malaysia, some asylum seekers and refugees
report having been pressured (even forced) by immigration officials
to opt for voluntary deportation; of these deportations, some have
resulted in *refoulement*. Most detainees, however, with the intervention of the UNHCR, are permitted to remain in immigration
detention depots, some after serving prison sentences for immigration offenses. This, however, is a difficult and stressful experience.
Most detainees in immigration detention depots complain of over-
crowding, intermittent water supply, insufficient access to medical attention, extreme temperature changes, insufficient and poor-quality
food, insufficient clothing and toiletries, being bitten by mosquitoes
and/or lice, and verbal and physical abuse by guards as discipline
is maintained.[23] This detention is made far worse by its *indeterminacy*: asylum seekers and refugees do not know how long they will
be detained. Most hope for the UNHCR to obtain permission for
their swift release. In most cases, however, the UNHCR has to show
proof that the refugee detained has received a confirmed place for
resettlement in a third country before the Immigration Department
concedes to the person's release—a process that can take several
months, if not years. When they first enter detention, some persons
lose heart after finding out that others of their same ethnic group
have been detained for more than two years, and thus opt for deportation despite being aware of the serious risks of doing so.[24]

Although the Immigration Act allows the minister to exempt any
person or class of persons from the Act, this clause has not been
invoked categorically in relation to asylum seekers, refugees, and
stateless persons. Trafficked victims are also not given automatic
exemption; there have been cases of trafficked victims detained for
long periods in immigration depots and prisons, some charged for
immigration offenses or for engaging in prostitution (SUHAKAM
2004; Wong and Saat 2002).

The regular arrests of illegal immigrants are supplemented periodically with crackdowns. What this means, in effect, is that im-

migration operations already conducted on a regular basis are given greater publicity and intensified. The crackdowns are usually preceded by an amnesty period during which undocumented immigrants are allowed to leave Malaysia without being arrested.[25] Rapid and repeated expulsion before and during crackdowns have left impoverished groups of deportees stranded at the Thai border and at those Indonesian ports where human smugglers and traffickers operate. In 2002, for example, mass deportations resulted in a humanitarian crisis in Nunukan, an Indonesian island off the coast of Sabah, where more than 137,000 deportees struggled with shortages of water and food, poor sanitation, and inadequate shelter.[26] This situation led to the deaths of over sixty people.[27] At the start of the 2005 crackdown, Nunukan again became severely overcrowded, "with migrants packed together like sardines," reportedly leading to thousands becoming ill with malaria and tuberculosis.[28] The crackdowns have, therefore, been repeatedly criticized by civil society groups. Not only have they caused deaths, led to human rights abuses,[29] and created humanitarian crises, they have not helped Malaysia to control illegal migration and have triggered severe labor shortages that have led to substantial financial losses.

REPRODUCTION OF BORDERS: NEGOTIATION, CONFLICT, AND AMITY

Such strict enforcement of Malaysia's borders in relation to asylum seekers, refugees, and stateless persons produces, unsurprisingly, constant tension between the UNHCR and the Malaysian state. In this section, I look at the negotiations in two specific cases: the high-profile arrests of asylum seekers outside the gates of the UNHCR in August 2003, and the crackdown on illegal immigrants that began in March 2005 (and continues to point of writing). These events illustrate several facets of border negotiation, and I describe them in turn.

As stated earlier, the UNHCR was trying, from 2002 onward, to cope with unexpected increases in asylum applications. Not only had the Malaysian government tightened its immigration laws and launched a crackdown on illegal immigrants in 2002, prompting undocumented immigrants to seek the assistance of the UNHCR, but thousands of Acehnese from Indonesia were coming to the Peninsula to escape the martial law declared in the Aceh Province of Indonesia on May 19, 2003, a military response to the activities of the Free Aceh Movement (Gerakan Acheh Merdeka, or GAM). From a total of around 100 to 150 applications per day, the numbers of Acehnese approaching the UNHCR alone began to reach almost 700 per day. In an effort to streamline its operations, the UNHCR office designated one specific day a week, Tuesday, for registration. Large crowds began to swell outside the UNHCR, drawing the attention of nearby residents and local police.

Before the UNHCR opened on that morning of August 19, local police officers set up roadblocks on the roads leading to its gates. Checking all who approached the UNHCR, they detained over 400 asylum seekers.[30] Some were eventually released, but 239 remained detained, most of these Acehnese.[31] In response, the UNHCR met with the home and foreign affairs ministries to negotiate for their release. They were given access to the 239 detainees, all of whom were immediately issued with temporary protection documentation. In order not to exacerbate the situation, the UNHCR decided to shut down its operations on August 20, after a police van set up surveillance of its premises. When asked about the arrests a day after they occurred, Home Minister Abdullah Ahmad Badawi (currently the country's prime minister) announced that the detainees would be treated as illegal immigrants and deported. However, his position wavered momentarily, and this statement was retracted the following day when he stated that Malaysia would reconsider the deportation of the detainees. "We are considering the possibility of giving them temporary stay but this is still uncertain. It needs serious consideration," Badawi was quoted; "We have to get a report from the police about the status of those who are arrested, who they are."[32] Three days after the arrests, immigration officers announced that 120 of 232 individuals still detained had opted for "voluntary repatriation."

Several days later, Minister for Foreign Affairs Syed Hamid Albar announced, "if Malaysia supports the asylum seekers, this will encourage other people to enter Malaysia. We [stand] together with the government of Indonesia and have agreed to prevent more and more Acehnese citizens coming to Malaysia." He further commented, "It is not right for UNHCR to register the Acehnese as possible refugees when they are not refugees. They are Acehnese people who have entered Malaysia without valid travel documents," and confirmed that

Malaysia would be deporting them.[33] On the same day, in Jakarta, Indonesia's security minister, Susilo Bambang Yudhoyono (currently president), expressed similar views, saying that Malaysia should reject asylum seekers from Aceh, a policy that he linked to Malaysia's respect for Indonesia's territorial integrity and to nonsupport of the GAM movement in Aceh.[34]

On the following Tuesday, police conducted similar operations and arrested 50 asylum seekers, of whom 30 were Burmese, 18 Indonesians (Acehnese), 1 Bangladeshi, and 1 Thai. At a press conference later in the afternoon, the officer in charge said that this was part of ongoing "normal crime prevention to look after the whole district."[35] These fresh arrests surprised the UNHCR in Kuala Lumpur, whose officials had been led to believe that they would be allowed to continue interviewing asylum seekers that week.[36] After meeting with (then) Indonesian president Megawati Sukarnoputri for annual consultative talks, (then) Malaysian prime minister Mahathir Mohamad announced that Malaysia would not grant the Acehnese asylum and that both Malaysia and Indonesia were finding ways of controlling the influx of illegal immigrants more effectively.[37] "We do not allow illegals or Acehnese to seek political asylum in Malaysia," Mahathir said at a news conference; "They are illegals and they will be caught and put in detention centres and deported." Indonesian foreign minister Hassan Wirajuda later confirmed that "both leaders agree that they (the Acehnese) will be sent back and Indonesia will cooperate to facilitate their deportation." He added: "The presence of Acehnese in Malaysia has nothing to do with the military operation in Aceh. They are not refugees. There is no excuse for them to come to Malaysia."[38] Despite the personal appeal of senior officials, including Ruud Lubbers, the UNHCR was unable to secure agreements to stop further deportations. Most were eventually pressured to go back; only a small number remained in detention until resettlement solutions surfaced. For months afterward, many Acehnese were afraid to approach the UNHCR to seek assistance.

The August arrests demonstrate the realpolitik behind the protection of asylum seekers and refugees. Malaysia was cautious about extending asylum to the Acehnese; it did not want to encourage further refugee flows to the Peninsula and was keen to stay out of Indonesia's domestic politics. Indonesia denied that there was any

cause for Acehnese to flee Indonesia; it was suppressing information about military activities in the Province of Aceh during martial
law.[39] Both these states refused to recognize that the Acehnese were
valid asylum seekers, and cast them as no different from other illegal
immigrants. What the administrations did not anticipate, however,
was the symbolic nature of the arrests. The fact that they occurred
outside the UNHCR in a public manner created ripples in the international community. It demonstrated visibly that Malaysia had vio-
lated the right to seek asylum as stated in the Universal Declaration
of Human Rights. This drew global attention to Malaysia's actions;
it placed Malaysia and its treatment of asylum seekers, refugees, and
stateless persons in the spotlight.

These high-profile arrests also reinforced an internal change
within the UNHCR. The UNHCR was already expanding in size
and scope of activities to deal with rapidly increasing asylum claims.
Shortly after the August arrests, the UNHCR accelerated changes
to its Malaysian operations.[40] A new representative, Volker Turk,
was appointed to head the Kuala Lumpur office. Under his leader-
ship, the UNHCR was restructured and repositioned, increasing its
capacity to fulfill its mandate of protection and assistance. From
having previously adopted a relatively meek and submissive role, it
began to engage proactively with the Malaysian government, with
civil society groups, and with the media on the issue of protection.
Operationally, it increased and amended its registration and refugee
status determination processes. It also defended the identity documents it produced, by intervening when registered persons of concern were arrested. It made appeals for their release, arranged for
legal representation in court, organized medical services for those in
detention, and processed more cases for resettlement. It lobbied for
change through every step of the law enforcement system.

Over time, the UNHCR became a more credible and public actor
in protecting asylum seekers, refugees, and stateless persons. They
found a measure of success when negotiating with the police, managing to convince them to recognize UNHCR documentation and
to release persons of concern. Immigration officials, however, were
more adamant that immigration laws be followed to the letter. They
did not recognize the identity documents issued by the UNHCR, and
refused to release persons of concern until resettlement places were
confirmed, thereby prolonging detention and increasing the chances

of refugees opting for deportation under duress. RELA members, also empowered with the right to make arrests during immigration raids, were less predictable; successful negotiation with them depended in large part on the amenability of district heads. Some resisted any form of negotiation, preferring instead to transfer cases directly to immigration officials. Cooperation with the UNHCR thus varied both across and within these enforcement agencies.

OPS TEGAS: THE 2005 CRACKDOWN ON ILLEGAL IMMIGRANTS

Several months after the August 2003 arrests, the Malaysian government announced that it would commence yet another crackdown on undocumented immigrants, one in which it planned to deport 1.2 million "illegals" by the end of 2005. Human rights advocacy networks and the UNHCR were concerned about the impact this would have on asylum seekers, refugees, and stateless persons, as the Malaysian state threatened that it would not spare any undocumented immigrants from imprisonment, whipping, and deportation.[41] As with previous crackdowns, human rights advocates were apprehensive that such operations would increase the vulnerability of immigrants at the point of arrest, during detention/imprisonment, and upon expulsion.

At the request of the governments of Indonesia and the Philippines, who were concerned about the potential implications of the mass operations on their citizens both in Malaysia and at home, the crackdown was postponed several times. Malaysia instituted an amnesty period during which undocumented immigrants were encouraged, through the media and through mock raids, to return home to obtain the right documents and to reenter Malaysia legally.[42] The amnesty was extended several times, once in response to the Asian tsunami on December 26, 2004, which caused overwhelming loss in the Aceh Province.[43] During the amnesty, between 385,000 and 400,000 immigrants left Malaysia, most from Indonesia. The amnesty inadvertently created a polarization between economic and forced migrants, the latter either unable or unwilling to leave Malaysia despite intense fears of reprisal during the impending crackdown.[44]

In an unexpected gesture in October 2004, the Malaysian government announced that it would recognize Rohingyas—a group of stateless persons from Myanmar registered by the UNHCR—as refugees, and allow them to work legally.[45] This news was met with great relief, as many Rohingyas had already suffered repeated deportations and whippings in the years they resided in Malaysia. The amnesty period allowed the UNHCR more time to prepare its response to the crackdown. It reinforced its 24-hour hotlines in order to respond promptly to arrests, organized briefings for police chiefs and senior members of RELA on refugee determination and documentation,[46] and continued public appeals for Malaysia to spare those under its protection. Aware that many asylum seekers, refugees, and stateless persons were still undocumented (and therefore even more vulnerable), the UNHCR supplemented its registration procedures, sending buses to areas with high concentrations of refugees to bring them in for documentation.[47] These efforts, however, were insufficient to cater to the thousands spread out across the Peninsula. Hundreds still flocked to the gates of the UNHCR, sleeping night after night, waiting with mounting anxiety to be documented. Unable to cope, the office turned them away.

The operations began on March 1, 2005, with intense media coverage. Journalists were invited to accompany law enforcement agents on their raids; Malaysia wanted to demonstrate to the world that it would observe human rights standards during arrests. Over four thousand immigrants were arrested in the first month alone.[48] Early into the crackdown, the government seemed to reconsider its position on other groups of asylum seekers and refugees. In response to the tsunami, (then) home minister Azmi Khalid announced that Acehnese, specifically, would be spared from crackdown,[49] as would other individuals possessing UNHCR documents.[50] The government's official position however, seemed to flip-flop. Azmi Khalid's statement was contradicted by Deputy Prime Minister Najib Tun Razak, who stated that all UNHCR document holders *would* be arrested. He said, "We will take action against anyone who is here illegally. There is no exemption on this including those who are carrying letters, genuine or otherwise, from the UNHCR. If the UNHCR wishes to appeal after these people are arrested, then it is up to them. But it is up to us whether we accept the appeal or not."[51] These incongruous statements indicated a lack of a clear policy and a wavering stance of the government in relation to persons of concern to the UNHCR. They also generated significant confusion, not only among those who held UNHCR documents but also among law enforcement officials.[52]

By May 5, 6,678 undocumented immigrants had been arrested, with 2,708 charged and 2,686 sentenced to imprisonment.[53] After three months of arrests, prisons and detention centres became over-crowded, holding 1.5 times their capacity.[54] By August, over 9,000 immigrants remained detained, awaiting resolution of court cases that dragged on because of delays in the legal process.[55] The UNHCR became burdened with a rapidly increasing caseload of arrests. This time, however, because of its preparations for the crackdown, it was better able to negotiate for the release of its persons of concern. Over a six-month period, the UNHCR intervened successfully for more than 1,000 persons of concern.[56] However, as of August 10, 957 persons of concern, most of whom had not yet been documented by the UNHCR at the point of their arrest, remained in detention, with an estimated 160 persons of concern prosecuted in immigration-related offences.[57]

Malaysia was also grappling with problems it did not anticipate. Expecting the swift return of workers after the amnesty period, pro-cedures for documentation and job-matching were unexpectedly slowed down by bureaucracy,[58] producing an acute labor shortage when the crackdown commenced. This led to losses of hundreds of millions of ringgit in different sectors of the economy; the oil palm plantation industry alone lost up to 70 million ringgit (US$18 mil-lion) per month.[59] It was estimated that over 420,000 workers were needed to replace those who left.[60] Unexpectedly, and in direct con-tradiction to its earlier position, Malaysian officials changed policies, publicizing that the country would accept Indonesians on tourist visas into Malaysia, wherein they could apply for work permits.[61] It also turned to other countries for its supply of labor, announcing that it would bring new workers in from Pakistan, India, Myanmar, Nepal, and Vietnam; it would accept 100,000 from Pakistan alone.[62]

In response, the UNHCR urged Malaysia to accept refugees as legal migrant workers.[63] "The ideal solution is for all refugees to be granted temporary stay permits," said UNHCR representative Volker Turk. "Not only will it address the country's labor shortage need, it is also in the humanitarian, economic, and security interest of the Malaysian Government."[64] This suggestion was met favorably by Minister Mohamed Nazri Aziz from the prime minister's depart-ment. "Since we have refugees in the country and most of them are unemployed, why not use them to resolve the labor shortage?" he said. Suggesting that the measure be temporary, he added, "All we need to do is provide them with some kind of identification and get them to fill the jobs."[65] A month later, this decision was confirmed by Azmi Khalid, who stated: "We know that in Malaysia there are refugees registered with the UN refugee agency. Since they are in Malaysia, we will allow them to work. They will be issued with a temporary work permit."[66] This unexpected announcement was welcomed by the UNHCR and transnational civil society groups.

Interviews with Acehnese refugees, however, indicate that immigration officials began a process of documenting Acehnese for work permits at the end of August 2005. These renewable permits allow them to work and to send their children to school. Acehnese held in detention centers have been given these permits and released. It is unclear whether the decision to issue these permits is related to the signing of a peace agreement between leaders of the Free Aceh Movement (GAM) and the government of Indonesia on August 15, 2005, but the timing is convenient: Malaysia cannot be accused of interfering in Indonesia's domestic politics by legalizing Acehnese refugees. Aside from these recent developments, it is yet unclear how or when the Malaysian government will legalize asylum seekers, refugees, and stateless persons of other ethnic and religious groups.

In these two case studies, we see realpolitik at play in the management of Malaysia's borders. Interstate relations are significant; states whose citizens are implicated in Malaysia's border controls have appealed for either their strict application (such as in the August 2003 arrests) or for some flexibility (in the 2005 crackdown), depending on their own political and domestic agendas. The UNHCR, too, has constantly attempted to influence Malaysia's policies on border control. To do so, it has had to reinvent and reposition itself, altering its practices to advocate for protection more effectively. During these events, Malaysia has had its own interests to look after. To fulfill these, it has implemented ad hoc shifts in policy that have left many uncertain about how border practices implicate asylum seekers, refugees, and stateless persons.

PERFORMANCES OF IDENTITY: DISRUPTION AND SUBVERSION

Identity production, creation, and reification are central to the operation of borders. Identities and borders are mutually constitutive; identities are the raison d'être of territorial borders, and borders reinforce the salience of identities. Territorial borders operate on a phantasm that a controllable inside can be delineated from an uncontrollable outside. Control on the inside is premised on the state's ability to know its population. It pursues such knowledge with fanaticism; it institutes practices of documentation, categorization, and aggregation in order to make individuals *legible*; it assigns them *identities*. Undocumented immigrants, as unknown individuals, are seen as a threat. In the case of Malaysia, this threat is managed by punishment and excision: throwing the persons out and warning them to stay outside until they capitulate to being registered and known according to state classifications on the inside. Malaysia's formal (and rigid) classification scheme does not include "asylum seeker," "refugee," or "stateless person" as options. All of these politico-legal identities are subsumed under "immigrant," and immigrants are either legal or illegal depending on the document they hold. It is on the basis of this formal classification scheme that law enforcement agents decide who belongs in the territory of Malaysia and who should be cast out.

In this section, I reflect on disruptions to the functioning of Malaysia's rigid identity classification scheme. The first is a formal act: the UNHCR's attempt to rework this scheme by inserting subcategories that allow a delineation of forced migrants in need of international protection from illegal immigrants. This is inherently an act of identity creation. For this to be successful, the identity has to be *performed* and *supported* so as to produce meaningful distinction. The second disruption is more informal; it occurs when law enforcement agents do not apply the classification scheme. I suggest three causes for this failure; first, self-interest and personal gain through extortion and abuse of powers; second, consideration of the immigrant's other identities (ethnicity, gender, religion, age) based on how their bodies are read; third, genuine uncertainty about what Malaysia's position is on immigrants holding UNHCR documents. These disruptions are part of producing the condition of (il)legality for asylum seekers, refugees, and stateless persons in Malaysia.

Registration and refugee status determination are core activities of the UNHCR in Malaysia, processes that involve interviewing, sorting, categorizing, and documenting persons of concern. A crucial prelude to registration is determining the country of origin of asylum seekers, as well as their ethnic identity. Once the UNHCR has established that an individual is likely to be a genuine asylum seeker, it issues documents to the person. The appearance of these documents has been an important element of protection.

Beginning in 2003, the UNHCR issued temporary protection documentation to Acehnese and Rohingyas. These documents were letters, printed with black ink on white paper, that identified the person as being in need of international protection until circumstances in their country of origin changed. Asylum seekers from all other ethnic groups and countries of origin were given a letter of similar appearance, indicating that they were under consideration for refugee status. Recognized refugees were given a blue letter to carry, a letter with a color photo of the bearer affixed. These letters were to be renewed every six months.

These documents lacked credibility when they were first issued; law enforcement officials frequently tore them up after examining them. They often either told the bearers of the documents that the letters did not have any value *(tak laku)* in Malaysia or accused them of forging the documents. There were multiple reasons for this reaction. The UNHCR's rapidly expanding registration activities created a sudden increase in the number of documents circulated; these documents were viewed with suspicion, and there were no official top-down directives by the Malaysian government to back their validity in Malaysia. Further, many law enforcement officials did not know what the function of the UNHCR was and did not understand what the documents signified; some thought that there were spatial limits to their validity—that they were valid in Kuala Lumpur, for example, but not in Penang. These documents also lacked *face validity*; they did not appear official; they looked more like correspondence than identity documents. Forgeries were also starting to appear.

The UNHCR had to work hard to change these perceptions. At the end of 2004, they modified their documentation processes. Adopting different technology, they began to issue plastic tamper-proof cards with color photo identification, gradually phasing out the white temporary protection letters to Acehnese and Rohingyas and blue refugee status letters to mandate refugees. These cards, which appeared more "professional," were given better recognition by law enforcement agents; those who held the cards reported that they were released more often than when they held the letters.

As stated earlier, the UNHCR also embarked on other activities to strengthen the distinction between their persons of concern and other illegal immigrants. It engaged more frequently with journalists, using the mass media to sensitize the public about its role and about the circumstances faced by those under its protection. Perhaps more significantly, the UNHCR intervened when persons of concern were arrested, thus giving credibility to the documents it issued. Its constant involvement when persons of concern were arrested and its ability to resettle refugees demonstrated to law enforcers that material, and not just discursive, distinction existed for those who held UNHCR documents. The identities have had to be performed with tangible results to create meaning.

Law enforcement officials play a crucial role in interpreting identity and determining the operation of Malaysia's borders in relation to immigrants. As stated earlier, immigration and police officers do not need to have warrants to check for identity documents; they are empowered to question anyone they please. The first reason for subverting Malaysia's classification schemes is self-interest; law enforcement officials often threaten undocumented immigrants with arrest to extort money from them. The amount taken ranges from whatever the immigrants hold when they are confronted (such as cell phones and available monies) to two or three months' salary.[67] Asylum seekers, refugees, and stateless persons have often had to provide bail money for friends and family to secure their release.

The decision whether or not to check immigrants for identity documents depends on how they appear. Bodies are read by law enforcement officials, who guess the origin and legal status of the individual. Those that "look foreign" are frequently asked to show identity documents, while those that "look Malaysian" are not stopped and questioned in the same way. Reports from refugees also suggest that other dimensions of difference (gender, age, religion, and ethnicity) have some influence on how a person is treated at the point of arrest. After the immense destruction caused by the tsunami in December 2004, Acehnese, as *mangsa tsunami* (victims of the tsunami), found more sympathy among law enforcement officers, who would question them about what happened in their homelands and then let them go. Women, children, and babies were sometimes pitied by civilian groups during raids and told to stay quiet while arrests were proceeding. These personal acts of consideration, however, are uncommon; most undocumented immigrants are not spared from arrest.

Those that carry UNHCR documents are cautious of how they conduct themselves with officers of the law. As (il)legals who *may* be given exception if the law enforcement official pleases, they believe that the performance of meekness and respect increases their chances of avoiding arrest. As an Acehnese refugee told me, "We cannot show attitude to the officers. If we are arrogant, if we look them in the eye, if we act too bold, they get angry with us. They shout 'Who do you think you are? You think the UN can protect you? They don't own Malaysia.' We have to be humble, [we have to] duck our heads and be very nice to them."[68] Those who hold documents also try to give reasons for their presence in Malaysia; they explain the conditions under which they fled their homelands and why they sought protection from the UNHCR. Facility with local languages becomes an added asset in these dialogues. The most effective form of securing release (without the aid of the UNHCR), however, remains the ability to pay money.

The continually shifting position of the Malaysian government also creates uncertainties for law enforcement officials. When the announcement was made at the end of 2004 that Rohingyas would be recognized as refugees, Rohingyas stopped by the police would discuss with them the government's position on their status. Similarly, when ministers announced in 2005 that those who held UNHCR documents would be exempt from arrest, Acehnese, Rohingyas, and Chins registered with the UNHCR would argue (politely) that they were legal in Malaysia. Over time, an increasing number of people who held UNHCR documents were not arrested.

(IL)LEGALITY: AN UNSETTLED POSITION

It is inaccurate to say that those documented by the UNHCR are definitely legal. Neither are they clearly illegal. Those who possess identity documents from the UNHCR occupy an indeterminate space, an unsettled socio-legal location on which the operation of Malaysia's borders is unclear. Malaysia recognizes the role of the UNHCR in its territories. In general, Malaysia adheres to the international customary law of non-*refoulement*.[69] It allows UNHCR officers to visit detainees, to intervene on their behalf, and to process them for resettlement. In the past two years, Malaysia has been slowly responding to the requests of the UNHCR to legalize forced migrants, conceding in this regard in order to address acute labor shortages. However, these are partial accommodations in terms of time and scope; they are temporary measures that can be withdrawn at any time; and (so far) they have been extended only to particular ethnic/religious groups. No changes to domestic laws have been made, and no formal legal recognition given to the persons of concern on the basis of being asylum seekers, refugees, or stateless persons, identities attached to specific rights in international law.

Living in indeterminacy is extremely disquieting for those who hold UNHCR documents; personal security remains their primary concern in Malaysia. They do not know whether or not they will be arrested, or, if so, how long they will have to remain in detention. They are not sure what their prospects are for resettlement. The UNHCR has been able to create some space for them to be exempted from normal rules concerning undocumented immigrants. It has done this by unsettling Malaysia's regimes of truth concerning illegal immigrants, by disrupting the premises of its immigration practices, and by challenging Malaysia's legitimacy in punishing and expelling asylum seekers, refugees, and stateless persons. Until Malaysia concedes to more durable solutions, however, the fate of asylum seekers, refugees, and stateless persons remains uncertain.

NOTES

The observations and primary material included in this article is based on work conducted as the refugee affairs coordinator of the National Human Rights Society (HAKAM) in Malaysia. This article was written while undertaking a Southeast Asia Visiting Fellowship at the Refugee Studies Centre at the University of Oxford. I am thankful for the institutional support of both these organizations, and for comments on an earlier draft from Dr. Eva-Lotta Hedman. The views expressed here, however, remain my own.

1. A notable exception is that of deportations to Thailand, a matter I elaborate on later.

2. Singapore was included briefly, but became an independent state in 1965.

3. This stood at 10,541.7 million people in the second quarter of 2005, with a low unemployment rate of 3.1 percent. Department of Statistics Malaysia, "Key Statistics," September 29, 2005. http://www.statistics.gov .my/english/frameset_keystats.php (accessed November 15, 2005).

4. Economic immigrants from Flores (Indonesia) have told me that

they lost large sums of money to agents who were supposed to provide them with necessary documents but absconded, leaving them with debts they could only pay by working illegally in Malaysia (Flores immigrants, group interview by author, Selangor, June 22, 2005).

5. LKBN Antara, "Malaysia to Resume Deportation of Illegal Workers," September 22, 2004.

6. V. Vasudevan, "RM9.4 Million to Upkeep 14 Immigrants Depots Last Year," *Malay Mail*, November 30, 2004. The department was responsible for four more immigration depots in Sabah and Sarawak.

7. Agence France Presse, "18,000 Illegal Immigrants Whipped in Malaysia," August 16, 2004.

8. UNHCR, 2004, *Country Operations Plan: Executive Committee Summary. Country: Malaysia, Planning Year: 2005*.

9. UNHCR, *Statistics Provided to NGOs* (UNHCR, 2004: *Active Cases Breakdown* as of December 31, 2004).

10. UNHCR, *Statistics Provided to NGOs* (UNHCR, 2005, *Active Cases Breakdown*, as of July 31, 2005.)

11. Ibid. The largest numbers of these are from other Asian countries: Sri Lanka (331 persons), Cambodia (343), Nepal (186), and Bangledesh (172). There are also asylum seekers from Middle Eastern and African countries.

12. These procedures have been altered several times for different groups of asylum seekers over the past two years, as the UNHCR struggles to find the most efficacious method of registering them safely.

13. Malaysia is, however, a state party to the Convention on the Rights of the Child, and is obligated to observe Article 22, which relates to refugee children specifically. Although it initially expressed a reservation to this article, this was subsequently withdrawn, following protest by other governments. For further elaboration of Malaysia's obligations, see Amer (2005).

14. Immigration Act of 1959/63, Sections 5 and 6.

15. Ibid., Section 6(4).

16. Ibid., Section 6(3).

17. Ibid., Section 15(1).

18. Ibid., Section 15(4).

19. Ibid., Section 51(3a).

20. This is in sharp contrast to Malaysian citizens, who have to be produced before a magistrate within twenty-four hours and cannot be detained for longer without his/her authorization (Immigration Act of 1959/63, Section 51(5a)).

21. In 2004, the UNHCR estimated an average monthly case load of 250 to 350 persons of concern in detention (UNHCR-NGO meeting, July 24, 2004). This, however, does not inform us of the numbers of new entrants per month, or of the numbers who left detention by opting for deportation, under duress. NGOs were only able to track several incidents of arrests. See,

for example, "Urgent Alert—Malaysia: Burmese Asylum Seekers Targeted in a Series of Arrests," SUARAM, April 14, 2004; "Malaysia: Urgent Alert—23 Burmese Detained and Face Charges of Illegal Assembly," SUARAM, May 19, 2004; "Malaysia: Urgent Alert—Mass Arrest of Achehnese Asylum Seekers at Bukit Jambul, Penang," HAKAM, June 18, 2004; "Urgent Alert—Acehnese Refugee Children Arrested," HAKAM, July 17, 2004; "Urgent Alert—Mass Arrest of Asylum Seekers at Selayang," July 27. Also see Malaysiakini, "NGO Upset by Arrest of Acehnese Refugees," June 2, 2004.

22. Asylum seekers and refugees from Myanmar inform me that, prior to 2003, they were merely left at the Thai border, typically at Golok River. In 2003, however, the Malaysian immigration authorities began to hand them over to Thai immigration authorities, who would continue their deportation through Ranong to Myanmar.

23. These conditions have been independently verified by the National Human Rights Commission of Malaysia (SUHAKAM) and NGOs. See for example, SUHAKAM, "Keadaan di Depot Tahanan Semenyih Perlu di Ambil Perhatian," September 8, 2004; SUHAKAM, "Prisons Department Should Take Over Management of More Immigration Depots," September 22, 2003.

24. Acehnese refugees, interview by author, Kuala Lumpur, August 1, 2004; Chin refugees, interview by author, March 28, 2005; Burmese asylum seekers, interview by author, July 16, 2005. See also *Tenaganita*, "Press Statement: Release All Refugees Held in Detention Camps in the Country and Stop Deportation of Refugees," September 2, 2005.

25. During the five-month amnesty period in 2002, for example, 450,000 opted to go home ("Deportations Put on Hold," *New Straits Times*, August 18, 2004); in the amnesty period of 2004 to 2005, between 385,000 and 400,000 immigrants left Malaysia.

26. "VP Denies Govt Not Paying Enough Attention to Workers' Plight," *Asia Pulse*, September 4, 2002.

27. Bernama, "Conditions in Indon Illegal Workers' Shelter in Nunukan Worsens," September 3, 2002. See also "Veep to Visit Indon Workers in Nunukan, E Kalimantan," *Asia Pulse*, September 1, 2002. However, the Indonesian government disputed these numbers, claiming only twenty-nine had died (Organization of Asia-Pacific News Agencies, September 3, 2002): "Only 29 Indon Workers Died in Nunukan, Says Minister."

28. "Indonesians Expelled from Malaysia in 'Terrible' Condition: Red Cross," Agence France Presse, March 3, 2005.

29. The Philippines government protested Malaysia's treatment of its citizens during the 2002 crackdown, after reports surfaced of the deaths of twelve children and the sexual abuse of young women by police and prison guards in detention camps. Mark Baker, "Filipino Outrage as Deaths of 12 Child Detainees Revealed," *Sydney Morning Herald*, September 6, 2002.

30. A UNHCR official later confirmed that one detainee was a recognized refugee, fifty-one held protection documents, and another forty-five carried tokens requiring them to return to the UNHCR office for registration. The remainder were approaching the UNHCR for the first time. "Police Presence Forces UNHCR to Cease Operations," http://www .malaysiakini.com/news/16770, August 20, 2003.

31. "Police Swoop on Acehnese Immigrants, 240 Arrested," http:// www.malaysiakini.com/news/16752, August 19, 2003.

32. "Malaysia Rethinks Deportation of 250 Acehnese," UN Wire, August 21, 2003. See also "Gov't Mulls Temporary Stay for 250 Detained Acehnese," Agence France Press, August 21, 2003.

33. Indoleft News, "Asylum Seekers from Aceh Will Still Be Deported," August 26, 2003. See also "Malaysia Takes Hard Line on Acehnese Refugee Bid," Agence France Press, August 25, 2003.

34. "Indonesia's Top Minister Urges M'sia to Reject Acehnese Asylum Seekers," Agence France Press, August 25, 2003.

35. "50 More Asylum Seekers Arrested in KL," http://www.malaysiakini .com/news/16839, August 26, 2003.

36. Ibid.

37. "No Asylum for Acehnese, Says PM," http://www.malaysiakini .com/news/16883, August 28, 2003.

38. "Malaysia and Indonesia Agree to Step Up Cooperation on Terrorism," Agence France Press, August 28, 2003.

39. For information on events in Aceh during martial law, see Eva-Lotta E. Hedman, ed., Aceh under Martial Law: Conflict, Violence, and Displacement, 2005, RSC Working Paper no. 24, Queen Elizabeth House, Department of International Development, University of Oxford. See online at http://www.rsc.ox.ac.uk/PDFs/workingpaper24.pdf.

40. Country Operations Plan: Executive Committee Summary, Country: Malaysia, Planning Year: 2005, UNHCR, 2004.

41. "Malaysia Vows to Punish Illegal Migrants before Deportation: Report," Agence France Presse, August 11, 2004.

42. Farrah Naz Karim, "Deportation Put on Hold," New Straits Times, August 19, 2004.

43. The tsunami caused the death of over 126,600 people (with 93,600 individuals missing) and triggered the displacement of more than 514,000 (Government of Indonesia, "Master Plan for the Rehabilitation and Reconstruction of the Regions and Communities of the Province of Naggroe Aceh Darussalam and the Islands of Nias, Province of North Sumatra," Relief Web, April 12, 2005). Three weeks after the tsunami, the UNHCR released an information note appealing to governments to suspend involuntary returns to countries affected by the tsunami, including the province of Aceh in Indonesia (UNHCR, "Information Note: Request for the Suspen-

sion of Forced Returns to Areas Affected by the Tsunami," January 12, 2005). See also Nah and Bunnell (2005) for a commentary on the effects of the tsunami on Acehnese refugees in Malaysia.

44. There were also "economic migrants" who did not go back during the amnesty, either because they could not afford to pay for the travel back and the processing fee for reentry or because their wages had been withheld from them (Flores immigrants, interview by author, Putrajaya, July 31, 2005).

45. Claudia Theophilus, "Daily Says Malaysian Minister Confirms Refugee Status of Burma's Rohingyas," http://www.malaysiakini.com, November 2, 2004. The decision by the government to legalize these refugees is likely a result of lobbying efforts by the UNHCR and civil society groups (including Rohingyas), as well as of recommendations made by the National Human Rights Commission (SUHAKAM).

46. "Update on UNHCR Operations in Asia and the Pacific," Executive Committee, UNHCR, 2005.

47. "UN Combs Malaysian Jungles for Refugees ahead of Crackdown," Agence France Presse, December 1, 2004.

48. "Ops Tegas: More Than 4,000 Detained," *Bernama Daily Malaysian News,* March 30, 2005.

49. Hamidah Atan and Ahmad Fairuz Othman, "Illegals from Aceh to Be Spared Detention," *New Straits Times,* March 3, 2005.

50. "Malaysia Arrests 563 Illegal Immigrants: To Intensify Crackdown—Minister," *AFX News,* March 2, 2005.

51. "Illegal Immigrants: None Will Be Spared from Ops Tegas," Malaysia *Star,* March 4, 2005.

52. Disagreement also occurs between law enforcement agents. In one incident, a Mon refugee holding UNHCR documentation was detained during a joint operation near his home. The RELA member who stopped him argued at length with an immigration official as to whether he should be detained. He was subsequently released (Mon refugee, interview by the author, July 12, 2005).

53. "Ops Tegas: Is It Moving in the Right Direction?" *Bernama Daily Malaysian News,* May 11, 2005.

54. "Malaysian Jails Overcrowded after Migrant Crackdown," Agence France Presse, May 25, 2005. The Prisons Department confirmed that the fifty-two prisons and detention centers in Malaysia, meant to house 30,000 inmates, were holding 45,000 people, most of whom were illegal immigrants.

55. "Report: Malaysia Will Try and Whip Illegal Aliens despite Swelling Backlog in Detention," Associated Press Newswires, August 14, 2005.

56. Information provided at a UNHCR-NGO meeting on August 24, 2005. Also reported in UNHCR Executive Committee, "Update on UNHCR Operations in Asia and the Pacific," 2005.

57. UNHCR Executive Committee, "Update on UNHCR Operations in Asia and the Pacific," 2005.

58. "Slow Return of Workers to Malaysia as Indonesia Waits for Job Orders," Organization of Asia-Pacific News Agencies, March 23, 2005.

59. "Expulsion of Illegal Workers Hits Malaysian Economy," Agence France Presse, May 11, 2005.

60. Annie Freeda Cruez, "Acute Labour Pains," *New Straits Times,* March 17, 2005.

61. "Rules Eased for Foreign Labour," SBS World News Headline Stories, May 27, 2005; Farrah Naz Karim, "Entry Eased for Former Illegals," *New Straits Times,* May 26, 2005.

62. Jasbant Singh, "Malaysia Seeks Migrant Workers from India, Other Nations to Curb Labor Woes," Associated Press Newswires, April 4, 2005. "Malaysia Widens Recruitment for Foreign Workers," Agence France Presse, April 4, 2005.

63. "UN Urges Malaysia to Allow Refugees to Work amid Labor Shortage," Agence France Presse, June 13, 2005.

64. Jessica Lim, "All They Want Is . . . REFUGE," *New Sunday Times,* June 19, 2005.

65. Suat Ling Chok, Jason Gerald, and Ranjeetha Pakiam, "Refugees May Be Given Jobs," *New Straits Times,* June 22, 2005.

66. "Malaysia to Allow Thousands of Refugees to Work to Solve Labor Crunch," Agence France Presse, July 5, 2005.

67. The high levels of corruption among law enforcement officials have already been reported. For a summary of issues related to the police specifically, see the *Report of the Royal Commission to Enhance the Operation and Management of Royal Malaysia Police (2005), Kuala Lumpur.* In relation to corruption and extortion of undocumented immigrants, see Jones (2000, 1996) and Human Rights Watch (2004, 2000).

68. Acehnese refugee, interview by author, Klang Valley, August 18, 2005.

69. That is, those documented by the UNHCR are not forcefully deported, even if they are detained for long periods in detention centers and prisons.

WORKS CITED

Amer, Hamzah Arshad. 2005. "Malaysia's Forgotten Children." *Aliran Monthly* 25, no. 5. http://www.aliran.com/monthly/2005a/5b.html (accessed November 15, 2005).

Brochmann, Grete. 1999. "Mechanisms of Control." In *Mechanisms of*

Immigration Control: A Comparative Analysis of European Regulation Policies, ed. Grete Brochmann and Tomas Hammar. Oxford, U.K.: Berg.

Bunnell, Tim. 2002. "(Re)Positioning Malaysia: High-Tech Networks and the Multicultural Rescripting of National Identity." *Political Geography* 21, no. 1: 105–24.

Hugo, Graeme. 1991. "Recent International Migration Trends in Asia: Some Implications for Australia." In *Immigration, Population, and Sustainable Environments*, ed. J. W. Smith. Adelaide: Flinders Press.

———. 2000. "Living in Limbo: Burmese Rohingyas in Malaysia." http://www.hrw.org/reports/2000/malaysia/ (accessed November 15, 2005).

———. 2002. "Introduction." *Migration and the Labour Market in Asia: Recent Trends and Policies*. Paris: OECD, 7–16.

Human Rights Watch. 2004. "Aceh under Martial Law: Problems Faced by Acehnese Refugees in Malaysia." http://hrw.org/reports/2004/malaysia0404/index.htm (accessed November 15, 2005).

International Labor Organization. 1996. "Hope and Tragedy for Migrants in Malaysia." *Asia Pacific Magazine* 1: 23–27. http://coombs.anu.edu.au/SpecialProj/APM/TXT/jones-s-01-96.html (accessed November 15, 2005).

———. 1998. *The Social Impact of the Asian Financial Crisis*. Bangkok: Regional Office for Asia and the Pacific of the International Labor Organization.

Iredale, Robyn, Kalika N. Doloswala, Tasneem Siddiqui, and Riwanto Tirtosudarmo. 2003. "International Labour Migration in Asia: Trends, Characteristics, Policy, and Interstate Cooperation." In *Labour Migration in Asia: Trends, Challenges, and Policy Responses in Countries of Origin*. Geneva: International Organisation for Migration.

Jones, Sidney. 2000. "Making Money Off Migrants: The Indonesian Exodus to Malaysia." Sydney, Australia: Center for Asia Pacific Social Transformation Studies, University of Wollongong.

Nah, Alice M., and Tim Bunnell. 2005. "Ripples of Hope: Acehnese Refugees in Post-Tsunami Malaysia." *Singapore Journal of Tropical Geography* 26, no. 2: 249–56.

Spaan, Ernst. 1994. "Taikongs and Calos: The Role of Middlemen and Brokers in Javanese International Migration." *International Migration Review* 28, no. 1: 93.

SUHAKAM. 2004. *Economic, Social, and Cultural Rights Report: Trafficking in Women and Children*. http://www.suhakam.org.my/en/document_resource/details.asp?id=73 (accessed November 15, 2005).

Wong, Diana, and Gusni Saat. 2002. "Trafficking of Filipino Women to Malaysia: Examining the Experiences and Perspectives of Victims, Governmental and NGO Experts, United Nations Global Programme against Trafficking in Human Beings." United Nations Office on Drugs and Crime. http://www.unodc.org/pdf/crime/human_trafficking/Exec_Summary_IKMAS.pdf (accessed November 15, 2005).

3

The Foreigner in the Security Continuum: Judicial Resistance in the United Kingdom

ELSPETH GUILD

In examining the development of the practice and theory of surveillance at distance, Didier Bigo finds a transnational field of security where internal and external security become a continuum
in which the worlds of the police and military find themselves in
competition (Bigo 2005, 129–60). In this chapter, I build on his
work in particular as regards the construction of the enemy who is
both external and internal at the same time and thus requires the
compensating measures of the state in the development of a security continuum. However, I will examine this continuum from the
perspective of legal challenges to the merging of what have been
two distinct fields. The field of law, the interpretation of which is
the domain of a powerful group of experts, lawyers and judges, has
much to lose from the development of the continuum. In Europe,
the United Kingdom in particular, courts have traditionally been
reluctant to claim jurisdiction over the legality of state action in the
field of foreign affairs. The issue of national security has also been
viewed by the U.K. courts as part of a world subject only to a light
legal scrutiny. But the merging of the worlds has coincided with an
increasing judicial reconsideration of the role of law in the area. This
is particularly so where the security professionals' internal-external
continuum starts to affect fields inside the state within which these

other professionals of security (this time the security of rule of law), the judges, have traditionally held a strong grip on legality.

Where a continuum is established between internal and external security, state authorities have sought to argue that the legal rules that apply to external security should apply, and that the courts should acknowledge their lack of competence in fields involving foreign and security policy. The reaction of the courts has not been uniformly in favor of such a renunciation of judicial control of administrative action. The tool deployed has been the duty of the courts to apply international human rights standards, in particular the European Convention on Human Rights. This duty was imposed on the courts by Parliament itself only in 1998, with the Human Rights Act. This legislative move has transformed the nature of judicial control in the United Kingdom through the insertion, at the heart of the jurisdiction of the courts, of a meta-level. It is beyond the control of the national yet clearly defined in law—unlike international obligations, which suffer from too great a distance between the national judicial authority and the international norm and (quasi-) enforcement mechanisms. One response to Bigo's security continuum is a judicial continuum that counters the apparent removal of fields of action from the supervision of the courts, which continuum depends on the meta-level of European legal norms that the state cannot control (at least, not directly).

JUDICIAL DEFERENCE AND PRE-SEPTEMBER 11, 2001, JUDICIAL APPROACHES

On February 9, 1993, Ur Rehman, a national of Pakistan, arrived in the United Kingdom, having been granted a visa as a minister of religion to work at a mosque in Oldham.[1] His father was also a minister of religion in Halifax, and both his father and mother were British citizens. Two of his children were born in the United Kingdom. He and his family remained resident in Oldham where he worked. In December 1998, he applied for indefinite leave to remain (a long-stay residence status). This was refused and deportation proceedings were commenced against him. The grounds for the refusal (as quoted in the judgment) were: "the Secretary of State is satisfied, on the basis of information he has received from confidential sources that you are involved with an Islamic terrorist organization. . . . He is satisfied that in the light of your association with the [organiza-

tion] it is undesirable to permit you to remain and that your continued presence in this country represents a danger to national security." The matter was ultimately determined by the House of Lords
(the final judicial instance in the United Kingdom).[2]

The charges against Rehman bear mention, not least in the
light of the autumn 2001 bombing campaign against the Taliban
in Afghanistan: recruitment of British Muslims to undergo militant
training; fund-raising for a terrorism organization; sponsoring individuals for militant training camps; responsibility for the existence
in the United Kingdom of British Muslims returned from the militant camps who had been indoctrinated with extremist beliefs or
given weapons training.

The interest of the individual foreigner is security of residence;
against this interest is the interest of the state to assure national
security. The balance found between these interests expresses the
division between the foreigner and the citizen. The interest of the
citizen is such that his or her security of residence on the territory
cannot be brought into play.[3] The right of entry and residence on
the territory for citizens is established not only in national law but
also in numerous human rights conventions, not least the European
Convention on Human Rights.[4] The interest of foreigners in security
of residence is much more tenuous, a field of negotiation in national
law between the benefits of integration of all persons resident on the
territory in the interests of social harmony and tolerance, and the
division between those who belong and those who are excludable
(Nascimbene 2001, 139).

This dividing line between the individual as a foreigner and
the state has been the subject of an increasing number of judgments of the European Court of Human Rights from 1993 onward
(Groenendijk 2001, 15). The development by that court of the concept of an integrated alien against whom expulsion can only be justified on very substantial grounds has been resisted by a number of
Council of Europe countries, as evidenced by the continuing stream
of cases pending before the European Court of Human Rights. The
possibility that the Court of Human Rights may be moving toward
a position of prohibition of expulsion of long-resident foreigners, as
promoted by Professor Henry Schermers in his partly concurring,
partly dissenting opinion in *Lamguindaz*,[5] seems to be receding,
though the position of enhanced protection remains. The balancing

of the interests of the individual and the state even in the light of substantial criminal activity by the foreigner does not, according to the Human Rights Court, necessarily come down in favor of the state security interest.[6]

What happens when the state's interests are enhanced by reason of national security arguments? How is the balance with regard to the interests of the individual changed? When the label of *terrorism* is added, a number of results flow. Where the charge of terrorism is against a national of the state, it arises in respect to a criminal offense. The offense is specific and regulated by the constitutional settlement between the rulers and the ruled, which is contained in criminal law. The Terrorism Act of 2000 was controversial in Parliament not least because of the changes to the balance of the rights of the defense and the rights of the state that flow from the addition of the characteristic of "terrorism" to what would otherwise be a criminal offense subject to the normal rules.

When the individual subject to the suspicion of terrorism is a foreigner, another option is open to the state: expulsion. Here the proceedings are of an administrative kind and not subject to the need for a criminal charge to which to add the terrorism sobriquet. Being a foreigner is sufficient to allow expulsion measures to be taken. As Lord Slynn put the issue in *Rehman*:

> Here the liberty of the person and the practice of his family to remain in this country is at stake and when specific acts which have already occurred are relied on, fairness requires that they should be proved to the civil standard of proof. But this is not the whole exercise. The Secretary of State, in deciding whether it is conducive to the public good that a person should be deported, is entitled to have regard to all the information in his possession about the actual and potential activities and the connections of the person concerned. He is entitled to have regard to the precautionary and preventive principles rather than to wait until directly harmful activities have taken place, the individual in the meantime remaining in this country. In doing so he is not merely finding facts but forming an executive judgement or assessment. There must be material on which proportionately and reasonably he can conclude that there is a real possibility of activities harmful to national security but he does not have to be satisfied, nor on appeal to show, that all the material before him is proved, and his conclusion is justified, to a "high civil degree of probability." (paragraph 22)

The logic at work is that of borders. This is so in two senses. First, borders are a protection: once someone is put on the other side of a border, the state or collectivity is safer. Scond, borders are the dividing line of legal orders. Only those fully within the legal order, i.e., citizens, can enjoy the benefits of the legal order for which the border forms the outer limit. Even a "high civil standard of probability" to protect the security of residence of the foreigner is rejected in favor of an ordinary civil standard of proof. The safety of borders around legal orders is for those who cannot be expelled. The civil liberties protections of citizens may be diminished where the allegation of terrorism is added to a charge of a criminal offense, but the standard of proof is not similarly diminished. The underlying framework of criminal law provides a resistant structure against which the terrorism label operates. The integrity of borders lies also in their effectiveness at providing separation. By placing an individual on the other side of a border, it is not necessarily self-evident that a state or community's security is increased.

The difference in the level of legal protection of the status of a foreigner, as opposed to that of a national, where alleged to be engaging in acts likely to compromise national security is found not least in the standard of proof that the state must satisfy. As Lord Hoffmann points out in *Rehman:* "The civil standard of proof always means more likely than not. The only higher degree of probability required by the law is the criminal standard" (paragraph 55). Of course, foreigners may be subject to criminal charges involving questions of terrorism, in which case the standard of proof is the same for them and for nationals of the state. In light of the decision of the first instance court in *Rehman* on the facts (which I shall discuss later), it is evident that the state would have had grave difficulties seeking to satisfy the higher criminal standard of proof even if appropriate charges could have been found for a criminal prosecution of Rehman. The preference to rely on the logic of borders and exclusion lies not least in the disapplication of the civil liberties protections that would otherwise apply.

The United Kingdom has some experience with the issue of the borders of the legal order and national security as regards foreigners. In 1996 the European Court of Human Rights handed down judgment against the United Kingdom as regards the proposed expulsion of Karamjit Singh Chahal, an Indian national, to his country of origin on national security grounds ([1996] 23 EHRR 413). Chahal, who was resident in the United Kingdom, had sought political asylum in the United Kingdom on the grounds that, if he were returned to India, as a suspect of Kashmiri terrorism, he would be subjected to torture. The secretary of state rejected his claim and sought to deport him on the basis that this would be conducive to the public good on national security grounds. In domestic law, no appeal, except in an advisory procedure, was permitted against a decision of deportation based on national security. Because there was no judicial remedy at the national level (only an advisory mechanism), Chahal had no domestic venue in which to put his claim to protection against expulsion notwithstanding the allegation of a threat to national security. Accordingly, the meta-level was engaged. Chahal petitioned the European Court of Human Rights (ECtHR), claiming that his proposed expulsion to India would result in a substantial risk that he would be subjected to torture contrary to Article 3 of the European Convention on Human Rights (ECHR). The finding of this supranational human rights court on the terrorism issue and the right of access to a judge would prove decisive also in the *Rehman* case four years later.

The European Court of Human Rights held that the prohibition on torture contained in Article 3 of the European Convention on Human Rights was absolute and in no case could a signatory state return an individual to a country where there was a serious risk that he or she would suffer torture. This prohibition is absolute and applies even where a question of national security is raised. Second, as regards the effectiveness of the remedy of an advisory procedure, Lord Slynn points out in *Rehman:* "This [advisory procedure] however, was held by the European Court of Human Rights . . . not to provide an effective remedy within section 13 of the [European Convention on Human Rights]. Accordingly, the [Special Immigration Appeals] Commission was set up by the 1997 Act and by subsection 2(1)(c) a person was given a right of appeal to the Commission" (paragraph 9). Lord Hoffmann, also commenting on the *Chahal* judgment, noted that the European Court of Human Rights had also held that "if [an individual] was detained pending deportation, he was entitled under Article 5(4) [European Convention on Human Rights] to the determination of an independent tribunal as to whether his detention was lawful. The European court rejected the United Kingdom

Government's argument that considerations of national security or international relations made it impossible to accord such a right of appeal. The court . . . commended the procedure established by the Canadian Immigration Act 1976, under which the confidentiality of secret sources could be maintained by disclosing it only to a special security-cleared advocate appointed to represent the deportee who could cross examine witnesses in the absence of the appellant" (paragraph 36). Such a system was established by the Special Immigration Appeals Commission Act of 1997.

Rehman appealed to the Special Immigration Appeals Commission against the secretary of state's intention to deport him on grounds of national security in accordance with the legislation introduced following the European Court of Human Rights decision in *Chahal*. The commission reviewed the evidence and information provided to it in accordance with its procedural rules (which do not permit sensitive information to be made available to the appellant). It held:

1. Recruitment. We are not satisfied that the appellant has been shown to have recruited British Muslims to undergo militant training as alleged;

2. We are not satisfied that the appellant has been shown to have engaged in fund-raising for the [terrorist organization] as alleged;

3. We are not satisfied that the appellant has been shown to have knowingly sponsored individuals for militant training camps as alleged;

4. We are not satisfied that the evidence demonstrates the existence in the United Kingdom of returnees, originally recruited by the appellant, who during the course of that training overseas have been indoctrinated with extremist beliefs or given weapons training, and who as a result allow them to create a threat to the United Kingdom's national security in the future. (paragraph 4)

The importance of rights of appeal to a judicial instance is evident here. The balancing of security interests between the state and the individual in this most sensitive of areas must be reviewed externally and independently. The assessment of the state as to the requirements of national security is not necessarily shared by judicial instances. In *Rehman* for the first time the issue of the content of national security was addressed. The government argued that the definition of national security and what could constitute a threat to it was a matter for the Home Secretary to decide (in paragraph 3).

The commission rejected this argument, holding that the definition was a question of law that it had jurisdiction to decide. This point was one of the key legal questions on which the state appealed the commission's decision.

What then is national security? Before one can determine how it can be protected, its contours must be understood. There is no statutory definition of national security. But being a danger to national security is a ground for deportation. In *Rehman,* counsel for Rehman argued that national security must be understood within the meaning of the job with which the security services have been charged under the Security Services Act 1989. The logic here is that the duties of the security services to protect national security must be coterminous with the state's power to take measures on grounds of national security. Section 1(2) Security Services Act 1989 states the duties are "the protection of national security and in particular, its protection against threats from espionage, terrorism and sabotage, from the activities of agents of foreign powers and from actions intended to overthrow or undermine parliamentary democracy by political, industrial or violent means" (paragraph 14).

Although not expressly rejected, this argument was left unanswered. Instead, Lord Hoffmann stated: "There is no difficulty about what 'national security' means. It is security of the United Kingdom and its people" (paragraph 50). Here a question of substantial interest to me is posed (and is never addressed in the judgment): who are the United Kingdom's people? When and to what extent can Rehman become or be one of the United Kingdom's people? The people are those entitled to the protection of the borders, including as expressed in the legal order. Chahal was excluded most fully from the social settlement of the legal order, in that an allegation of national security risk against him did not give rise even to the rudimentary elements of judicial control of administrative action. This exclusion was held incompatible, inter alia, with Article 13 of the European Convention on Human Rights.

European human rights norms imposed on the United Kingdom a transformation of the concept of "its people." The United Kingdom was not entitled to exclude entirely from judicial scrutiny, on grounds of national security, a class of persons on the basis of their nationality. The inclusion of even a limited form of judicial scrutiny, though the commission, results in a substantially different appreciation of who "its people" are. The assessment of national security includes, albeit at a limited level, Rehman and his interest in security of residence. He has slipped inside the border of the UK legal order. In the words of Lord Slynn, "It seems to me that the appellant is entitled to say that 'the interests of national security' cannot be used to justify any reason the Secretary of State has for wishing to deport an individual from the United Kingdom."

Rehman has arrived at the edges of the "UK's people." Lord Hoffmann went on to state that, "On the other hand, the question of whether something is 'in the interests' of national security is not a question of law. It is a matter of judgment and policy. Under the constitution of the United Kingdom and most other countries, decisions as to whether something is or is not in the interests of national security are not a matter for judicial decision. They are entrusted to the executive" (paragraph 50). Lord Slynn seems to accept more readily that Rehman is somewhere within the legal borders: "There must be some possibility of risk or danger to the security or well being of the nation which the Secretary of State considers makes it desirable for the public good that the individual should be deported." The next question then is how direct or indirect the threat must be.

The commission adopted a narrow approach to this question: "A person may be said to offend against national security if he engages in, promotes, or encourages violent activity which is targeted at the United Kingdom, its system of government or its people. This includes activities directed against the overthrow or destabilisation of a foreign government if that foreign government is likely to take reprisals against the United Kingdom which affect the security of the United Kingdom or of its nationals. National security extends also to situations where United Kingdom citizens are targeted, wherever they may be" (paragraph 2). Here the borders are clearly those of citizenship. The security of Rehman or other integrated foreigners in the United Kingdom is not the subject of national security except as a by-product of the more general security of the United Kingdom. Abroad, the United Kingdom's national security is engaged if British citizens are targeted, though there is no responsibility for integrated foreigners. The assumption is that the latter remain the responsibility of their state of nationality.

The issue of the narrow or wider conception of the risk to

national security formed the subject of substantial discussion, in the House of Lord's judgment. In the words of Lord Slynn:

> It seems to me that, in contemporary world conditions, action against a foreign state may be capable indirectly of affecting the security of the United Kingdom. The means open to terrorists both in attacking another state and attacking international or global activity by the community of nations, whatever the objectives of the terrorist, may well be capable of reflecting on the safety and well-being of the United Kingdom or its citizens. . . . I accept that there must be a real possibility of an adverse effect on the United Kingdom for what is done by the individual under inquiry but I do not accept that it has to be direct or immediate. Whether there is such a real possibility is a matter which has to be weighed up by the Secretary of State and balanced against the possible injustice to that individual if a deportation order is made. (paragraph 17)

The balance between the state and the individual, where the interests of national security and security of residence conflict, has now been established as the threshold of a real possibility of an adverse effect on the state. Even though there is permitted a wide assessment of the meaning of an adverse effect permitting action against foreign states, nonetheless the test that must be satisfied is one of a real possibility. Rehman's interest in security of residence cannot be extinguished at a level lower than a real possibility of an adverse effect, even though he is not, or perhaps not yet, directly an intended subject of the protection of national security. However, the borders within which the security interest may be assessed have been enlarged beyond those of the United Kingdom or indeed of its citizens. They now encompass indirect threats, attacks on other states, or the international or global activity of the community of nations.

Two of the judges in *Rehman* made specific reference to the attacks in the United States of America of September 11, 2001 (Lords Steyn and Hoffmann) (paragraphs 29 and 62). Both stated that, although they had reached their decisions before the events, those attacks confirmed their opinions. In both cases, they indicate that the judiciary must respect the decision of the government regarding the evaluation of threats to national security. Lord Slynn's measured approach to the balance of interests and the role of the judiciary in achieving that balance does not refer to terrorist threats or acts outside the allegations relevant to the case itself.

AFTER SEPTEMBER 11, 2001:
FOREIGNERS, JUDGES, AND EXCEPTIONS

The final UK judgment on the indefinite detention of foreigners (which I described in some detail in Guild 2003, 491–515) transforms the nature of judicial control over the security continuum yet again.[7] The challenge arises from part 4 of the Anti-Terrorism, Crime and Security Act of 2001, passed on December 14, 2001, as the United Kingdom's main response to the September 11, 2001, attacks in the United States. Part 4 of the act permits the secretary of state to designate foreign nationals as suspected international terrorists. The result of such a designation is that they may be detained until they choose to leave the United Kingdom, a rather impossible choice for those of them who fear torture and persecution in their countries of origin. Nine men (eight detained in December 2001 and one in February 2002) appealed against their detention. One was released on bail with strict conditions in April 2004, and another was released without conditions in September 2004. Two left the United Kingdom voluntarily—one to Morocco on December 22, 2001, and one to France (of which he was a national) on March 13, 2002. In the months following the House of Lords' judgment of December 2004, the rest were released subject to new and very strong surveillance measures, entitled control orders, made possible under new legislation passed to replace the 2001 act. However, all were retained under Immigration Act powers this time. (I will return to this matter at the end of this chapter.)

The men, mainly from North Africa (none from Iraq or Afghanistan), challenged their detention on the grounds that the statute under which they had been detained failed to comply with the European Convention on Human Rights' guarantee of the right of liberty of the person in conjunction with the right to nondiscrimination. At first instance, before the Special Immigration Appeal Commission (SIAC), their appeal was allowed on the basis that the legislation discriminated, contrary to the duty of nondiscrimination (Article 14 of the European Convention on Human Rights), in conjunction with the right to liberty (Article 5 of the European Convention on Human Rights), from which the United Kingdom had derogated for the purposes of passing the legislation. The main thrust of the SIAC judgment was that, as such legislation could not be applied to British citizens, it was inconsistent with the European

Convention on Human Rights to apply it exclusively to foreigners, many of whom, it appeared, were suspected of being international terrorists as a result of their association with British citizens. The Court of Appeal very rapidly reversed this finding and held that the position of foreigners and that of citizens could not be considered comparable for the purposes of the application of the rule against discrimination.

The House of Lords, again relying on Article 14 of the European Convention on Human Rights (in conjunction with Article 5), found that the legislation offended against the European Convention on Human Rights and issued a declaration of incompatibility.[8] The European Convention on Human Rights protects the right of liberty of the person in Article 5. That article does provide for an exception to the strict rule of liberty of the person where the detention is for the purpose of expulsion or to prevent an unauthorized entry into the state. However, as the men involved could not be expelled on account of the risk of torture or persecution, the exception was not effective. It was for this reason that the UK government had derogated from Article 5 of the European Convention on Human Rights for the purposes of passing the legislation. The procedure for derogation is set out in Article 15 of the European Convention on Human Rights, and is only permitted in respect of certain provisions of the European Convention on Human Rights (for instance, the prohibition on torture cannot be the subject of a derogation). According to Article 15, derogation is only possible "in time of war or other public emergency threatening the life of the nation." The measures that a state may take in such circumstances to derogate must not exceed the extent strictly required by the exigencies of the situation, and provided that such measures are not inconsistent with the state's other obligations under international law.

The majority of House of Lords' judges in the case accepted that it was for the UK government to determine when a state of public emergency threatening the life of the nation exists. Judicial control over this assessment would be light. In the lead judgment, Lord Bingham gave three reasons for refusing to interfere with the decision on the state of emergency: (1) the SIAC saw confidential material not available to the other courts, and on that basis was satisfied that the government was justified in declaring a state of emergency, and so the House of Lords should not lightly interfere with that finding; (2) the

European Court of Human Rights permits a wide margin of appreciation on this issue to national authorities (and the judge reviewed all the European Court of Human Rights jurisprudence on the point); (3) the political nature of the decision must be considered:

> The more purely political (in a broad or narrow sense) a question is, the more appropriate it will be for political resolution and the less likely it is to be an appropriate matter for judicial decision. The smaller, therefore, will be the potential role of the court. It is the function of political and not judicial bodies to resolve political questions. Conversely, the greater the legal content of any issue, the greater the potential role of the court, because under our constitution and subject to the sovereign power of Parliament it is the function of the courts and not of political bodies to resolve legal questions.

In the end the United Kingdom withdrew the derogation under Article 15, effective March 16, 2005, by notification to the Council of Europe, as the provisions of part 4 of the 2001 act ceased to have effect from March 14, 2005, when the legislation was replaced with new measures (which I shall discuss later).

However, the judges agreed that the measure taken (indefinite detention) was not proportionate to the risk. Faced with the charge of usurping the political, Lord Bingham stated: "But the function of independent judges charged to interpret and apply the law is universally recognised as a cardinal feature of the modern democratic state, a cornerstone of the rule of law itself. The Attorney General is fully entitled to insist on the proper limits of judicial authority, but he is wrong to stigmatise judicial decision making as in some way undemocratic."

One final comment on the question of the legality of the declaration of a state of emergency came from Lord Scott:

> The Secretary of State is unfortunate in the timing of the judicial examination in these proceedings of the "public emergency" that he postulates. It is certainly true that the judiciary must in general defer to the executive's assessment of what constitutes a threat to national security or to "the life of the nation." But judicial memories are no shorter than those of the public and the public have not forgotten the faulty intelligence assessments on the basis of which United Kingdom forces were sent to take part, and are still taking part, in the hostilities in Iraq. For my part I do not doubt that there is a terrorist threat to this country and I do not doubt that great vigilance is necessary,

not only on the part of the security forces but also on the part of individual members of the public, to guard against terrorist attacks. But I do have very great doubt whether the "public emergency" is one that justifies the description of "threatening the life of the nation." Nonetheless, I would, for my part, be prepared to allow the Secretary of State the benefit of the doubt on this point and accept that the threshold criterion of article 15 is satisfied.

The majority of the judges were in agreement that, though the determination of the state of emergency should only be subject to a light judicial scrutiny, the measures taken (which were effectively to institute indefinite detention for foreigner nationals) did not satisfy the "strictly required" test of Article 15. Thus, although the state may declare a state of emergency without a strong judicial control over the reasoning, the measures that the state takes as a result of that declaration of a state of emergency will be subject to a much stricter test regarding their necessity. Undoubtedly, the seriousness of the emergency is in fact judged by the severity of the measures that the judges consider acceptable or not under the "strictly required" test.

The benchmark against which the state's right to declare the state of emergency, and to take measures as a result of it, is external to the control of the state: here, the benchmark is the European Convention on Human Rights and its interpretation by the European Court of Human Rights. The United Kingdom itself is not capable of changing the European Convention on Human Rights or its interpretation. As a result of the embedding of the European Convention on Human Rights both at the national level and at the EU level, it is no longer possible for the United Kingdom to escape the effect of the European Convention on Human Rights and its interpretation both (internationally) by the European Court of Human Rights and by its own courts. Further, a dialogue has begun between the national court, here the House of Lords, and the European Court of Human Rights on the nature, meaning, and effect of a state of emergency and measures taken as a result of such a declaration by a state. This dialogue seems unlikely to result in the European Court of Human Rights reversing the House of Lords' judgment in favor of greater state control. The state of exception within which the security continuum designates the enemy within the state is subjected to judicial scrutiny though the application of supranational rules at the national level. The medium through which this process is occurring in the United Kingdom is the body of the foreigner.

The Meta-Level: Respecting the Borders of Security?

Finally, I shall discuss an admissibility decision of the European Court of Human Rights that challenges the security continuum in another way. Here the supranational court considers the responsibility of a state regarding the actions of its military that take the security continuum across the border of the state itself into a neighboring state. In order to safeguard security within its borders, the state has authorized the use of the military. But in order effectively to secure state security, the military consider it necessary to continue their activities on the far side of the state border. The state in question is Turkey and the far side of the border is Iraqi Kurdistan. Six Kurdish women of Iraqi citizenship who had lived all their lives in Iraq brought an action against Turkey for the torture and unlawful killing of their husbands and sons in 1995. The description of the facts of the case by the European Court of Human Rights has a rather biblical tone:

> On the morning of 2 April 1995, Ismail Hassan Sherif, Ahmad Fatah Hassan, Abdula Teli Hussein, Abdulkadir Izat Khan Hassan, Abdulrahman Mohammad Sherriff, Guli Zekri Guli and Sarabast Abdulkadir Izatthe, together with the first, third, fourth and fifth applicants, left the village to take their flocks of sheep to the hills. The second and sixth applicants remained in the village to take care of their children.
>
> 1. After the party of eleven shepherds (the first, third, fourth and fifth applicants and Ismail Hassan Sherif, Ahmad Fatah Hassan, Abdula Teli Hussein, Abdulkadir Izat Khan Hassan, Abdulrahman Mohammad Sherriff, Guli Zekri Guli and Sarabast Abdulkadir Izzat) had walked for fifteen minutes in the direction of Spna, with the four women walking in front of the seven men, they met Turkish soldiers. The latter started to shout abuse at the eleven shepherds, hitting them with their rifle butts, kicking them and slapping them on the face. They separated the women from the men. They told the women to return to the village and then took the men away. The four applicants returned to the village and told the other villagers what had happened.[9]

When the men's bodies are finally found, they have been mutilated and the men are dead. The women seek redress against the Turkish state for the actions of its military in security operations in northern Iraq. As the European Court of Human Rights acknowledge, "The Turkish security forces carried out fourteen major cross-border operations between January 1994 and November 1998. The largest operation, called 'Çelik (steel) operation' and carried out with the participation of seventy to eighty thousand troops accompanied by tanks, armored vehicles, aircraft and helicopters, lasted almost six weeks between 19 March and 2 May 1995. The Turkish troops penetrated 40 [to] 50 kilometres southwards into Iraq and 385 kilometres to the east." Nevertheless, it found the case inadmissible on the basis that it was not satisfied (notwithstanding what the lay observer might consider overwhelming evidence to the contrary, including video film footage of the events) that the Turkish armed forces were responsible for the human rights abuses that took place and resulted in the deaths of the husbands and sons of the applicants. As the Court put it, "On the basis of all the material in its possession, the Court considers that it has not been established to the required standard of proof that the Turkish armed forces conducted operations in the area in question, and, more precisely, in the hills above the village of Azadi where, according to the applicants' statements, the victims were at that time."

Nevertheless, an important step is taken in this judgment toward the responsibility of states in international human rights law for the actions of their security forces even when those are acting outside the borders of the state. The Court had to consider whether the actions of the Turkish armed forces outside the borders of the state still engaged the human rights obligations of Turkey. This issue revolves around the meaning of Article 1 of the European Convention on Human Rights, which requires signatory states to comply with the commitments of, and guarantee the human rights set out in, the convention to all persons within their jurisdiction. Jurisdiction is not necessarily coterminus with state borders. States often claim jurisdiction over events and persons outside their borders—for example, over their own nationals' actions abroad (one recent and fairly controversial example is criminal liability for certain sexual behavior in foreign countries whether or not that behavior is contrary to the law of the state where it takes place). However, states guard carefully their right to control the delimitation of their jurisdiction; thus the interpretation of Article 1 of the European Convention on Human

Rights, is undoubtedly one of the more controversial issues arising in this case. The European Court of Human Rights found:

2. According to the relevant principles of international law, a State's responsibility may be engaged where, as a consequence of military action—whether lawful or unlawful—that State in practice exercises effective control of an area situated outside its national territory. The obligation to secure, in such an area, the rights and freedoms set out in the Convention derives from the fact of such control, whether it be exercised directly, through its armed forces, or through a subordinate local administration. (ibid., §52)

3. It is not necessary to determine whether a Contracting Party actually exercises detailed control over the policies and actions of the authorities in the area situated outside its national territory, since even overall control of the area may engage the responsibility of the Contracting Party concerned. (ibid., 2235–36, §56)

4. Moreover, a State may also be held accountable for violation of the Convention rights and freedoms of persons who are in the territory of another State but who are found to be under the former State's authority and control through its agents operating—whether lawfully or unlawfully—in the latter State (see, mutatis mutandis, *M. v. Denmark*, application no. 17392/90, Commission decision of 14 October 1992, DR 73, p. 193; *Illich Sanchez Ramirez v. France*, application no. 28780/95, Commission decision of 24 June 1996, DR 86, p. 155; *Coard et al. v. the United States*, the Inter-American Commission of Human Rights decision of 29 September 1999, Report No. 109/99, case No. 10.951, §§ 37, 39, 41 and 43; and the views adopted by the Human Rights Committee on 29 July 1981 in the cases of *Lopez Burgos v. Uruguay* and *Celiberti de Casariego v. Uruguay*, nos. 52/1979 and 56/1979, at §§ 12.3 and 10.3 respectively). Accountability in such situations stems from the fact that Article 1 of the Convention cannot be interpreted so as to allow a State party to perpetrate violations of the Convention on the territory of another State, which it could not perpetrate on its own territory. (ibid.)

In effect, what the court is saying is that states take their human rights obligations with them when they act abroad even when such actions are in the interests of state security. Further, the standard applicable is the same that applies within the state. As the court put it, if a state is not permitted to act in a certain way within its borders, then it is also prohibited from carrying out such an act outside its borders. The European Court of Human Rights carefully supports this finding by reference to the findings of a very wide range of international tribunals, courts, and human rights bodies. The authority of the finding is effectively linked to the force of rule of law as an international interlinking framework within which the meaning of security and the legitimacy of acts in the name of security are controlled by law.

Following January 2005

Twelve months before the ruling of the House of Lords against indefinite detention of foreigners, the UK authorities had already had a report from the Privy Counsellors recommending that the powers be repealed for want of legality. A consultation process was commenced that resulted in a wide-ranging report published by the government.[10] The response of the UK authorities to the House of Lords judgment was to pass new legislation in March 2005, the Prevention of Terrorism Act of 2005, which was passed in eighteen days (in comparison with the emergence legislation of 2001, which took thirty-two days to get through Parliament). The key element of the new legislation as regards the fight against terrorism was the creation of a power for the UK authorities to apply "control orders." The legislation withdrew part 4 of the 2001 act, and with it the power of indefinite detention.

Under the new legislation, the control orders take two forms—derogating control orders, and nonderogating control orders (s.1). Derogating control orders can only be made on the express authority of the secretary of state for the Home Department, where he or she has authorized a prior designation order that is a statutory instrument designating the United Kingdom's derogation from the European Convention on Human Rights in the specific case. So far, there is no derogation in place with the Council of Europe under the new legislation. A derogating order can only be made where the Secretary of State applies to a court and obtains an order that confirms the order. The court must be satisfied on a balance of probabilities that the individual has been involved in "terrorism-related activity." Nonderogating orders do not require prior notification to the Council of Europe of the invocation of the Article 15 right. Nonetheless, unless there are exceptional circumstances, the UK authorities are still required to obtain a court order to validate a nonderogating control order. Here, the test of which the court must be satisfied is that the UK authorities have "reasonable grounds for suspecting" the individual's involvement in "terrorism-related activity." A court must issue the order unless it is satisfied that the request is "obviously flawed."

A control order can include obligations that the UK authorities consider "necessary for purposes connected with preventing or restricting involvement by that individual in terrorism-related activity" (s.1(a)–(o)). These include use of specified articles or substances; use of specified services or specified facilities; a person's work or other occupation, or specifics in respect of the person's business; association or communications with specified persons, or with other persons generally. Positive obligations include giving access to specified persons into a place of residence or other premises to which the individual has access; allowing specified persons to search that place or any such premises (at any time); complying with a demand to provide information to a specified person in accordance with the demand. Examples of the use of these powers could include requiring a person to remain in only one room of his or her house or flat, and subjecting it to random and regular searches; limiting telephone access or contact with any other person, including family.

Although control orders were initially made against the men who had been under indefinite detention on their release in January 2005, the men were all rapidly rearrested, this time under the detention powers of the Immigration Act of 1971, with a view toward expulsion. Most of them have been released pursuant to control orders. However, there is a further complication, as the men cannot be expelled to their countries of origin as a result of the reasonable risk that they would suffer torture there, contrary to Article 3 of the European Convention on Human Rights. The UK authorities are currently seeking assurances from the countries of origin that, if returned there, the individuals would not be subject to torture or treatment contrary to Article 3.[11] At this juncture, five key issues arise. (1) First, will the third state even discuss providing an assurance? (2) If so, will the wording of the assurance be legally binding? (3) Does the department giving the assurance have the authority to ensure that it is observed? (4) How will the assurance be monitored? (Mainly foreseen is that well-respected nongovernmental bodies in the country of origin will agree to do the monitoring, but this plan is in doubt (5). What responsibility rests with the UK government in the event that the assurance is not respected?

This issue of assurances engages the same questions regarding the interaction of the national and supranational levels as does the detention of foreigners. Here the relationship of the judges at national and supranational levels is reversed. It was the European Court of Human Rights that in 1996 held, in respect to the proposed expulsion of an Indian national to India on national security grounds, that the UK authorities were barred by Article 3 of the European Convention on Human Rights from returning the man, notwithstanding diplomatic assurances given by the Indian authorities that they would not subject him to torture, that the United Kingdom would be in violation of Article 3 of the European Convention on Human Rights if it returned the man, since the Indian government would not be able to ensure that members of the security forces would respect the assurances.[12] The UK authorities once again sought to obtain and rely upon diplomatic assurances in 2004, regarding an Egyptian national.

Youssef, an Egyptian national, had sought asylum in the United Kingdom. Although the UK authorities accepted that he had a well-founded fear of persecution in Egypt for his political opinions (the definition of a refugee), his application had been rejected because the authorities considered that he had been involved in terrorist acts. He was detained pending expulsion to Egypt. However, because of Article 3 of the European Convention on Human Rights, he could not be expelled to Egypt. He claimed that he was being falsely imprisoned on this account. Between his detention in September 1998 and his release in July 1999, Youssef (and the three others detained with him) made no less than three applications for release (habeas corpus), all except the last ultimately rejected on the basis of evidence from the UK authorities that the assurances would be forthcoming shortly and thus the return of the men to Egypt would become possible. Judgment on the false imprisonment claim was handed down on July 30, 2004.[13]

In the judgment, the judge includes the long correspondence with the Egyptian authorities, including the personal intervention by the United Kingdom's prime minister in the matter. At one point, the home secretary wrote, in a note to the prime minister, as published in the judgment:

I am clear that, without any assurances, the men would face an Article 3 risk if they were returned to Egypt. As we have already ruled out the possibility of removing the men to anywhere other than Egypt this means that there is no longer a basis for detaining them under immigration powers. I will therefore have no option other than to agree to their very early release. In my letter of 25 May [1999], I did, however, make clear that I would provide you with a report before any action was taken to release the men. I am doing that now. If you decide to write to President Mubarak in the terms advised by FCO (i.e., making general points but not raising the issue of assurances) we will need to make arrangements to release the men as a matter of urgency. I will therefore be grateful if your officials could let mine know, if possible, within the next forty-eight hours, how you would prefer to proceed. Although the habeas corpus hearing I mentioned in my last letter was adjourned *sine die* we may need to explain our actions to a court at a future date. We are, in any event, required to account for our actions since the habeas hearing to the representatives of one of the four by Monday of next week at the latest. (paragraph 34)

The answer of the prime minister to this clear statement of the international human rights obligations and the consequences of continuing detention was by letter dated June 14, 1999, and again was published in the judgment:

The Prime Minister has reflected further on this difficult issue. He is also aware of the strong advice from our Embassy in Cairo, yourselves and SIS that we should not revert to President Mubarak to seek a full set of assurances from the Egyptians.

However, the Prime Minister is not content simply to accept that we have no option but to release the four individuals. He believes that we should use whatever assurances the Egyptians are willing to offer, to build a case to initiate the deportation procedure and to take our chance in the courts. If the courts rule that the assurances we have are inadequate, then at least it would be the courts, not the government, who would be responsible for releasing the four from detention. The Prime Minister's view is that we should now revert to the Egyptians to seek just one assurance, namely that the four individuals, if deported to Egypt, would not be subjected to torture. Given that torture is banned under Egyptian law, it should not be difficult for the Egyptians to give such an undertaking. He understands that additional material will need to be provided to have a chance of persuading our courts that the assurance is valid. One possibility would be for HMG [Her Majesty's Government] to say that we believed that, if the Egyptian government gave such an assurance, they would be sufficiently motivated to comply with it. We would need some independent expert witness to back that up.

The back and forth with the Egyptian authorities continued until the arrival of the deadline of a further court hearing on the man's habeas corpus application, where the UK authorities would have to provide some excuse and evidence of progress regarding the assurances. At this point, the UK authorities preferred to release the men rather than continue. The consequences of the international relations of the United Kingdom and Egypt were usefully summarized in yet more of the UK ministerial correspondence published in the judgment, as follows: "The position is very difficult; particularly as it is far from clear what Number 10 [the prime minister's residence] believe will be gained from pursuing the matter further. All the evidence from FCO [Foreign and Commonwealth Office] is that the Egyptians are not interested in pursuing the idea of assurances (regardless of the nature of the assurances being requested); and that losing the cases in the courts here would not assist our bilateral relationship" (paragraph 48).

What bears note from this case is the degree to which the courts, both national and supranational, through their interpretation of the obligation to protect the individual against torture, transform the nature of politics. If one considers the time scale from the decision in *Chahal* (1996) that individuals can never be returned to a country where there is a serious risk that they will be tortured, to the matter of Youssef (1999), it is impressive how quickly even the most powerful leader of a state, in the case of the United Kingdom, the prime minister, finds himself constrained by supranational human rights obligations in specific cases where he has invested political capital in his government's capacity to act without consideration for such obligations.

Finally, the United Kingdom's engagement with antiterrorism measures did not finish with the Prevention of Terrorism Act of 2005 (in respect to which, at the moment, only one person is subject to a nonderogating control order).[14] On November 3, what has now become the Terrorism Act of 2005 passed (by one vote) its second reading in the House of Commons. Two provisions of the act were particularly contentious: (1) the power of the police to detain terrorist suspects for up to ninety days without charge (s23(5)); and the creation of an offense of glorifying terrorism. Parliament reduced the period of precharge detention to twenty-eight days. However, in February 2007 the UK home secretary announced that he would make a new attempt to extend the maximum period beyond twenty-eight days.[15]

CONCLUSION

In this chapter, I have examined the response of courts, first in the United Kingdom and then at the European meta-level, to the development of an internal-external security continuum as Bigo has described it. At the outset, I noted that in the development of the continuum, the security services have sought to define their field as one subject only to the weaker judicial supervisory controls that apply to external security matters and national security. The response from the judges in the three cases that I have examined here indicates an increasing skepticism regarding the claims to exceptionalism by the security forces. Relying increasingly on the strengthening web of interlocking judgments at the national and supranational levels, the judges seek to reestablish judicial oversight over the field of security. In the first two cases, which deal with the foreigner within the state as the object of exceptional measures based on national security, the judges have refused to concede their territory of supervision of the application of supranational rights of the individual to the needs of national security. In the third case, the security concerns of the state lead it to act not only within its borders but also outside its borders to secure security within; the supranational court, although exempting the state on somewhat questionable grounds of an evidential shortcoming regarding the authors of the human rights abuses, nonetheless finds as a matter of law that the European human rights norms that apply within the state, as a result of the European Human Rights Convention, also apply to the actions of the state when its agents act outside its borders or indeed even outside the territory of the member states of the Council of Europe. The court specifically finds that there is no difference in degree or nature of the duty to protect fundamental rights between when the state acts within and when it acts outside its borders, so long as it has control over the territory within which it is acting.

The actions of Bigo's professionals of security, merging their fields and entering into new forms of competition and convergence, are mirrored by those of the professionals of legal security, who are also in the process of enunciating and interpreting the rule of law in the changing framework of security activity. The inside and outside of the European legal world are also in a process of redefinition, the end result of which is far from clear. What is evident, however, is the mimetic action of these two fields, both of which are renegotiating their relation with democracy.

NOTES

1. I have examined this in more depth in "The Face of Securitas: Redefining the Relationship of Security and Foreigners in Europe," in *Law and Administration in Europe,* ed. P. Craig and R. Rawlings (Oxford: Oxford University Press, 2003).

2. *R v Secretary of State for the Home Department ex parte Rehman,* (AP) UKHL 47, October 11, 2001.

3. The limitation on rights of entry and residence for certain types of British nationals, notably British overseas citizens and British nationals (overseas), remains problematic. The European Court of Justice, however, has declined to enter the discussion; see C-192/99 *Kaur* judgment 20.02.01.

4. Protocol 4, European Convention on Human Rights, contains the right of admission of nationals to their territory of nationality. The United Kingdom has not signed or ratified this protocol. Nonetheless, the European Court of Justice has made specific reference to it in a number of judgments, notably C-370/90 *Singh,* [1992], ECR I-4265.

5. "I am not so sure, however, whether international law concerning the expulsion of aliens is not changing fundamentally as a result of growing concerns for human rights and of a perceived need for solidarity among States in the face in increasing interstate relations. By admitting aliens to their territories, States inevitably accept at least some measure of responsibility. This responsibility is even more important in the case of children educated in their territory. For any society, individuals like the present applicant are a burden. Even independent of human rights considerations, I doubt whether modern international law permits a State which has educated children of admitted aliens to expel these children when they become a burden. Shifting this burden to the State of origin of the parent is no longer so clearly acceptable under modern international law. It is at least subject to doubt whether a host country has the right to return those immigrants who prove to be unsatisfactory." Series A No. 258-C, June 28, 1993.

6. European Court of Human Rights Application 54273/00 *Boultif v Switzerland*, 2.08.01.

7. *A (FC) and others (FC) v Secretary of State for the Home Department*; *X (FC) and another (FC) v Secretary of State for the Home Department*, [2004], UKHL 56.

8. This is the procedure set out in the Human Rights Act of 1998, which incorporated the European Convention on Human Rights into national law.

9. *ISSA and Others v. Turkey*, European Court of Human Rights, November 16, 2004.

10. *Counter-Terrorism Powers: Reconciling Security and Liberty in an Open Society*, Cm 6147 (London: Home Office, February 2004).

11. For an overview of the latest situation worldwide regarding this type of assurance, see Human Rights Watch, *Developments Regarding Diplomatic Assurances since April 2004*, http://hrw.org/reports/2005/eca0405/5.htm (accessed April 2005).

12. *Chahal v UK*, European Court of Human Rights Reports, 1996-V.

13. *Hani El Sayed Sabaei Youssef v Home Office*, [2004], EWHC 1884 (QB).

14. Kier Starmer, QC, Justice Human Rights Conference, October 26, 2005.

15. "Fresh Bid to Extend Detention Period for Terror Suspects," *Canadian*, February 1, 2007.

WORKS CITED

Bigo, Didier. 2005. "Gére les transhumances: La surveillance à distance dans le champ transnational de la sécurité." In M. C. Granjon, *Penser avec Michel Foucault: Théorie critique et pratiques politiques*. Paris: Karthala.

Groenendijk K., E. Guild, and R. Barzilay. 2001. *The Legal Status of Third-Country Nationals Who Are Long-Term Residents in a Member State of the European Union*. Luxembourg: European Commission.

Guild, E. 2003. "Exceptionalism and Transnationalism: UK Judicial Control of the Detention of Foreign 'International Terrorists.'" *Alternatives* (Special English Language Issue of *Cultures and Conflicts*) 28, no. 4: 491–515.

Nascimbene, B. 2001. *Expulsion and Detention of Aliens in the European Union Countries*. Milan: Guiffre.

4

Ambivalent Categories: Hill Tribes and Illegal Migrants in Thailand

MIKA TOYOTA

The border is not a neutral line of separation; borders not only demarcate boundaries between nation-states, they also make the distinction between belonging and nonbelonging to the state. Most works on *territorialization* look at the border in relation to international boundaries, but this chapter will focus on the border as an internal phenomenon; in particular, it will focus on the way the border defines belonging from not belonging to the nation-state.

Legally, the territorial sovereign state regulates persons within its territory through the institution of citizenship. But citizenship is not necessarily available to all, or not necessarily on equal terms. At the individual level, there will be exclusions as well as inclusions, sometimes on a very arbitrary basis. This chapter will examine the workings of this in relation to the hill tribe people in Thailand. The concept *hill tribe*[1] mainly denotes people residing in the highland zone bordering China, Burma, Laos, and Thailand, and whom anthropologists refer to as *upland people* or *highlanders*. According to a recent (2002) official tribal population survey, the number of hill tribe persons was 914,755, which amounted to around 1.4 percent of the total population of Thailand (61.81 million).

Two questions to be asked are, first, why the Thai state denies citizenship rights to more than 370,000 hill tribe people who reside within its territory, thus making them noncitizens, and, second, how this discrimination is justified by the state. In other words, this chapter explores how this borderspace not only serves to mark off Thailand *externally* from other sovereignties but also is defined *internally* by the Thai state as a space of exception (Agemben 2005) where the residents are scrutinized and controlled precisely through their legal exclusion.

Denial of citizenship means that hill tribe people are legally alienated and deprived of basic rights, such as the right to own land and access to essential health care. It means, too, that they are not eligible for school certification, which in turn makes it difficult for them to find proper jobs, and hence many resort to illegal activities to make a livelihood for themselves and their families. Further, if hill tribe people wish to travel outside their home districts, permission from the local authority is required, even for everyday activities such as meeting friends or family members or looking for jobs. If caught without such permission, they can be arrested and punished. Finally, their stateless status justifies acts of expulsion from the country altogether by state authorities on the specious grounds of their being "illegal migrants."

The political construction of the *border* in ways that allow it to make distinctions between inclusion and exclusion is a means of legitimating the structure of territorial power and its embodiment in the state. But simply demarcating territorial boundaries is not sufficient to establish the legitimacy of the nation-state. The state needs in addition to be seen to have objective classifications of belonging, and this involves categorizing the population and clearly marking off legitimate citizens, against whom noncitizens may be discerned, the identity of the included only becoming meaningful when contrasted with the excluded. Thai identity, therefore, to a significant degree is defined through creating the "non-Thai"; this exclusion affirms and legitimates sovereign power. But what is a matter of power and sovereignty at the national level can become a question of life and death when brought down to the individual level.

The predicament of hill tribes in Thailand has been mostly analyzed within the framework of the exercise of a power of domination—in other words, the hierarchical power of the majority over the minority or the center-periphery relationship (McKinnon 1989; McCaskill and Kempe 1997). But although this framework can account for how hill tribes are marginalized, assimilated, or "domesticated" using a fixed dichotomy between the state, as an oppressor, and hill tribes as the oppressed, it does not explain the underpinning mechanisms of how and why denial of citizenship to the hill tribes in Thailand is rationalized by the sovereign state in the first place. This is a problem that fundamentally goes back to the rationale and very existence of the modern nation-state itself.

In his use of the concept of *governmentality*, Foucault distinguishes between power and domination (Foucault 1991). Domination refers to asymmetrical relationships of power in which the subordinated have little room to maneuver. His concept of governmentality instead accounts for the systematization of the power relationship. States of domination are the effects, not the primary source, of the indirect technologies of government. These effects go beyond the spontaneous exercise of power over others. Foucault's concept of governmentality provides an analytical framework to study technologies of power through an analysis of the political rationality underpinning them. In this way, Foucault gives us a more comprehensive account of the mechanism of the legitimization of domination, and of the inherent contradiction and hidden geographies of the modern nation-state. His model also allows us to conceptualize the politics of borderscapes as an ongoing process and to illuminate the contested process of justifying and asserting the border. As Foucault notes (1982), modern sovereign states and modern individuals codetermine each other's emergence. In this vein, the aim of this chapter is to illuminate such intersections of discourses by investigating why and how the border space between highland and lowland has been repeatedly territorialized and reterritorialized through discourses in the various stages of state building.

Many liberal scholars, such as John McKinnon and Chupinit Kesmanee, hope and are expecting that the segregation of Thai proper from hill tribes will disappear once external factors are improved. Some assume that the segregation is a matter of prejudice and discrimination based on ignorance of the other, and thus should improve when people acquire mutual knowledge and respect. Others blame lack of state capacity, such as incomplete democratization and/ or bureaucratic corruption, and believe that once decent democratic government is established the problem should be solved.

However, I believe that the segregation of Thai and hill tribes is an enduring feature inherent in the governmentality of the Thai state. If this is so, then improvement to democratic standards or even the use of scientific mechanisms such as DNA analysis of individuals to ascertain "objectively" who has the right to belong to the nation-state and who may legitimately be excluded will not lead to a fundamental solution. Past experience shows that, even after the number of hill tribes granted Thai citizenship was increased, the gulf between the two groups did not diminish. So why did such differences emerge in the first place, why are they so persistent, and how is this borderscape rationalized?

This chapter is divided into four sections, analyzing four discourses that shape the borderscape. First I look at why making the symbolic distinction between *upland* and *lowland* was necessary in the process of building the modern Thai nation-state. Second, I examine the purpose of constructing the *hill tribes* category during the late 1950s, and its meaning in Thai society in the 1960s and 1970s. Third, I examine the impact of environmental conservation policies in mapping the forest boundaries in the 1980s and 1990s, and how these led to villagers being driven from their land and forbidden to farm or even enter their own homeland. And, last, I discuss how hill tribes themselves have appropriated the discourse to fight back against the state's stigmatization of them as "illegal migrants."[2]

THE DISCOURSES OF BORDER MAKING BETWEEN UPLAND AND LOWLAND

Although notions of upland people (primitive) versus lowland people (civilized) existed, and although physical elevation has long been associated with ethnic difference throughout Southeast Asia (Reid 1993, 5), all of this does not necessarily mean that there existed a clear political boundary between highland people and lowland people based on geographic elevation. The actual political space traditionally was quite blurred. In premodern mainland Southeast Asia, the critical element of sovereignty was the people, not the territorial entity. The borders of center-oriented "galactic polities" of the traditional state were "porous and indistinct" (Tambiah 1976). It is said that the peoples of these margins used to be under the "indirect rule" (Marlowe 1969) of the Thai authority. For example, the highland local authority of the Karen and Lua used to pay tribute to the princes of Chiang Mai. In return for this, the princes recognized the legitimacy of the local authorities and extended them their protection while permitting them to take control over land and people and thus enjoy a degree of semiautonomy. In this way, at the overlapping margins of Siam and its adjacent kingdoms, the coexistence of multiple loyalties to several overlords of the peripheral minorities was common and was accepted by the ruling state (Thongchai 1994, 97). Through this tributary relationship, the peripheral population was linked with the major lowland kingdom in a loose, symbolic relationship (Keyes 1979; Lehman 1979). Loyalty at the border area had always been fluid and fluctuated according to shifts in power within the autonomous tributary relationship. As Jørgensen notes, during the early colonial encounter, Thai rulers appointed feudal chiefs *(chao muang)* among such frontier peoples as the Mon, Lawa, and Karen (Jørgensen 1979, 84). The relationship was not always one of subordination to the dominant group but, rather, of interdependence. It was only during the reign of King Vajiravudh (1910–1925) that usage of Thai language became a marker of "Thai-ness."

However, the emergence of Siam as a buffer state in the Franco-British encounters in colonial Southeast Asia, and the use of Western-style political mapping techniques gave substance to the notion of a territorially bounded nation-state. This fundamentally altered the structure of the highland/lowland relationship. Frontier people were forced to give up the practice of multiple loyalties. From the perspective of the modern state, with its need for clear-cut allegiances, the issue of belonging or not belonging, with clear territorial boundaries, is crucial to state integrity and security.

Bangkok officials classified the population at the periphery as *Khon Pa* (the "wild people"). This implied that they shared little or nothing with their fellow Thai. They were depicted as "strange," "filthy," "wild," and "uncivilized," in contrast with the civilized Thai (the realm of *pa* implies the dangerous "wild frontier"). In the process of creating "Thai-ness," the dichotomy between *muang* (the center) and *pa* (highland) was constructed and became a tool to identify "Thai-ness" in contrast with the "wild [non-Thai] others" (Stott 1991; Thongchai 1993, 2000). In effect, a Darwinian theory of human evolutionary development came to serve as the basis of the distinction between *muang* and *pa*. *Pa* was seen as the historical past of the *muang,* and was represented as "backward" and thus an object of contempt for the Bangkok elite. This virtual Darwinism justified Bangkok officials in looking down on non-Thai. They were perceived as primitives of the forest, isolated remnants living "in the absence of the later civilizing influences," retaining features of "the original inhabitants" *(khon dangdoem)* (Pracha Khadikit 1885, 164, cited by Laungaramsri 2003). Since the number of *Khon Pa* was fairly insignificant, the Thai central government could be politically fairly indifferent to them and, characteristically, a relationship of noninterference prevailed.

This, however, does not mean that interaction between highland and lowland peoples was nonexistent. The long-established relations with lowland Thai-speaking people continued in the forms of mixed marriages, adoption, day-to-day trade, labor exchanges, etc. As Jonsson notes, historically, trading with the highlanders for forest products was essential to the running of the lowland (1998, 20). Moreover, although the genealogical common-ancestor/clanship system used to be seen as a source of ethnic identity (Kammerer 1988), recent detailed genealogical studies reveal that other ethnic groups were an integral part. For example, the Akha genealogical system includes Thai and Chinese descendants (von Geusau 2000). Alting von Geusau's genealogical study of the Akha clearly demonstrates the dynamic nature and adaptability of their ethnic identification; von Geusau's work has done much to show how flexible and open "Akha-ness" can be, as instanced by the following:

> Several originally non-Akha groups entered the Akha ethnic alliance system; . . . these include poor marginalized Thai and Chinese, mountain people such as the Lahu, and forest people such as the Wa. These became Akha through attaching themselves to the ancestor system and accepting Akha customary law. The Akha call this *padaw-eu,* or adoption of a group or person into the Akha alliance system by intermarriage or, in the past, as *jakh'a* (bonded servant); this latter did not happen in a class context, however, but in a family context, leading to integration. There are particular places in the genealogical system where a group or person can attach himself/herself. (Alting von Geusau 2000, 134)

In this way "the ethnic sub-groups could change affiliation and become members of a different ethnic system" (von Geusau 2000, 122).

Although the dialectical oppositions of up-slope and down-slope were developed in Akha oral history (von Geusau 1983), this does not imply that highland people never lived in the lowlands.

While Kammerer proclaims that "cultural identity . . . presupposes a sense of territoriality" (Kammerer 1988, 263), I argue instead that discourses were consciously developed in the Akhas' oral history through their being pushed into highland areas by more powerful peoples at a certain time in their history. But not all moved to the highlands; some moved to other places. Thus it is also reasonable to suppose that some so-called highland people might not have actually lived in the highlands or not for generations. For example, not all Akha were shifting cultivators; some earned their living as traders (Toyota 2000). Reynolds notes that, historically, *pa* did not always carry the connotation of *wild*: "In the Thai inscription of King Ramakhamheng (Inscription I), *pa* is the term for 'groves of coconut, jackfruit, areca, and tamarind, surely sustenance for the nearby muang and not its nemesis. In northern Zhuang, one of the Thai languages, *pa* is scrub land where cows graze. In this Thai language, a 'real' jungle with wild animals and tall trees would be *dong,* not *pa*" (2003, 117). He thus suggests that "Muang and pa are best seen as the outer limits of a continuum along which stretch gradations of wilderness, from jungle to scrub land to rice field."

In fact, these people have been in a constant state of flux, moving not only across national boundaries, but also across ethnic boundaries, and so blurring highland/lowland territoriality. In opposition to this fluid reality, the sedentary framework introduced by administrators fixes upland people within the marginal highland domains, at the same time laying the basis of the idea that the civilized lowlanders are the core of the nation-state.

SECURITY DISCOURSES ON "HILL TRIBES"

It was in the late 1950s that the marginal highland population at the edge of the Thai nation-state became of concern to the Thai government. Following the emergence of the People's Republic of China (PRC) in 1949, and the associated Indochina conflict, the Thai government, under the influence of the United States, became increasingly concerned about communism creeping over the border. Subsequently, the *Khon Pa* at the border zone were no longer simply seen as wild others but became "threats and problems" to Thai nationhood (Toyota 1999, 239).

The Border Patrol Police (BPP) was established in 1953. Then, in 1955, the U.S. Operation Mission (USOM) came into being to provide

substantial financial support to establish a highlands Thai-language
school program run by the Border Patrol Police. Through this program, some highland people were trained as village guards to form
border security volunteer teams in conjunction with the Communist
Suppression Operation Command (CSOC) (Tapp 1989, 32; 1990,
154). As part of these programs, photographs of the Thai king were
distributed to the border villages and instructional speeches on Thai
nationalism were delivered to raise patriotic awareness among villagers (Kunstadter 1967). The Thai king became symbolically a bridge,
integrating highland people into the Thai nation-state. Loyalty to
the king was to mean loyalty to the Thai nation. The Thai king's personal patronage was emphasized in the Royal Highland Development
Project, justified by a projection of the highland people as "innocent,
helpless and pitiful": in need of royal protection.

In 1959, the official identification "hill tribes" (chao khao in
Thai), which includes nine ethnic highland minorities, was established (Vienne 1989, 36). According to McKinnon (1989, 307), the
term chao khao was derived from a British colonial term used in
Burma, where highland people were called hill tribes. Thai officials
translated this English term "hill tribes," into the Thai chao khao
(chao translates as "people," khao as "hill") to refer to the non-
Thai-speaking population of the highland periphery who had yet to
be assimilated into the Thai nation-state. Although there are other
people living in the hill areas, such as Yunnanese Chinese (often
called Chin-Ho) and Shan people (Thai-speaking people from the
Shan state of Burma), they were not included in the category hill
tribes in spite of the fact that they had established trade links and
intermarried with highland people (Toyota 2000). This indicates
that the term hill tribes does not simply refer to the minority people
who live in the highlands, but has specific political implications in
terms of making a distinction between those who can be included
in the classification Thai citizen and those who cannot.

The creation of the official category hill tribes intensified the
Pa (non-Thai)/Muang (Thai) ideology with its rigid geographical
territoriality of hill/valley. In this way, in the process of confirming the boundary of the integrated Thai nation-state, the category
hill tribe came to be applied to the area where historically ethnic
identifications had been ambiguous and porous. In the drive to secure a territorially bounded modern Thai nation-state and secure

national integration, the ambiguity of transferable identities was no longer to be allowed. The impact of the creation of the category *hill tribes* has been threefold: first, the practice of physically moving back and forth across the national borders and lowland/highland boundary becomes a problem from the government's perspective (which emphasized sovereign control over borders and crossings); second, the symbolic mobility across ethnic boundaries could no longer be accepted; third, the politicization of space, that is, the lowland/highland division, became a marker in differentiating Thai from non-Thai citizen, with non-Thai citizens perceived as threats to national security.

In the 1960s, several, mostly American, institutions provided significant financial support for research into identifying *hill tribe* populations in Thailand. At the request of the Thai government, the United Nations assisted the first socioeconomic survey of the hill tribes in northern Thailand, between October 1961 and March 1962 (Bhruksasri 1989, 14). This survey was initiated by an Australian anthropologist, Hans Mannorff. Another well-known example of U.S. donor research was conducted by Cornell University for USAID in northern Thailand in 1963 (Diamond 1993; Price 2003; Wakin 1992/2004). Although anthropologists might not have intended to support the Cold War structures, and did use the research fund for their own ends, they followed the specific agendas and categories established by the funding agencies. Within these specific regimes of development and modernization, the so-called scientific categorization of ethnicity, delineating clear boundaries, resulted in the outlining of a distinct ethnic group based on which *hill tribes* development policies were formulated.

Representational ethnic difference was constructed from the new knowledge. Canonical works on the peoples and ethnicity of the region, such as *Ethnic Groups of Mainland Southeast* (1964), *Southeast Asian Tribes, Minorities, and Nations* (1967), have classified ethnic differences based on language groups, social organization, religion, etc., and *Farmers in the Forest* (1978) further broke these people down on the basis of their association with particular geographic elevations. These early works by Western academics addressed the fluid nature of identity boundaries and the mobility of these people between upland and lowland (Leach 1954; Keyes 1979). Kunstadter and Chapman even caution readers that they found almost as much variation in land use within the same ethnic categories as between them. Such ambiguous and fluid elements of ethnic boundaries are ignored, however, when these works are used as references for policy implementation. Instead ethnic differences are simplified and essentialized; for example, wet rice cultivation is said to represent Thai, and shifting agriculture to represent hill tribes' modes of farming.

Although several development programs to improve social welfare among hill tribes—Thai elementary school education, primary health care service, and occupational training—were introduced to encourage and support hill tribes' integration into Thai society, hill tribes have never been viewed as truly Thai citizens, and citizenship rights have been begrudged them. For example, in 1956, when the government conducted a survey to register "all" households in the country, the hill tribe population was excluded. This reflects a mixture of prejudice, concern, and confusion as to how to place hill tribes within the Thai state. From the security angle, the Thai authority had to consider individuals within the Thai territory as citizens who should be under the control and protection of the state, but at the same time the very concept of hill tribes as outsiders of the lowland Thai realm acknowledged them as non-Thai. This results in a peculiarly ambiguous legal status, which in effect makes these people subjects of the nation without citizenship. I have discussed in detail elsewhere the bewildering array of identity cards in use in the highlands and borderlands of Thailand (Toyota 2005), but I wish to mention briefly the legal conditions arising.

The 1965 Nationality Act granted Thai citizenship to people belonging to ethnic minority groups who were born in the kingdom, provided that both their parents were Thai nationals. Withdrawal or cancellation of citizenship was possible when a parent was proved to be an alien. The preconditions required for obtaining Thai nationality were first instituted by the Ministry of Interior's Regulation on Consideration for Granting Thai Nationality (to the hill tribes), issued in 1974. However, many members of hill tribes could not prove their families had lived in Thailand for any length of time, and thus were regarded as illegal migrants. Citizenship in Thailand is, in principle, determined not by place of birth but by the citizenship status of a person's parents. Restrictions on citizenship are stated in several laws, including the Citizenship Act. Following the end of the Indochina war in 1975, more refugees, both highlanders and lowlanders, from the neighboring countries came into Thailand. Keyes notes: "Their presence justified the continuation of policies that precluded illegal migrants from becoming Thai" (Keyes 2002, 1181). In 1976, a cabinet memorandum called for the acceleration of the registration of ethnic minorities who had entered Thailand prior to 1975, with the ultimate aim of enabling them to become Thai citizens. Meanwhile Thai authorities provided a "Pink Card" to political refugees who had arrived from Burma before 1976—Mon, Karenni, Tai-yai, Lawa, etc. The distinction between refugees and those who entered Thailand after 1975 and are thus not entitled to citizenship remains in effect.

These complex legal restrictions have been keeping the majority of hill tribe people from holding Thai citizenship. Before applying for citizenship, a Thai birth certificate is required, to prove the applicant's identity as belonging to the hill tribes in Thailand. However, quite a number of these persons never had birth registration, although they were born within Thai territory. In some cases, parents did not know where to go to register, or did not know that they should register, their children, or did not know how to fill in the registration form since they could not read or write Thai. These problems with citizenship qualification have impeded the process of citizenship approval. Without sufficient legal procedures and the requisite papers, many hill tribe people living in Thailand even for the second or third generation have been stuck in an endless process.

THE DISCOURSES OF ENVIRONMENTAL THREATS AND MAPPING THE FOREST

Territorialization of the forest began as far back as 1896, when the Royal Forestry department was established and issued "the declaration that all unoccupied land within the national boundaries was state forest under the jurisdiction of the Royal Forestry Department" (Vandergeest and Peluso 1995, 408). Along with the establishment of the territorially bounded modern Thai nation-state's control over the forest resources, about 75 percent of the total land area was claimed by the Royal Forest Department (RFD) in 1896 (Vandergeest 1996, 161). However, this did not legally prevent local people from using forest resources for domestic needs. Government policies encouraged the clearing and cultivating of new land for the

production of rice, including new land out of undemarcated forest, as there were no laws specifically preventing the local villagers
from doing so. Although the state declared forest ownership to be
Bangkok's, access to forest products remained controlled by influen-
tial local people in the upper northern part of Thailand rather than
by the central administrators.

Then, after 1932, the second stage of the demarcation between
reserve and permanent forest was initiated, when the monarchy was
replaced with a government composed of bureaucrats and military
officers (Vandergeest and Peluso 1995, 409). In theory, the Protection
and Reservation of Forest Act of 1938 provided for this demarcation. Clearing and burning were prohibited in "protected" forests,
local inhabitants were forbidden to graze animals there, and per-
missions were required to extract any forest products (Vandergeest
and Peluso 1995, 409). In practice, however, territorial control was
neither of interest to, nor feasible for, the central government. Most
forests were defined as unoccupied land, and "the territorial boundaries of the forest remained ambiguous, changing and unenforce-
able" (Vandergeest and Peluso 1995, 409). Although a series of new
laws, such as the 1960 Wildlife Conservation and Protection Act,
the 1961 National Park Act, and the 1964 National Forest Reserve
Act, were enacted in the 1960s, and by 1985 the area mapped as reserve forest had reached 42 percent of national territory, commercial
exploitation of forest resources was not stopped. As a result, rapid
deforestation took place in the upper north of Thailand in the 1960s
and 1970s. It is officially estimated that in the early 1950s almost
two-thirds of the country was still covered with forest; by the early
1980s, however, forests covered less than one-third of the nation
(Buergin 2003, 48).

In 1985, the RFD reclassified the forest reserves into *conservation
forest* and *economic forest*. By the middle of the 1980s, deforestation
was perceived as a problem for the first time by Thailand's wider
public. This perception was partly due to the influence of growing
international awareness of global environmental issues, and partly
due to the heavy floods and landslides in the south in November
1988.

The idea that shifting cultivation was a dangerous form of agriculture owed much to international opinion. It was not until the
emergence of international objections that shifting cultivation came

to be perceived locally as a problem and a prime cause of forest destruction. As Kunstadter notes (1978, 3), shifting cultivation used to
be practiced by both Thai and hill people in both the lowlands and
the highlands of the region. However, since the FAO argued in 1967
that "shifting cultivation created harmful effects on the number of
trees in the forest and caused ecological destruction" (Launagramsri
1997, 30), such cultivation has come to be viewed as indeed de-
structive and harmful. Further, in this negative discourse, shifting
cultivation has been exclusively associated with the hill tribes, as
their typical mode of economy. As a result, hill tribes who live in
the forest area, even those hill tribes who no longer practice shifting
cultivation, are directly blamed for destroying the natural resources
of the country.

Territorial sovereignty claims authority over not only people but
also the resources within the boundaries it defines for the allocation
and realization of access rights. "Territorialization is about controlling what people do and their access to natural resources within those
boundaries" (Vandergeest 1995, 388). The concept of the Protected
Area System (PAS) became a new instrument of forest conservation
(the Thai Forestry Sector Master Plan of 1993); this new "functional
territorialization" shifted from resource control to the "surveillance
of boundaries and the simple prohibition of most activities within
these boundaries" (Vandergeest and Peluso 1995, 410). Scientific criteria such as soil type, slope, and vegetation have become the basis
for laws prohibiting and prescribing specific activities in these areas
(Vandergeest and Peluso 1995, 408). This zoning approach gives
priority to conservation forestry by ejecting and prohibiting human
settlement in the affected areas.

The FAO's evaluation on shifting cultivation, accompanied by
the associated discourse of hill tribes as shifting cultivators, justified the Hill Tribe Resettlement Project. The resettlement policy,
which aimed to remove hill tribe people from these forest areas and
to protect the watersheds against encroachment by hill tribes, were
perceived as the most important tasks of the RFD. The program was
set up with the help of the military and thus led to "militarization
of forest conservation policy" (Vandergeest 1996, 171), showing the
militarization of forest space under the name of forest conservation.
In particular, the program intensified during the military rule in
1991 and 1992, and continued even after the civilian government

came in after May 1992. As the Protected Areas were extended, from about 10 percent to more than 17 percent in 1999, the survival of hill tribes people in the forest area became more and more problematic. Their land use was restricted, they were charged with being "illegal encroachers," and a forced resettlement policy was imposed.

In this process, hill tribes as non-Thai Others has revived in the RFD's discourse, projecting them as threatening the welfare of the state by destroying the national forests. In May 1998, the director general of the RFD signed an agreement with the supreme commander of the army, specifying the cooperation of the RFD and the army to protect Thailand's remaining forests. In this agreement, the army was given far-reaching authority as well as financial support for operations in forest areas (*Nation*, May 9, 1998). From April 18 to May 12, 1999, under this operation a pilot project involving an alliance between the military and the Royal Forest Department was created. This involved soldiers and forest rangers going to the Karen villages in the wildlife sanctuary and demanding that they cease to grow rice, a demand accompanied by demolishing huts and destroying personal belongings (*Bangkok Post*, May 13, 15, and 16, 1999). A cabinet resolution of June 30, 1998, stated that those who failed to prove that they had lived in the forest before the Forestry Department declared the area a conservation zone would be forced to move out. Since then, a way of life and practices are depicted as illegal by the state authorities and are blocked by territorial borders, immigration controls, and other forms of legal restriction. An eruption of the military into the daily lives of hill tribes people is now observable. Emerging nationalistic sentiment to protect the forest against "forest destroyers" justifies the military's tough handling of hill tribes. For example, the director general of the RFD laments that the territory of Thailand is gradually being given away to non-Thai (*Nation*, September 18, 2000).

HILL TRIBES OR ILLEGAL MIGRANTS?

As the influx of refugees, irregular migrant workers, and trafficked people from neighboring countries increased in the 1970–1990s, the issue of granting citizenship to hill tribe people became problematic for Thai officials. The concept *hill tribes* could no longer simply represent non-Thai Others. Who belonged to the categories *hill tribes*,

illegal migrant workers, or *refugees* had to be established at the individual level. This situation made Thai officials even more restrictive in granting citizenship—although admittedly it was not an easy
task to distinguish and identify hill tribe people from those refugees
or illegal immigrants from neighboring countries. The Immigration
Police classified the illegal workers into three groups: first, those
suspected of trafficking workers into Thailand; second, those entering and exiting the country frequently, causing a disturbance;
third, those coming to Thailand illegally to take up employment.
Unregistered, vulnerable hill tribe people, however, could easily end
up in any category. In particular, the fact that many foreign workers
from Burma were ethnic minorities sharing close ethnic and kinship
networks with the highland minority in Thailand made a clear distinction almost impossible, and consequently both groups were put
in the same category of illegal migrants.

Unauthorized workers were often treated as a reserve of flexible
labor, being used to ensure low-cost labor provision in the agricul-
tural sector, the fisheries industry, domestic service, and the sex
industry. The "miracle" economic development in Thailand of the
1980–1990s could not have been achieved without the cheap labor
provided by illegal foreign migrants from neighboring countries
and also by the hill tribe people. Nevertheless, these workers were
outside the protection of labor workplace safety, health, minimum
wage, and other standards, and are easily deportable. With the rapid
expansion of the foreign labor force all over Thailand, the government decided in 1995 to implement a regularization policy to bring
them under some form of control. Illegal migrant workers became
visible in Thai society when the estimated number became available: the number of illegal workers was seen to have increased from
525,000 (1994) to 987,000 (1998).

The issue of identifying and classifying non-Thai people in
Thailand became pressing when the Thai government started seriously dealing with foreign labor problems. After the economic crisis of the late 1990s, when the average unemployment rate jumped
from 1.5 percent (1997) to 4 percent (1998) and 4.1 percent (1999)
(Chalamwong 2001, 306), law enforcement against illegal workers
was stepped up. The estimated figure of undocumented workers in
1998 was 932,200, and it was equivalent to almost 70 percent of
Thai unemployed. "It was suggested that if the government could

get all of the illegal immigrants out of the country, the employment situation of Thais would improve considerably" (Chalamwong 2001, 306). Given their lack of legal recognition, Thailand's illegal migrants became a target of deportation. As a result of measures taken in 1999, arrests and deportation of illegal migrants stepped up. According to the statistics of the National Security Council (NSC), 319,629 were arrested in 1999 and 444,636 in 2000. In the year 2000 alone, more than 1,000 employers who continued to hire undocumented workers after the granting period expired were arrested and sentenced.

In the process of categorization of peoples, it has not been uncommon for some families members to be divided into different categories—for example, a father classed as an illegal migrant worker, mother as a refugee, son as a Thai citizen, daughter as of the hill tribe, grandmother perhaps never granted any status. In the process of "solving the problem of illegal migrants," those categorized as non-Thai people in Thailand have been severely put upon. This has brought about the further marginalizing of those already marginalized. The livelihoods of the irregulars among the marginalized minority people of the Burma-Thai borderlands have been criminalized, and the numbers of these persons defined as stateless or without citizenship increased. The issue of right to citizenship, to work, to study, and to settlement now intrudes into the concerns of people who in the past were free from such exactions and pressures. Distinctions that never existed historically between those living somewhere legally and those not have come into being. Persons without official acceptance are technically illegal and will be harassed, fined, and generally bullied by the authorities. The modern state is forcing distinctions on legal/technical grounds onto peoples who were previously one, and so has created divisions that have no historical validity.

In their depiction as "non-Thai" people, the public images imposed both on foreign workers and on hill tribes are identical. According to this, these people are (1) the source of contagious diseases such as HIV/AIDs, (2) the cause of increased crime, (3) the makers of stateless babies. Further, without the legal entitlement of Thai citizens, they are excluded from basic human rights such as (1) appropriate access to public health services, (2) educational attainment, (3) land rights, (4) occupational options (employers pay lower wages to hill tribe ID holders [or nonholders] than to Thai citizens, and certain professions are open only to Thai nationals), and (5) freedom of mobility, as both hill tribe ID holders and registered illegal workers are prohibited from leaving the district in which they are registered without permission. If caught at one of the many police checkpoints outside their district without a pass, they can be arrested and detained, and face both fines and imprisonment, and sometimes deportation. It is estimated that at least two to three million illegal migrants from Burma alone are currently working in Thailand (2002). But the strategy of eliminating illegal migrants severely disrupts the daily life not only of foreign migrant workers but also of those hill tribe people still waiting for legal recognition from the Thai authority.

The insecurity of hill tribe life increased in the late 1990s. On the one hand, these people were forbidden to farm or even to enter their home villages, due to the forest management policies, which means they had no choice but to leave their home villages. On the other hand, they were at the same time under tighter control over their mobility, should they wish to leave the home district to seek employment elsewhere.

To further their struggle for collective rights, leaders in the hill tribes community saw that the abstract concept of *hill tribes* forced upon them had to be turned to their advantage by being strategically utilized in a collective mobilization. For this reason, the discourses of indigenous rights and community forest were appropriated to proclaim the peoples' right to existence within a specific territorialized space. At the same time, these people sought to represent their traditional life in terms tuned to the theme of ecological conservation: not forest destroyers but guardians of the forest and preservers of a traditional lifestyle, in contrast with the meretricious modernity of the Thai lowland cities, was the image they sought to convey, more generally, by agency of NGOs, academics, the media, and directly to the general public.

It is ironic that in this way the standard account of distinction between Thai modern life versus hill tribes' traditional life has been further intensified. Romanticizing "native culture" as "the sacred," as spiritual and traditional, can be a double-edged sword. The main problem is that it traps users in the representation of themselves as the isolated noble savage, something far from the actual practice of hill tribe life today. Indeed, contrary to such imaginary projections, most hill tribe people are involved in some kind of sedentary market-related economic practices, such as growing flowers, carrots, cabbages, and potatoes, or short rotational agriculture for the commercial market. Further, many of the hill tribe younger generation no longer live in the highlands, which they left behind to attend school or find work. Even though it may be strategically useful to emphasize that shifting cultivation is ecologically friendly and that hill tribe lifestyle is harmonious with nature, there is the danger of the reproduction of the an impermeable symbolic binary of *highland* and *lowland*, and of the social hierarchy attached to it.

The real needs of hill tribes are served not so much by the demand for recognition of ethnic difference, for example in the form of indigenous rights, but rather by securing substantive citizenship rights and social equality on the same terms as other Thais. Without questioning the ways in which the essentialized category *hill tribes* was constructed, the current movement to preserve and/or essentialize traditional ways of life of hill tribes may in the long run do more harm than good.

CONCLUSION

In this chapter, I suggest that Foucault's notion of governmentality is important to understanding the process of shaping the modern Thai nation-state. Boundarying practices have taken place not only between states, but also within states; this produces equally sharp demarcations at the popular level. This chapter seeks to shed light on the discourses in which the rationalization of exercising state power and making such distinctions has been justified.

The borderscape of the modern state is based on what we call an abstraction of knowledge and technologies. Abstract space is linear space, which can be cut up into discrete parts. It is represented as uniform and homogeneous units. Although it cannot adequately represent the diversity, complexity, and dynamics of daily reality, modern mapping is a key technique in the government of the modern nation-state in conceptualizing abstract space and in legitimating boundaries.

This chapter illustrates the ways in which the new racialized domains of the Thai internal borderscape emerged. Various techniques of knowledge abstraction justified such reconfiguration. These included a Darwinian theory of human evolutionary development making distinctions between *civilized* and *uncivilized*; the "scientific" classification of ethnic groups based on an essentialized representational ethnic difference; and mapping technology that territorializes not only physical space but also social space.

In the first section of this chapter, the institutionalization of distinctions between upland and lowland was examined; the symbolic boundaries between *primitive* and *civilized* were already part of the traditional ideology of the Buddhist polity, but the introduction of a Darwinist notion of human evolution provided the basis for a new sharper justification. In the second section, I investigated the rationale of constructing the *hill tribes* category; this category was made possible through the emergence of the idea of closed, exclusive, presumed scientific ethnic classifications, which were employed to categorize peoples within the state boundaries. In the third section, the role of technologies of spatial abstraction, of mapping and its role in the territorialization of space, was explored; maps became instruments enabling state agencies not only to create and classify territories but to legitimatize territorial surveillance. In the final section, I considered the way that new discourses of indigenous rights provide the basis for an abstracted self-imaginary among hill tribe peoples allowing for the reinforcement of their claims to social and economic rights.

The chapter also suggests that the configuration of the Thai internal borderscape is not a one-way process, with power emanating from the state over the subjects. Demarcating borders within the state is legitimized by a plurality of actors, from state administrators, international aid agencies, media, environmental groups, NGOs, and academics to hill tribes themselves.

NOTES

1. Those officially categorized as "hill tribe" (*chao khao* in Thai) by the Thai authority normally consist of nine ethnic groups: Karen, Meo (Miao, Hmong), Lahu, Lisu, Yao (Mien), Akha, Lua (Lawa), H'Tin, and Khamu. According to a recent official Tribal Population Survey, the number of hill tribe members was 914,755 (2002), around 1.4 percent of the total population of Thailand (61.81 million). Human rights groups estimated that 600,000 to one million hill tribe persons are not officially recognized as citizens (*Nation,* July 16, 2004). This figure reflects considerable deliberate official underestimation and/or lack of knowledge.

2. A study by the Population and Social Research Institute, Mahidhol University, revealed that there were at least 2.4 million illegal residents in Thailand (*Nation*, July 1, 2004). According to UN statistics, almost one million children between the age of six and twelve are not in school in Thailand.

WORKS CITED

Abu-lughod, Lila. 1991. "Writing against Culture." In *Recapturing Anthropology: Working in the Present*, ed. Richard G. Fox. Santa Fe, N.M.: School of American Research Press, 137–62.

Agar, Michael H. 1980. *The Professional Stranger: An Informal Introduction to Ethnography*. New York: Academic Press.

Agemben, Giorgio. 2005. *State of Exception*, trans. Kevin Attell. Chicago: University of Chicago Press.

Aguettant, Joseph L. 1996. "Impact of Population Registration on Hilltribe Development in Thailand." *Asia-Pacific Population Journal* 11, no. 4: 47–72.

Alting von Geusau, Leo. 2000. "Akha Internal History: Marginalization and the Ethnic Alliance System." In *Civility and Savagery: Social Identity in Tai States*, ed. Andrew Turton. Surrey: Curzon, 122–58.

Bao, Jiemin. 1999. "Reconfiguring Chineseness in Thailand: Articulating Ethnicity along Sex/Gender and Class Lines." In *Genders and Sexualities in Modern Thailand*, ed. Peter A. Jackson and Nerida M. Cook. Chiang Mai: Silkworm Books, 63–77.

Barry, Brian. 2000. *Culture and Equality*. Cambridge, U.K.: Polity.

Bhruksasri, Wanat. 1989. "Government Policy: Highland Ethnic Minorities." In *Hill Tribes Today: Problems in Change*, ed. John McKinnon and B. Vienne. Bangkok: White Lotus–Orstorm, 5–31.

Bremen, Jan van, and Akitoshi Shimizu, eds. 1999. *Anthropology and Colonialism in Asia: Comparative and Historical Colonialism*. Anthropology of Asia. Richmond, Surrey: Curzon.

Buergin, Reiner. 2003. "Trapped in Environmental Discourses and Politics of Exclusion: Karen in the Thung Yai Naresuan Wildlife Sanctuary in the Context of Forest and Hill Tribe Policies in Thailand." In *Living at the Edge of Thai Society: The Karen in the Highland of Northern Thailand*, ed. Claudio O Delang. London and New York: Routledge Curzon, 43–63.

Castles, S., and A. Davidson. 2000. *Citizenship and Migration: Globalization and the Politics of Belonging*. Basingstoke, U.K.: Macmillan.

Chalamwong, Yongyuth. 2001. "Recent Trends in Migration Flows and

Policies in Thailand." OECD Proceedings, International Migration in Asia: Trends and Policies. OECD.

———. 2002. "Thailand." In Migration and the Labour Market in Asia: Recent Trends and Policies. Paris: Organization for Economic Cooperation and Development (OECD).

Coedes, G., trans. 1925. Documents sur l'histoire politique et religieuse du Laos Occidental, including a translation of the Jinakalamalini (Chronicle of Chieng Mai). BEFEO 25: 1–189.

Darling, Frank C. 1965. Thailand and the United States. Washington, D.C.: Public Affairs Press.

Delang, Claudio O, ed. 2003. Living at the Edge of Thai Society: The Karen in the Highland of Northern Thailand. London and New York: Routledge Curzon.

Diamond, Sigmund. 1992. Compromised Campus: The Collaboration of Universities with the Intelligence Community, 1945–1955. New York: Oxford University.

Dikötter, Frank. 1990. "Group Definition and the Idea of 'Race' in Modern China." Ethnic and Racial Studies 13: 420–32.

Fineman, Daniel. 1997. A Special Relationship: The United States and Military Government in Thailand, 1947–1958. Honolulu: University of Hawaii Press.

Foucault, Michel. 1982. "The Subject and the Power." In Michel Foucault: Beyond Structuralism and Hermeneutics, ed. Hubert Dreyfus and Paul Rabinow. Brighton: Harvester, 208–26.

———. 1991. "Governmentality." In The Foucault Effect: Studies in Governmentality, ed. Graham Burchell, Colin Gordon, and Peter Miller. London: Harvester Wheatsheaf, 87–104.

Geddes, William R. 1967. "The Tribal Research Centre, Thailand: An Account of Plans and Activities." In Southeast Asian Tribes, Minorities, and Nations, ed. Peter Kunstadter. 2 vols. Princeton: Princeton University Press, 553–81.

Giersch, Charles Patterson. 2000. "The Sipsong Panna Tai and the Limits of Qing Conquest in Yunnan." Chinese Historians 10, no. 17: 71–92.

———. 2001. "A Motley Throng: Social Change on Southwest China's Early Modern Frontier, 1700–1880." Journal of Asian Studies 60, no. 1: 67–94.

Hanks et al., eds. 1964. Ethnographic Notes on North Thailand. Data paper no. 58, Southeast Asia Program. Ithaca, N.Y.: Cornell University.

Hastrup, Kirsten, and Karen Olwig. 1997. "Introduction." In Siting Culture: The Shifting Anthropological Object, ed. Karen Olwig and Kirsten Hastrup. London and New York: Routledge.

Hills, Ann Maxwell. 1998. Merchants and Migrants: Ethnicity and Trade *among Yunnanese Chinese in Southeast Asia*. New Haven, Conn.: Yale University Press.

Hugo, Graham. 1999. "Undocumented International Migration in South-East Asia." In *Asian Migration: Pacific Rim Dynamics*, ed. Yen-Fen Tseng, C. Bulbeck, Lan-Hung Nora Chiang, and Jung-Chung Hsu. Interdisciplinary Group for Australian Studies, Monograph No. 1: 73–97. National Taiwan University, Taipei.

Jonsson, Hjorleifur. 1998. "Forest Products and Peoples: Upland Groups, Thai Polities, and Regional Space." *Sojourn* 13, no. 1: 1–37.

Jørgensen, Anders Baltzer. 1979. "Forest People in a World of Expansion." *Transactions of the Finnish Anthropological Society*, no. 2.

Kammerer, Cornelia Ann. 1989. "Territorial Imperatives: Akha Ethnic Identity and Thailand's National Integration." In *Hill Tribes Today: Problems in Change*, ed. John McKinnon and B. Vienne. Bangkok: White Lotus–Orstorm, 259–301.

Kasian Tejapira. 1997. "Imagined Uncommunity: The *Lookjin* Middle Class and Thai Official Nationalism." In *Essential Outsiders: Chinese and Jews in the Modern Transformation of Southeast Asia and Central Europe*, ed. Daniel Chirot and Anthony Reid. Seattle: University of Washington Press, 75–98.

Kesmanee, Chupinit. 1994. "Dubious Development Concepts in the Thai Highlands: The *Chao Khao* in Transition." *Law and Society Review* 28, no. 3: 673–86.

Keyes, Charles, ed. 1979. *Ethnic Adaptation and Identity: The Karen on the Thai Frontier with Burma*. Philadelphia: Institute for the Study of Human Issues.

———. 2002. "Presidential Address: The Peoples of Asia—Science and Politics in the Classification of Ethnic Groups in Thailand, China, and Vietnam." *Journal of Asian Studies* 61, no. 4: 1163–1203.

Kelly, David, and Anthony Reid, eds. 1998. *Asian Freedoms: The Idea of Freedom in East and Southeast Asia*. New York: Cambridge University Press.

Khadikit, Pracha. 1885. "Wa Duay Khon Pa Ru Kha Fai Nua" (On the Wildman or the Kha in the North). In *Wachirayanwiloet*. Bangkok: Klum Pitak Chiwit Pua Matubhumi (Mother Earth Life Protection Group).

Kirsch, A. T. 1990. "The Quest for Tai in Tai Context." *Crossroads* 5, no. 1: 69–79.

Kukathas, Chandran. 1992. "Are There Any Cultural Rights?" *Political Theory* 20, no. 1: 105–39.

Kunstadter, Peter. 1967. *Southeast Asian Tribes, Minorities, and Nations*. Princeton, N. J.: Princeton University Press.

Kunstadter, Peter, and E. C. Chapman. 1978. "Problems of Shifting Cultivation and Economic Development in Northern Thailand." In *Farmers in the Forest: Economic Development and Marginal Agriculture in Northern Thailand,* ed. by Peter Kunstadter, E. C. Chapman, and S. Sabhasri, 3–23.

Kymlicka, William. 1989. *Liberalism, Community, and Culture.* Oxford: Clarendon.

———. 1995. *Multicultural Citizenship.* Oxford: Clarendon.

———. 2001. *Politics in the Vernacular: Nationalism, Multiculturalism, and Citizenship.* Oxford: Oxford University Press.

———, ed. 1995. *The Rights of Minority Cultures.* Oxford: Oxford University Press.

Kymlicka, William, and Wayne Norman, eds. 2000. *Citizenship in Diverse Societies.* Oxford: Oxford University Press.

Laungaramsri, Pinkaew. 1997. "On the Discourse of Hill Tribes." Paper presented at Workshop on Ethnic Minorities in a Changing Environment, Chiang Mai University. February.

———. 2003. "Constructing Marginality: The 'Hill Tribe' Karen and Their Shifting Locations within Thai State and Public Perspectives." In *Living at the Edge of Thai Society: The Karen in the Highland of Northern Thailand,* ed. Caudio O. Delang. London and New York: Routledge Curzon, 21–42.

Leach, E. 1954. *Political Systems of Highland Burma: A Study of Kachin Social Structure.* London School of Economics Monographs on Social Anthropology, no. 44. London: London School of Economics.

———. 1960. "The Frontier of 'Burma.'" *Comparative Studies in Society and History* 3: 49–68.

Lehman, F. K. 1979. "Who Are the Karen, and If So, Why? Karen Ethnohistory and Formal Theory of Ethnicity." In *Ethnic Adaption and Identity: The Karen on the Thai Frontier with Burma,* ed. Charles Keyes. Philadelphia: Institute for the Study of Human Issues, 215–49.

Manzo, Kathryn. 1995. *Creating Boundaries: The Politics of Race and Nation.* Boulder, Colo.: Lynne Rienner.

Marlowe, David H. 1969. "Upland-Lowland Relationship: The Case of the S'kaw Karen of Central Upland Western Chiang Mai." *Tribesmen and Peasants in North Thailand.* Proceedings of the First Symposium of the Tribal Research Centre, Chiang Mai, Thailand, 53–68.

Marx, Anthony. 1998. *Making Race and Nation: A Comparison of South Africa, the United States, and Brazil.* Cambridge: Cambridge University Press.

McCaskill, Don, and Kampe, Ken, eds. 1997. *Development or Domestication? Indigenous Peoples of Southeast Asia.* Chiang Mai: Silkworm Books.

McCoy, Alfred W. 1972. *The Politics of Heroin in Southeast Asia.* New York: Harper and Row.

McKinnon, John, and B. Vienne, eds. 1989. *Hill Tribes Today: Problems in Change*. Bangkok: White Lotus–Orstorm.

Moerman, Michael. 1965. "Ethnic Identification in a Complex Society: Who Are the Lue?" *American Anthropologist* 67: 1215–30.

Muscat, Robert J. 1990. *Thailand and the United States: Development, Security, and Foreign Aid*. New York: Columbia University Press.

O'Connor, R. 1990. "Siamese Tai in Tai Context: The Impact of a Ruling Center." *Crossroads* 5, no. 1: 1–21.

Pels, Peter, and Oscar Salemink. 1999. "Introduction: Locating the Colonial Subjects of Anthropology." In *Colonial Subjects: Essays on the Practical History of Anthropology*, ed. Peter Pels and Oscar Salemink. Ann Arbor: University of Michigan Press, 1–52.

Price, David H. 2003. "Subtle Means and Enticing Carrots: The Impact of Funding on American Cold War Anthropology." *Critique of Anthropology* 23, no. 4: 373–401.

Renard, Ronald. 1980a. "Kariang History of Karen-T'ai Relations from the Beginning to 1923." Unpublished PhD dissertation, University of Hawaii, Honolulu.

———. 1980b. "The Role of the Karens in Thai Society during the Early Bangkok Period, 1782–1873." *Contributions to Asian Studies* 15: 16–28.

———. 2000. "The Differential Integration of Hill People into the Thai State." In *Civility and Savagery: Social Identity in Tai States*, ed. Andrew Turton. Richmond: Curzon Press, 63–83.

———. 2001. *Opium Reduction in Thailand, 1970–2000: A Thirty-Year Journey*. United Nations International Drug Control Programme Regional Centre for East Asia and the Pacific, Bangkok, Thailand. Chiang Mai: Silkworm Books.

Reynolds, Craig J. 2003. "Review: Tai-land and Its Others." *South East Asia Research* 11, no. 1: 113–30.

Shapiro, Ian, and Will Kymlicka, eds. 1997. *Ethnicity and Group Rights*. New York: New York University Press.

Skinner, G. William. 1957. *Chinese Society in Thailand*. Ithaca, N.Y.: Cornell University Press.

Stott, Philip. 1991. "Mu'ang and Pa: Elite Views of Nature." In *Thai Constructions of Knowledge*, ed. Manas Chitakawem and Andrew Turton. London: School of Oriental and African Studies, University of London, 142–54.

Streckfuss, David. 1993. "The Mixed Colonial Legacy in Siam: Origins of Thai Racialist Thought." In *Autonomous Histories, Particular Truths: Essays in Honor of John R. W. Smail*, ed. Laurie J. Sears. Monograph no. 11, Center for Southeast Asian Studies. Madison: University of Wisconsin Press, 123–53.

Supaporn, Jarunpattana. 1980. "Phasi Fin Kap Naiyobai Kankhlang Khong Rathaban," Ph.S. 2367–2468 (Opium Revenue and Fiscal Policy of Thailand, 1824–1925). Unpublished MA thesis, Chulalongkorn University, Bangkok.

Tambiah, S. J. 1976. *World Conqueror and World Renouncer*. Cambridge: Cambridge University Press.

Tambiah, Stanley J. 1977. "The Galactic Polity: The Structure of Traditional Kingdoms in Southeast Asia." *Annals of the New York Academy of Science* 293: 69–97.

Tapp, Nicholas. 1986. "The Hmong of Thailand, Opium People of the Golden Triangle." Indigenous Peoples and Development Series Report no. 4. London: Anti-Slavery Society.

———. 1989. *Sovereignty and Rebellion: The White Hmong of Northern Thailand*. Singapore: Oxford University Press.

———. 2002. "In Defence of the Archaic: A Reconsideration of the 1950s Ethnic Classification Project in China." *Asian Ethnicity* 3, no. 1: 62–84.

Taran, Patrick A. 2001. "Human Rights of Migrants: Challenges of the New Decade." *International Migration* 38, no. 6: 7–51.

Taylor, Rupert. 1999. "Political Science Encounters 'Race' and 'Ethnicity.'" In *Ethnic and Racial Studies Today*, ed. Martin Bulmer and John Solomos. London and New York: Routledge, 115–23.

Thongchai, Winichakul. 1993. "The Other Within: Ethnography and Travel Literature from Bangkok Metropolis to Its Periphery in Late Nineteenth Century Siam." Paper presented at the fifth International Conference on Thai Studies. SOAS, University of London, July 4–10.

———. 1994. *Siam Mapped: A History of the Geo-body of a Nation*. Chiang Mai: Silkworm Books.

———. 2000a. "The Others Within: Travel and Ethno-Spatial Differentiation of Siamese Subjects, 1885–1910." In *Civility and Savagery: Social Identitiy in Tai States*, ed. Andrew Turton. Richmond: Curzon Press, 38–62.

———. 2000b. "The Quest for 'Siwilai': A Geographical Discourse of Civilizational Thinking in the Late Nineteenth- and Early Twentieth-Century Siam." *Journal of Asian Studies* 59, no. 3: 528–49.

Tong Chee Kiong and Chan Kwok Bun, eds. 2001. *Alternate Identities: The Chinese of Contemporary Thailand*. Singapore: Times Academic Press: Leiden E. J. Brill.

Toyota, Mika. 1999. "Cross Border Mobility and Multiple Identity Choices: The Urban Akha in Chiang Mai, Thailand." Unpublished PhD thesis, Hull University, United Kingdom.

———. 2000. "Cross Border Mobility and Social Networks: Akha Caravan Traders." In *Where China Meets Southeast Asia: Social and Cultural*

Change in the Border Regions, ed. Grant Evans et al. Singapore: ISEAS, 204–21.

———. 2003. "Contested Chinese Identities among Ethnic Minorities in the China, Burma, and Thai Borderlands." *Ethnic and Racial Studies* 26, no. 2: 301–20.

———. 2005. "Subjects of the Nation without Citizenship: The Case of 'Hill Tribes' in Thailand." In *Multiculturalism in Asia: Theoretical Perspectives,* ed. Will Kymlicka and He Baobang. Oxford University Press, 110–35.

———. Forthcoming. *The Akha: A Transnational Ethnic Minority in the Borderlands of Thailand, Burma, and China.* London and New York: Routledge Curzon.

Tully, James. 1995. *Strange Multiplicity: Constitutionalism in an Age of Diversity.* Cambridge: Cambridge University Press.

United States Department of the Army. 1970. *Ethnographic Study Series: Minorities Groups in Thailand,* Washington, D.C.: U.S. Government Printing Office.

Vandergeest, Peter. 1996. "Mapping Nature: Territorialization of Forest Rights in Thailand." *Society and Natural Resources* 9:159–175.

———. 2003. "Racialization and Citizenship in Thai Forest Politics." *Society and Natural Resources* 16:19–37.

Vandergeest, Peter, and Nancy Lee Peluso. 1995. "Territorialization and State Power in Thailand." *Theory and Society* 24, no. 3 :385-426.

Vienne, Bernard. 1989. "Facing Development in Highlands: A Challenge for Thai Society." In *Hill Tribes Today: Problems in Change,* ed. John McKinnon and Bernard Vienne. Bangkok: White Lotus–Ostrom, 33–60.

Wade, Geoff. 2000. "The Southern Chinese Border in History." In *Where China Meets Southeast Asia: Social and Cultural Change in the Border Regions.* Ed. Grant Evans et al. Singapore: ISEAS, 28–50.

Wakin, Erik. 1992/2004. *Anthropology Goes to War: Professional Ethics and Counterinsurgency in Thailand.* Center for Southeast Asia Studies Monograph no. 7. Madison: University of Wisconsin Press.

Walker, Andrew. 1999. *The Legends of the Golden Boat: Regulation, Trade, and Traders in the Borderlands of Laos, Thailand, Burma, and China.* Richmond, Surrey, and Curzon, Honolulu: University of Hawaii Press.

Walzar, Michael. 1983. *Spheres of Justice: A Defense of Pluralism and Equality.* Oxford: Blackwell.

Wijeyewardene, Gehan. 1991. "The Frontiers of Thailand." In *National Identity and Its Defenders: Thailand, 1939–1989,* ed. Craig J. Reymonds. Monash Papers on Southeast Asia no. 25, Centre of Southeast Asian Studies. Melbourne: Monash University Press.

Willett, Cynthia, ed. 1998. *Theorising Multiculturalism.* Oxford: Blackwell.

II

Borderpanic:
Representing Migrants and Borders

5

Danger Happens at the Border

EMMA HADDAD

The border can be understood as a dangerous place. Things that cross the border undermine the border's authority and have the capacity to "pollute" the inside that the border is trying to protect. To highlight this understanding of pollution, this chapter uses the concept of the refugee as one moving individual who operates at the border. Neither inside nor outside, the refugee moves across borders as an inherently polluting person who defies the order that the border would like to dictate. Europe has become fixated on keeping refugees away from its territorial borders via policies aimed at preventing their arrival. Such strategies can be seen as an attempt at pollution prevention or *source reduction,* which aims to place "polluting" refugees in areas of "protection" no longer dependent on state borders. To avoid the clean inside becoming polluted, danger must be kept as far away as possible.

DANGEROUS BORDERS

As previous chapters in this book have argued, the border is discursively identified as a site of danger (Douglas 1966). In this discourse, the border is the boundary between inside and outside: the inside is safe, outside there is danger. The discourse enacts a particular reality; danger is not merely a discursive skilfulness by which

119

the inside is made coherent. In his account of the Jacalteco Maya, Thompson conjures up this image vividly:

> If Jacaltecos are natives, it is to a bleeding land and history. Especially because of the violence of places they inhabit . . . conceptions of Jacalteco identities have come about in part through hardships caused by forces that have pushed them into geographically confined areas and have attempted to keep them there. Even so, they have not remained passive recipients of these borders, as one might contain livestock, but instead they have found ways of moving beyond them, even if there is pain involved. (Thompson 2001, ix)

The risk of danger at the border is high, since it is here that inside and outside merge: "One side is not simply the site of violence and *el otro lado,* the site of refuge. 'Here' bleeds into 'there,' and . . . the two sides often bleed together" (Thompson 2001, 179-80). The Jacalteco search for refuge on one side of the border from danger on the other: "Pools of red spill on the borderlands between Guatemala and Mexico; a line nearly devoid of practical use and meaning, becomes this night divider between life and death" (Thompson 2001, 11). This image portrays not only the danger, but also the power and importance of the border: "Were borders insignificant, there would be no blood, and there would be no crosses erected along them" (Thompson 2001, 19). In this way, the border also symbolizes the boundary between life and death: those who manage to flee have crossed to the other side, into Mexico, *al otro lado*; those left behind have crossed the border of death into the afterlife, they too *al otro lado* (Thompson 2001, 11). In this way, "divisions of time—the past, present, and future—bleed into one place, a field, a space, where all who have been and who are, are together for a while, where those who are living know they will be" (Thompson 2001, 166). Thus there is danger at the border, and death may be the final, dreadful outcome.

Where inside and outside merge, there is a danger of pollution. Pollution is a type of danger likely to occur wherever there are clear lines and boundaries (Douglas 1966, 113). Paradoxically, as Douglas points out, "it is part of our human condition to long for hard lines and clear concepts . . . [and so] when we have them we have to either face the fact that some realities elude them, or else blind ourselves to the inadequacy of the concepts" (Douglas 1966, 162). The danger from outside threatens to penetrate the safe inside. People think of their social environment as consisting of other people joined or separated by lines that must be respected (Douglas 1966, 138). Wherever these lines are blurred or precarious, pollution ideas may be introduced in an attempt to strengthen them: "Physical crossing of the social barrier is treated as a dangerous pollution. . . . The polluter becomes a doubly wicked object of reprobation, first because he crossed the line and second because he endangered others" (Douglas 1966, 139).

THE POLLUTING REFUGEE

The refugee is created at the border, where inside and outside meet. She cannot exist solely inside, nor can she exist solely outside. The border is that which ensures her existence. Were there no borders, there would be no refugees (Haddad 2003a). Indeed the refugee only exists inasmuch as modern political borders exist and attempt to organize peoples and territories among nation-states. Sovereignty sorts and classifies, and refugees are created in the process. The ordering of peoples is a basic tenet of the pluralist world and places a sharp distinction between being inside or outside. But this ordering of peoples only takes place insofar as it relates to individual states. The refugee does not belong to any individual state; she exists by definition between states and thus falls outside the reach of the "international community." Once borders are put up and territorial jurisdiction is defined, the refugee is forced between such borders by the very system that creates her. States exert their sovereign right to decide whom they will represent and protect. When the individual loses her attachment to a particular territory, she ceases to behave according to the ordering of peoples that the international states system would demand. Thus a breakdown domestically leads to a similar breakdown internationally; the refugee loses her relationship with the state both internally and externally. The refugee is therefore an anomaly both within and between states: she is not supposed to exist internally or externally.

If the refugee's position in the states system is that of an anomaly, it is because she is displaced, primarily, in the physical sense (Tuitt 1999, 106). As such, she is associated with motion. Language employed in discourse on the refugee mirrors this perception: we talk of flows of refugees, mass movements, and tides (Tuitt 1999, 108), people who are running, escaping. They are victims of causal factors that have an endemic mobilizing force, be these war, persecution, or famine. And as Goodwin-Gill points out, "flight" then constitutes "the only way to escape danger to life or extensive restrictions on human rights" (Goodwin-Gill 1996b, 4). Liquid images associated with uprooting and displacement contrast with the territorializing metaphors of identity: "roots, soils, trees, seeds are washed away in human flood-tides, waves, flows, streams, and rivers" (Malkki 1995a, 15). Thus the "sedentarist bias in dominant modes of imagining homes and homelands, identities and nationalities" becomes evident (Malkki 1995a, 15), and the emphasis put on territorial belonging is underlined.

In conceiving the refugee as a "moving entity" (Tuitt 1999, 107), we construct her as different, an irregularity in the life of an otherwise stable, sedentary society (Malkki 1995b, 508). Displacement occurs within what Malkki describes as the "national order of things" (Malkki 1995a, 2), in which having a fixed, stationary existence is the norm. By virtue of her "refugeeness," the refugee occupies a "problematic" and "liminal" position in the national order of things (Malkki 1995a, 1–2). As the refugee cannot be fixed within one set of borders, she acts to "blur" (Malkki 1995a, 8) or "haemorrhage" (Douglas in Malkki 1995a, 7) national boundaries. The national order of things is subverted and "time-honoured and necessary distinctions between nationals and foreigners" are challenged (Arendt 1966, 286). Thus the refugee becomes a "problem" that requires "specialised correctives and therapeutic interventions" (Malkki 1995a, 8). If the nation classifies, orders, and sorts people into national kinds and types (Malkki 1995a, 6), "refugeeness" can be seen as an aberration of categories, a "zone of pollution" (Malkki 1995a, 4):

> Transnational beings are particularly polluting, since they are neither one thing nor another; or may be both; or neither here nor there; or may even be nowhere (in terms of any recognized cultural topography), and are at the very least "betwixt and between" all recognized fixed points in the space-time of cultural classification. (Turner in Malkki 1995a, 7)

Since the refugee acts at the border, indeed can only exist if there are borders, she is a dangerous figure. On the threshold between inside and outside, she threatens to penetrate the border, precariously blurring the meaning of inside and outside, and projecting danger onto others. The refugee can therefore be seen as a *polluting person*; she has crossed, or threatens to cross, a line that should not have been crossed, and her displacement unleashes danger (Douglas 1966, 113). As Douglas explains,

> Danger lies in transitional states, simply because transition is neither one state nor the next, it is undefinable. The person who must pass from one to another is himself in danger and emanates danger to others. The danger is controlled by ritual which precisely separates him from his old status, segregates him for a time and then publicly declares his entry to his new status. (Douglas 1966, 96)

Thus if one imagines the refugee as fluid and between categories, she can be seen to constitute a threat to established boundaries, as a pollutant who can easily cross such boundaries. The refugee is a side effect of the creation of separate sovereign states. In other words, international legal norms imagine all peoples organized among territories and divided among states, but the refugee is an inevitable, if unintended, externality. It is as if the international system breathes in oxygen in the form of stable state-citizen relationships, and breathes out carbon dioxide in the form of refugees. To ensure protection from such waste matter, borders must be equipped with filters to purify and keep toxic material away. To ensure pollution levels do not reach drastically high levels, the borders must be fitted with systems to monitor, measure, and control.

According to Douglas, dirt is essentially disorder, something that offends against order. Notions of dirt and defilement are contrasted with notions of the positive structure of a society, which must not be negated or polluted (Douglas 1966, 159): "eliminating it is not a negative movement, but a positive effort to organise the environment" (Douglas 1966, 2). To impose order and stability, it is necessary, first, to recognize those things that are out of place and a threat to order. Their identity is constituted by their being unwanted. At this stage, such unwanted anomalies are dangerous: their old identity still clings to them, and the ability to clarify and order the inside is impaired by their presence. It is only after a cycle of pulverizing, dissolving, and rotting that the dirt is finally utterly disintegrated and undifferentiated, its identity completely removed, and the danger eradicated, since it now clearly belongs to a defined place—a rubbish heap (Douglas 1966, 160). In other words, we can identify a cycle: the anomaly is initially nondifferentiated; it then becomes dirt by being imagined as different and by being created by the imposition of order; during the process of differentiating, it threatens the distinction between inside and outside, order and disorder; the threat is finally absolved by returning the dirt to its true indiscriminant character (Douglas 1966, 161).

Dirt is a by-product of the creation of order, just as the refugee is a by-product of the creation of separate sovereign states and thus a source of disorder and instability. Instability is a security or pollution risk, and therefore needs to be corrected. Attempts to keep the refugee away from borders are not therefore negative action, but positive action to keep order within and thus protect the citizens in an organized, stable society. Before becoming a refugee the individual is like any other nondifferentiated individual. She is then displaced to a border where she begins to be imagined as different and a threat to the creation of a safe, stable place of order. To resolve the threat the refugee is kept out and, preferably, sent back from whence she came so as to regain her "indiscriminable" character by being reinstalled in a state-citizen relationship. The solution to the refugee problem is the reterritorialization of those individuals acting between states so as to put an end to their disorderly and potentially polluting movement at and between borders.

The "polluting" refugee can be compared to other social outcasts. Kathy Stuart's study of ritual pollution conflicts involving defiled trades in early modern Germany is illuminating in this respect. Certain tradespersons were known as *Unehrliche Leute* or "dishonourable people." Social status was expressed in terms of honor or dishonor. Society was honorable and was separated by a social boundary from specific groups of dishonorable people—for example, ethnic and religious minorities, criminals, and those performing certain trades, such as grave-diggers, millers, actors, and bailiffs (Stuart 1999, 2). The greater the degree of dishonor, the greater the exclusion from (honorable) society and "normal sociability" (Stuart 1999, 3). Executioners and skinners, for example, were to varying degree pelted with stones, barred from taverns, and denied an honorable burial (Stuart 1999, 3). In a similar vein, notes Stuart, the *burakumin* of Japan were the pariah group working as executioners, morticians, and night-soil fertilizers, while barber-surgeons, leather workers, and latrine cleaners are still found among the untouchables in India (Stuart 1999, 9). Indeed the Indian caste system is a clear example of the boundary between purity and impurity being exemplified by social class. Here status is determined by conceptions of the pure and the impure: "The whole system represents a body in which by the division of labor the head does the thinking and praying and the most despised parts carry away waste matter" (Douglas 1966, 123). Caste pollution is a symbolic system based on the image of the body. Its primary concern is the ordering of a social hierarchy. The higher the caste status, the purer it is and thus the more of a minority it will be. Accordingly, the anxiety about threats that can penetrate the bodily margins expresses greater dangers to the survival of the minority group as a political and cultural unit (Douglas 1966, 124).

Dishonor is in large part defined by its polluting quality. Speaking again of those in defiled trades, Stuart points out, "by coming into casual contact with dishonourable people or by violating certain ritualised codes of conduct, honorable citizens could themselves become dishonourable" (Stuart 1999, 3). Communication of dishonor corresponds to Douglas's definition of ritual pollution, which occurs *ex opere operato* ("by the act itself"). In other words, such pollution is effective regardless of the morals or intentions of the actor. Pollution is thus liable to be created inadvertently—as in the case of the refugee who has been forced against her will out of a sustaining political community and hence into the no-man's-land of *refugeeness*. Dishonor is a taboo, as is Douglas's ritual pollution, and in both cases the risk of contagion is evident. The polluting refugee encompasses a threat of contagion to the normal, sedentary citizen's identity and security, and so the citizen will therefore want to keep the refugee away by the strict control of borders where dangerous acts of pollution are most likely to occur.

For Douglas, there are four kinds of social pollution: the danger pressing on external boundaries; the danger from transgressing the internal lines of the system; the danger in the margins of the lines; and the danger from internal contradiction—that is, "when some of the basic postulates are denied by other basic postulates, so that at certain points the system seems to be at war with itself" (Douglas 1966, 121). The refugee is particularly dangerous since she is a form of all four types of social pollution. She is outside and inside and in the margins, or in-between, all at the same time. Further, she defies logic:

she is in the in-between of the state-citizen relationship, which by definition should not have an in-between; thus she also represents an internal or inherent contradiction to the very system of international states. She is that which should not exist, yet she exists in all the points where pollution is possible. The system wants to be rid of her, to eliminate the risk of pollution, yet she is inevitable, such that the system is at war with itself, fighting a losing battle.

That which is not part of society and subject to its laws is potentially against it—a danger. To remove the risk of pollution, the potential transgression must be separated, purified, demarcated, or punished. This will impose system on an inherently untidy experience: "It is only by exaggerating the difference between within and without . . . that a semblance of order is created" (Douglas 1966, 4). Contagion only makes sense in terms of an anomaly, an *other*, that has the potential to disrupt the unity or safety of the norm. Recognition of an anomaly leads to anxiety and then to suppression or avoidance (Douglas 1966, 5). The refugee is the anomaly who may transgress the border and pollute the clean inside. Anxiety over a potential source of pollution leads to a movement to suppress the source, to keep it out and avoid contagion—to, in other words, border control and manipulation of who is and is not included in the refugee category.

POLLUTED EUROPE

States claim they are being flooded with asylum seekers who bring crime and terrorism into their societies; asylum seekers are a pollutant. If the movement of such individuals is not controlled, if the pollutant and dirt is allowed in, the end result could be chaos, disorder, and a clash of civilizations (Ceyhan and Tsoukala 2002, 22). Where this movement is seemingly illegal or out of control, the insecurity increases. Accordingly, refugees are constructed as a high-level threat, a threat that is—according to Ceyhan and Tsoukala—articulated around four axes: the socioeconomic, the securitarian, the identitarian, and the political (Ceyhan and Tsoukala 2002, 24). Within this discourse of *danger*, several myths are perpetuated that construct the refugee as a potential source of pollution: the refugee causes unemployment and so pollutes the economy; the presence of the refugee indicates a loss of control over sovereign borders and so pollutes authority; the refugee weakens the national identity of the host society and so pollutes social cohesion; and the refugee brings disease and so pollutes the well-being of citizens. European societies are made to believe that there are precarious gaps in our borders, which if not closed can allow dirt and disease to seep through.

Yet this discourse suggests that states have always had complete control over their borders and that, before the onset of the so-called asylum crisis, borders were totally impermeable. Of course, just as sovereignty has never been absolute, so state control over borders has long been compromised by the flow of individuals in and out of states' territory, both legally and illegally.[1] But states like to endow their borders with a sense of the magical; borders are highly symbolic, demarcating the limits of political control and the boundaries between inside and outside, them and us: "Each discourse associating the control of migration flows to the reinforcement of border-control measures relies . . . on the myth of the existence of the sovereign state fully able to control its territory" (Ceyhan and Tsoukala 2002, 34). Borders represent sovereignty, designate national identities, and protect citizens against external threats of pollution; borders are accordingly charged with power. It is this power inherent in borders that keeps the securitarian discourse of a *European asylum crisis* alive. The European asylum crisis is constructed around several false truths. Not only is this a "new crisis," with "different" refugees coming from beyond Europe and numbers "bigger" than ever before, but further asylum seekers are upsetting the homogeneous national communities of European societies that existed in the past. The exclusion of these outsiders will allow for the reestablishment of pure, national communities.

This discourse is in direct relationship to the way refugee debates were framed and attempted to be solved in the interwar period when the issue first emerged, or was constructed, as a mass problem: giving every nation a national homeland protected in a state would solve, it was argued, the problem of persecution of minorities and ethnic conflict, all of which could be time bombs with the potential to decompress into refugee flows. Contemporary rhetoric harks back to this thinking, extolling the political, socioeconomic, and cultural benefits of societies "unpolluted" by refugees from foreign lands, societies that will be more peaceful, safe environments free from the threat of international crime and alien, undemocratic values. But looking back to the League of Nations era, we are reminded that attempts to create homogeneous nation-states as a condition of a stable international system failed miserably and refugee flows continued unabated. The political hyperbole and press myths that seek to justify the exclusion of refugees from European societies today on the bases of their potential to weaken national traditions and of the threat they pose to the very survival of national communities can be seen as the reproduction of a past that never existed. In the words of Huysmans, "arbitrarily defined threats are connected in a global discourse that produces artificial homogeneity" (Huysmans 1995, 56).

Where there is a danger of pollution, there is fear, and where there is fear, there is a need for security. Consequently, "the actors playing the security game desire to be free from that which scares them: the threat(s)" (Huysmans 1995, 54). At the same time, notes Huysmans, "the security story is a centring tale." Accordingly, the threatened identity is found in the center. This placement acts to create a periphery, "the outside environment, which is constituted from the position of the centre and where the threats are located" (Huysmans 1995, 55). Hence the actors fearing the threat are within the state-citizen relationship; the threatening refugee is excluded, outside. Ironically, this means that it is the actors inside who are seen to be in need of protection from the threat, not the vulnerable and marginalized refugee. If the center wants to survive, it has to control the periphery (Huysmans 1995, 59), and it has to do so *at* the periphery, leading to the institution of norms and legislative instruments to deal with refugees in an attempt to prevent pollution that could bring chaos. In short, the refugee (at the periphery) poses a security problem (to the inside or center), one which is fundamentally a potential pollution problem. Accordingly, borders must be sealed to prevent contamination of the pure inside.

The logic of security and the asylum crisis it produces, therefore, has come to characterize the normative understanding of the refugee problem in contemporary European politics, and rests on an understanding of the refugee as a disruption and a pollutant. Note the dichotomy within migration policy of the European Union itself: extra-EU migration is posited under a security ethos, while intra-EU movement of persons is to be found under a liberalization ethos (Kostakopoulou 2000, 506). Indeed the EU appears to have imitated individual state policies in trying to conjure up an image of the safe inside and the dangerous outside, an outside that has the potential to disrupt the homogeneous way of life and the "national interest" within. The Treaty on European Union has as its objective the creation of "an area of freedom, security and justice in which the free movement of persons is to be assured in conjunction with appropriate measures with respect to external border controls, immigration, asylum and the prevention and combating of crime" (Article 2, Treaty on European Union). Safety inside is not possible without measures to keep external threats under control. Accordingly, security becomes a given: the state, or in this case the EU, is the given norm, the refugee the given other, and security the given threat that exists in the relationship among them. Just as the refugee is silenced in this statist understanding of her, so the political process of critically articulating and defining what constitutes a security threat or problem is also rendered silent.

Inside, Europe is clean and secure; hence internal borders have been brought down and can remain open. But the outside is increasingly dangerous and polluted and needs to be clearly defined. Debates arise periodically in the United Kingdom regarding whether all individuals who enter the country should be screened to stop the apparent spread of infectious diseases. These debates are an example of this thinking brought to its logical conclusion. Politicians have proposed health tests of all new immigrants and asylum seekers before they are allowed to stay in the United Kingdom; asylum seekers would be detained until the tests had been carried out and the all-clear obtained.[2] These debates clearly overlook the fact that the biggest health threats are more likely to be to the asylum seekers themselves during their journey to Europe. Further, "the stress associated with being up-rooted, and the family disorganization that often accompanies it, are additionally erosive of [refugees'] health" (Carballo and Siem 1996, 35).

But these proposals also ignore the fact that traditional ways of coping with international health risks, such as tighter border controls, no longer work, as was seen with the rapid spread of the SARS virus during the first half of 2003, and seen again in current scares surrounding the avian flu epidemic. Screening, it would seem, is merely portrayed as "an easy enough and necessary way by which to raise a barrier to the spread of disease" (Goodwin-Gill 1996a, 64). The proposals once more reify the border as the impermeable concept it has never been. Such thinking also assumes that control of individuals, who might otherwise bring danger and cause pollution, must happen primarily at the border (Huysmans 2000, 759), thus reinforcing the idea of the border as a site of danger. Apart from the obvious undermining of international legal obligations not to reject asylum seekers on grounds of ill health, and the risk of such a policy having the opposite effect and actually driving HIV-infected immigrants underground, for example, this thinking has a clear focus on a constructed external danger that could pollute the inside if not controlled. As Carballo and Siem remark, "immigrant health policies reflect prevailing public attitudes and fears" (Carballo and Siem 1996, 33), and this is seen in the image of the diseased other who could contaminate the pure inside. In the words of Goodwin-Gill, "the power of illness combined with that [of the other] to generate arbitrary and emotional responses is nowhere more evident than in state measures to relate control over entry to their territory to HIV" (Goodwin-Gill 1996a, 55). The inside is pure, clean, and homogeneous, and pollution by refugees represents a type of contagion or disease. The refugee becomes a pollutant rather than a victim.

CONTAINMENT AND "SOURCE REDUCTION"

Pollution is contamination by the discharge of harmful substances. The metaphor of environmental pollution is a useful way of conceptualizing how European states conceive refugees and the supposed threat they carry, as well as the means by which this human pollution may be contained. International legislation aimed at preventing environmental pollution holds that, where possible, pollution should be prevented or reduced at its source. This is known as source reduction, which would act to reduce or eliminate the creation of pollutants before they have the chance to spread and contaminate clean things; *source reduction* means any practice that "reduces the amount of any hazardous substance, pollutant, or contaminant entering any waste stream or otherwise released into the environment . . . and reduces the hazards to public health and the environment associated with the release of such substances, pollutants, or contaminants." It is the best means of improving environmental protection, by avoiding the generation of waste and harmful emissions. Importantly, "source reduction makes the regulatory system more efficient by reducing the need for end-of-pipe environmental control by government." Reducing quantities of hazardous substances and increasing efficiency of operations means protecting human health, strengthening economic well-being, and preserving the environment. Further, disposal costs are reduced when the volume of waste is decreased.[3]

Europe has become fixated on keeping refugees far from its territorial borders. New ways of attempting to deal with the flow of asylum seekers are constantly being introduced, many with the aim of reducing numbers. Policies aimed at preventing the refugees' arrival in the first place include the "safe haven" concept, "in-country processing," and "safe country" lists. Pollution prevention is easier if caught at source and much more expensive to deal with at the "end of the pipe." Thus source reduction is imagined as a more economical way of dealing with the so-called refugee problem.

A whole multitude of terms has grown to describe the new concept of confining refugees to their country of origin: *safety zones, open relief centers, security zones, safe haven zones, safe corridor,* and *safety corridors* (Landgren 1995, 436). Such concepts can, at first sight, hint at the way humanitarian concerns take the form of intervention and act to override the sacred concept of sovereignty. The understanding of *safe areas* in humanitarian law, as Landgren notes, is that of "a location within the disputed country or territory, neutral and free of belligerent activity, to which humanitarian access is ensured" (Landgren 1995, 438). Yet more and more refugee flows stem from internal conflicts that have as their objective the forced removal or elimination of other ethnic groups. Safety areas were originally developed for the protection of civilians in wartime, yet finding safe areas within war zones is somewhat of a contradiction in terms in an age when 90 percent of war casualties are civilians (Landgren 1995, 437). Hence supposed solutions such as safe havens can be understood as attempts by European states to keep refugees at a distance. The nearer the refugee gets to the border, the greater the risk of pollution. Border controls and asylum determination systems are expensive, and repatriation of those not granted refugee status is often difficult. It may even be better to pay countries of refugee origin to keep putative refugees there. This payment may be in the form of help, such as aid, or of penalty, such as cutting off trade agreements. The outcome is the same: reduced refugee flows to Europe, and thus a reduced threat of pollution.

Hence, the international community now seeks protection on behalf of refugees more and more within the country of origin. Arguments put forward by European governments in favor of within-country protection justify it on the grounds that people prefer to stay close to similar national, religious, or ethnic groups; that it is no longer always the state that is the source of persecution, and the state may be able to provide sanctuary from a persecuting nonstate group in another part of the country; and that it is easier for refugees to go home, once the factors that caused the flight have been removed, if they remain nearby. Yet governments also consider that it is harder to repatriate refugees across long distances, especially after a significant period of time when a certain degree of integration and adjustment into the host community may have taken place. In other words, pollution is more likely to occur at the border and more difficult to clean up once it has taken place, so it is best tackled at the source.

Strategies of containment and in-country protection are an attempt to place refugees in areas of so-called protection that are no longer dependent on state borders, so as to keep danger and pollution as far away as possible: "A new paradigm is emerging by which refugee flows are prevented before asylum seekers cross an international border, the definitional trip-wire that heretofore has marked the threshold step in the world's response to refugees" (Frelick 1993, 5). Refugees are pushed farther and farther from Europe by the progressive extension of the *sovereign* borders of European states without an extension of their *territorial* borders. The result is that the borders of the state extend farther than ever, far from the state's territorial jurisdiction and its physical borders, and territory and sovereignty are decoupled. Protection is effectively deterritorialized. This development, too, points to the arbitrary nature of international borders as physical symbols of sustaining political communities—suggesting, instead, that the contents of substantive sovereignty could, in theory, be divorced from territory.

No longer the exilic bias whereby the refugee must flee her country of origin, cross an international border, and become a ward of the international community; now the emphasis is on ameliorating conditions back home to stop the refugee flow in the first place. However, apart from an obvious desire of European states to keep these so-called new refugees as far from their territory as possible, what this approach shows is, as Chimni stresses, a trend in internalist interpretations of the root causes of refugee creation (Chimni 1998). By blaming conditions in the country of origin for the creation of refugees, the international community can lay the burden of stopping the refugee flow on that country of origin: if the country of origin fails to comply with norms of good governance that would ensure the protection of its citizens, it is its own responsibility to improve the internal situation. This idea of good governance relies on a discourse, again, that sees refugee movements as the anomaly in an otherwise stable and sedentary international states system, in which belonging to a territory, with clearly demarcated borders, is the ideal condition. It is an idealization of *home* that links the individual with a space of territory within a political entity, reducing home to a matter of geopolitics. The border maintains a "politics of place" (Connolly cited in Warner 1994, 168) that never was.

The failure of refugee-producing countries to ensure the protection of their citizens means the creation of pollution, and states have a duty not to pollute other states or the international system as a whole. Hence, European states can attempt to keep refugees in their state or region of origin by calling for an improvement in the human rights situation there. But this move ignores two factors: first, that external factors play a large part in the creation of conditions that lead to refugee flows, not least because refugees are intricately wrapped up in the workings of the entire international system; and second, that the "international community" has a role to play in the international human rights regime, which implies a conception of negative sovereignty implicating the international system as a whole, not just conditions of positive sovereignty within states (Haddad 2003b).

In short, ideas relating to *regional protection* have the potential, if used carefully, to increase access to and levels of protection for would-be refugees close to their countries of origin. But if such ideas are exploited, it is not inconceivable to begin to see the export of asylum seekers to developing countries much as developed nations export their toxic waste. This exploitation is similar to the Australian model whereby Papua New Guinea and Nauru are paid by Australia to act as transit centers, affording Australia a clean, unpolluted territory. It is an ironic coincidence that environment and immigration policy are both discussed in depth at the European level with the assumption that such issues are no longer possible for individual states to deal with alone. Each issue transcends the borders of the state and thus any effective solution relies on a common, supranational approach. Global refugee flows are to be treated, or purified, in much the same way as environmental concerns, if Europe is to be kept free of pollution.

CONCLUSION

The concept of pollution can explain and highlight the fragility of territorial conceptions of community and identity. The border can be seen as an unsatisfactory and unstable line that fails to keep the inside in and the outside out. The refugee is one actor who threatens territorial concepts of community and identity by crossing the border precariously, bringing the danger of pollution to the clean, safe inside. Discourses about pollution underscore the nature of territoriality in international relations; by relating such discourses to the figure of the estranged refugee, the danger and instability of the border can be demonstrated. We are thus faced with the arbitrariness and limits of notions of political space, in addition to the restrictions of ethical and moral obligations.

An appreciation of the way in which European states recognize the "refugee problem" as a problem of pollution can help illustrate the refugee discourse in contemporary Europe. The construction of the so-called refugee problem as a security problem in Europe rests on an assumption that the refugee is a polluting person who has the potential to contaminate European states and societies. To prevent such pollution taking place, European states have undertaken various strategies of containment or source reduction. These aim to keep putative refugees in their country of origin, so that the risk of pollution to Europe is kept as small as possible. The humanitarian logic that should underpin refugee debates often appears to be lacking. The border becomes a blurred and unsatisfactory concept—something to keep the "polluting" refugee out, rather than to protect the citizen within. And, as Mills notes, the unsatisfactory border is continuously reinvented by policy and practice: "The conceptual abstraction is reified by such phenomena as border patrols and passports. Border patrols attempt to keep out the undesirables and passports help to regulate the temporarily desirables" (Mills 1996, 77).

The refugee is simply constructed as a dangerous and polluting person, such that the arguments justifying strict border controls are reinforced, and keeping the refugee as far as possible from state borders is assumed the optimum solution. The effects of this discourse of danger and pollution on individuals who have been forced across borders in search of protection are obvious, and a shift in the way "solutions" to the "problem" are played out in the international arena would, accordingly, seem called for. We must reimagine the border and our responses to border-crossing individuals.

NOTES

1. Note the popularist rhetoric used to denounce the presence of *illegal* asylum seekers, whereas the two terms are in fact misnomers: since everyone has the right to seek asylum, it is not literally possible to do so illegally.

2. For example, in August 2003 the British Conservative Party depicted London as "the TB capital of the western world" (*Guardian,* August 4, 2003). Tabloid coverage, such as the *Sun's* claim that immigrants are "polluted with terrorism and disease" and should be forcibly tested for HIV and hepatitis B, helped bolster this image (cited in the *Observer,* February 2, 2003).

3. See http://www.p2.org/about/nppr_p2.cfm.

WORKS CITED

Arendt, Hannah. 1966. *The Origins of Totalitarianism.* New York: Harcourt, Brace and World.

Carballo, Manuel, and Harald Siem. 1996. "Migration, Migration Policy, and AIDS." In *Crossing Borders: Migration, Ethnicity, and AIDS,* ed. Mary Haour-Knipe and Richard Rector. London: Taylor and Francis.

Ceyhan, Ayse, and Anastassia Tsoukala. 2002. "The Securitization of Migration in Western Societies: Ambivalent Discourses and Politics." *Alternatives* 27, supplement: 21–39.

Chimni, B. S. 1998. "The Geopolitics of Refugee Studies: A View from the South." *Journal of Refugee Studies* 11, no. 4: 350–74.

Douglas, Mary. 1966. *Purity and Danger: An Analysis of the Concepts of Pollution and Taboo.* London: Routledge.

Frelick, Bill. 1993. "Preventing Refugee Flows: Protection or Peril?" *World Refugee Survey, 1993.* Washington, D.C.: United States Committee for Refugees.

Goodwin-Gill, Guy. 1996a. "AIDS and HIV, Migrants and Refugees: International Legal and Human Rights Dimensions." In *Crossing Borders: Migration, Ethnicity, and AIDS*, ed. Mary Haour-Knipe and Richard Rector. London: Taylor and Francis.

———. 1996b. *The Refugee in International Law*. 2nd ed. Oxford, U.K.: Clarendon Press.

Gwynn, Robin D. 1985. *Huguenot Heritage: The History and Contribution of the Huguenots in Britain*. London: Routledge and Kegan Paul.

Haddad, Emma. 2003a. "The Refugee: The Individual between Sovereigns." *Global Society* 17, no. 3: 297–322.

———. 2003b. "Refugee Protection: A Clash of Values." *International Journal of Human Rights* 7, no. 3: 1–26.

Huysmans, Jef. 1995. "Migrants as a Security Problem: Dangers of 'Securitizing' Societal Issues." In *Migration and European Integration: The Dynamics of Inclusion and Exclusion*, ed. Robert Miles and Dietrich Thränhardt. London: Pinter Publishers.

———. 2000. "The European Union and the Securitization of Migration." *Journal of Common Market Studies* 38, no. 5: 751–77.

Kostakopoulou, Dora. 2000. "The 'Protective Union': Change and Continuity in Migration Law and Policy in Post-Amsterdam Europe." *Journal of Common Market Studies* 38, no. 3: 497–518.

Landgren, Karin. 1995. "Safety Zones and International Protection: A Dark Grey Area." *International Journal of Refugee Law* 8, no. 3: 416–32.

Malkki, Liisa H. 1995a. *Purity and Exile: Violence, Memory, and National Cosmology among Hutu Refugees in Tanzania*. Chicago: University of Chicago Press.

———. 1995b. "Refugees and Exile: From 'Refugee Studies' to the National Order of Things." *Annual Review of Anthropology* 24: 495–523.

Mills, Kurt. 1996. "Permeable Borders: Human Migration and Sovereignty." *Global Society* 10, no. 2: 77–106.

Stuart, Kathy. 1999. *Defiled Trades and Social Outcasts: Honor and Ritual Pollution in Early Modern Germany*. Cambridge: Cambridge University Press.

Thompson, Charles D. 2001. *Maya Identities and the Violence of Place: Borders Bleed*. Aldershot, U.K.: Ashgate Publishing.

Tuitt, Patricia. 1999. "Rethinking the Refugee Concept." In *Refugee Rights and Realities: Evolving International Concepts and Realities*, ed. Frances Nicholson and Patrick M. Twomey. New York: Cambridge University Press.

Warner, Daniel. 1994. "Voluntary Repatriation and the Meaning of Return to Home: A Critique of Liberal Mathematics." *Journal of Refugee Studies* 7, no. 2/3: 160–74.

6

Violence, Subversion, and Creativity in the Thai–Malaysian Borderland

ALEXANDER HORSTMANN

When the military arrived in a village in Narathiwat province in the week ending September 25, 2005, only women and children remained, holding a banner and saying: "You are the terrorist."

This courageous act followed a bloody confrontation of the Thai military and the frustrated and angry Malay youth, in which paramilitary forces suddenly appeared at the tea shop in the village to kill some Malay on the blacklist, and in which Malay youth retaliated by taking two soldiers hostage, later to murder them. Anticipating the arrival of the Thai military in force, 131 people escaped across the border to Malaysia. Meanwhile, some women in purdah and some children formed a human barrier to block the soldiers. The women's husbands all left the village in fear of joining 109 documented cases of disappeared people, mainly male, who were never seen again. The organization of Young Muslims in Yala tried hard to track every case of disappearance, but had to stop after being threatened by state authorities. Many women lost their husbands, leaving hundreds of children in an orphanage.

On the one hand, the female human barrier symbolized the current atmosphere of fear in which Pattani (or Patani, in Malay spelling) Malay people are estranged from the Thai state, which imposes its own notion of Thai national identity on Malay-speaking

Figure 6.1. Where is the justice? Photograph courtesy of Amporn Marddent.

Muslims. But the human barricade was also a showdown demonstrating the creativity of border people, which in this case involved the symbolic acts of unarmed women to avoid a full military assault on the village. By using the label *terrorist* for the state, people in Narathiwat were creatively changing the meaning of *terror* from its strong association with radical Islam and Islamic terrorist networks. Violence coming from a brutal state, which declared martial law, is countered by subversion in which people at the border use the border as a resource. By crossing the border, Malay people from Pattani can disappear into Malay villages that were originally set up by Pattani Malay migrants in northern Malaysia.

People in southern Thailand have suffered from unprecedented violence beginning in January 2004. Although few anthropologists working in the region expected this level of bloodshed, scholars repeatedly underlined the roots of structural violence in the southern border provinces (including in Chaiwat 1987). The reasons for the resurgence of the shocking violence in these days are to be sought in present sociopolitical constellations as well as in long-term tendencies of a violent borderland.[1]

As Thongchai points out, Pattani was colonized by Siam after the defeat of the Pattani Malay sultanate, and large sections of the local population were forcibly relocated to other regions of Thailand as prisoners of war (Thongchai 2002). From this time, Malay-speaking people of Pattani felt treated as inferior and second-class citizens, the estrangement being aggravated by poverty, violations of human rights, and chauvinistic behavior by the Thai government (Wan Kadir 1990). In many areas of the southern border provinces, the Thai state is perceived as a stranger, with officials coming predominantly from the Thai Buddhist provinces. On the other hand, chauvinistic Thai policies had the unintended effect (resembling what happened in the southern Philippines) of producing the imagination of a shared (in this case, Malay) culture among a heterogeneous and loosely organized population (Horstmann 2002). This semicolonial constellation, in addition to the supply of small weapons, sex work, and drug trade activities, produced a violent *borderscape* characterized by a permanent state of exception (see the introduction to this volume).[2]

My own research reveals the formation of a border society with a high potential for violence both within and from without the borderlands (Horstmann 2005). In the border regions, a specific political ecology is developing, with a large illegal economy as a basis of transactions. Illegal border transactions include illegal logging, especially in Yala, the methamphetamine trade, human trafficking, and smuggling (smuggling of small arms, logs, foodstuff, petrol, and consumer products). The border economy is sustained by coalitions of corrupt officials, the local mafia, military, the Border Patrol Police, and the Royal Thai Police. The illegal border economy is a multimillion dollar business, which provides large rents for complicit officials. The recent "war on drugs" by Prime Minister Thaksin Shinawatra brought vast pressure on government officials and possibly induced a huge upsurge of homicides as drug dealers in major positions were eager to silence small dealers by eliminating them. The government further dissolved the balance by replacing the military with fresh police forces from Bangkok, who abducted and killed more than twenty former separatist fighters who had surrendered and since become informers of the military. These informers were believed to be safe from state threat. Thaksin was eager to dissolve military forces associated with the Democratic Party; however, he made a crucial mistake by dissolving avenues for mediation between communal leaders and the government (McCargo 2005).

Malay-speaking Muslim people in the southern border provinces of Thailand have a long history of suffering. It is important to keep in mind the social memory villagers hold of former atrocities, as these experiences continue to shape their perception of the Thai government (see Chaiwat 2005).

In the Muslim provinces of Pattani, Yala, and Narathiwat, and in parts of Satun and Songkla where Malay-speaking Muslims and Thai-speaking Buddhists coexist, shared cultural institutions could not develop, partly because Thai policies included resettlement to unbalance the population in favor of the Buddhist majority. In contrast, in the Thai-speaking provinces of southern Thailand, the world of ancestor spirits is a powerful culture in which the hybridization of local cosmologies takes place, as follows. Some of the oldest Buddhist monasteries can be found in the region around Lake Songkla, monasteries that coexist with Islamic mosques in mixed neighbourhoods. In the Lake Songkla region, rituals with mixed religious elements can be observed, as both Buddhists and Muslims believe in common ancestors *(taa jai)*. In the Pattani region, pre-Islamic Hindu beliefs and Malay popular culture are reproduced among the Malay group, but Malays are careful not to blend Malay Muslim and Buddhist traditions. Chavivun Prachuabmoh argued that Malay women play an important role in maintaining boundaries to the Buddhist "other" (Horstmann 2004).

The Pattani Bay basin was the area of one of the earliest sultanates to introduce Islam. Sufi brotherhoods of the Naqshbandi Sufi order introduced these as early as the twelfth or thirteenth centuries. Although few studies exist on the localization of Islam in Pattani, the flexibility of Southeast Asian systems in integrating the Qur'an in local contexts should be underlined. However, from the seventeenth and eighteenth centuries, networks of Malay Indonesian and Middle Eastern *ulama* were gradually gaining influence (Azra 2004). In the 1980s, increasing pressure was exercised by *ulama* coming back from Egypt and Saudi Arabia, occupying key positions in traditional Islamic schools (in Malay, *pondoks*). Orthodox reformist thinking of the new leaders questioned many of the premises of the traditional pondoks. A new generation of students were attracted by the new orthodox knowledge offered by semiprivate schools and colleges that were lavishly financed by foundations from Brunei and Saudi Arabia (see Madmarn 2002).[3]

Dorairajoo argues that the suicidal operation of 107 Muslim lightly armed youth who died in the memorial Kru Se mosque after attacking police and military posts represented the beginning of a new and more dangerous round in the protracted conflict of the Thai state and the Malay Muslim minority in the Thai border provinces of Pattani, Yala, and Narathiwat (Dorairajoo 2004a). Although the separatist movement in the 1960s was essentially an elitist movement, the recent violence seems to have been instigated by a new group or groups of individuals not connected to the old separatist groups. Dorairajoo points out that the new violence may have been instigated by some pondok teachers within Thailand, able to exert greater authority over Muslim youth than were separatist organizations led by Malay elites who, operating outside of Thailand, have been lately out of touch with Thai–Malay youth. After the massacre at Kru Se mosque, the police identified as insurgents certain young people who then retreated into their homes after carrying out attacks not only against police but also against Buddhist and Muslim civilians. Historian Nithi Eisriwongse has argued in a recent essay that the spiritual preparation of Muslim fighters resembles the small people's revolt of the messianic Buddhist utopian movements in Isaan (Northeast Thailand), where peasants believed themselves invulnerable against bullets.

In fact, the violence in southern Thailand may have been triggered by many factors, including insurgent Islam; disillusioned young Malay Muslims; the infighting among competing police and military factions, million-dollar drug dealers, warlords, and mafia-like racketeers controlling smuggling and prostitution rings; and a government whose unscrupulous use of violence in the "war on drugs," involving human rights violations, has contributed to an overall atmosphere of fear. Underlying the conflict is a notion of structural violence, in which Pattani Muslims constantly perceive the administrative and military intervention of the Thai state as an attack on the foundations of Malay Islamic culture. The persistence of poverty, drug addiction, and infant mortality is perceived as due to the economic exploitation by strangers.

Kraus (1982) observed in the 1970s the emergence of a fundamentalist movement that might recruit from urban as well as rural elements. In my own work, I have argued that there was a very important shift in the 1990s away from ethnic identity toward religious, namely, Muslim, identity, as local Muslims become ever more entangled in greater circuits of Muslim pilgrimage, education, and global missionary (Da'wa) movements in the Islamic world. Unlike the inward-looking Thai Buddhists, for whom Thai identity is intimately bound to Thai nationalism and to the Buddhist order (Thai *sangha*), Muslims have become exposed to transnational organizations and networks and to much wider trends of religious (Islamic) resurgence, including violent forms and organizations (see Horstmann 2002 and 2004).

Further, Dorairajoo (2004) notes that the American-led "war against terrorism" has been interpreted by many Muslims as a war on Islam, and has given substantial fodder to the Islam-under-threat mentality. In southern Thailand, where there has been a history of state aggression against the area's Malay Muslims, the Islam/Muslims-under-threat mentality has great currency and has easily won converts. Young Muslims, being highly mobile and media-informed, are extremely agitated by graphic images or reports of Iraqi prisoners abused by American soldiers or that versions of the holy Qur'an were demolished and flushed down the toilet in the Guantanamo prison camp. Young Muslims are vulnerable to analogizing the suffering of Muslims in Bosnia, Palestine, or Afghanistan with their own situation in Pattani, no matter how disconnected such contexts may be.

The circulation of graphic images of Muslim suffering in the community media (radio, magazine, newspaper) in southern Thailand is like pouring oil onto the fire for young Muslims, mostly estranged from the Thai monarchy since they live in a collapsing society in which poverty and corruption are rampant. After the brutal crackdown of a peaceful demonstration in Takbai, of which videos have been circulated on both sides of the border, there has been a deepening division within the Muslim community, particularly among college students, many of whom feel unable to accept the brutal behavior by the military any longer. The call by the Thai queen for the armament of Thai Buddhist villagers in the border provinces further demolishes confidence in the institution of the Thai monarchy.

Saroja Dorairajoo (2004, 469) states, "The increased and unabated killing of policemen, civil servants, and police informants by motorcycle gunmen who are mostly Muslim youth, since the 28 April massacre, point to the success that this new organization has had in arousing young Thai–Malays to rebel against the state."

Conflicts between the police and the military may account for some of the violence. Members of the security forces (Border Patrol Police) are involved in the trafficking of people, narcotics, and small arms. Organized crime is rampant in the southern border provinces and is facilitated by systemic official corruption.

This impression was confirmed in a recent meeting of the author with representatives of the BERSATU (the United Front for the Independence of Pattani, or Bersatu—Malay for "united") alliance in political exile in Hamburg, Germany. The current insurgency has arisen from a new radical organization that can mobilize mainly young, frustrated Malay people, along with some splinter groups from BRN, PULO, or Mujahedin Patani. Although the representatives distance themselves from the terror acts of the insurgents, there is some fascination with the selfless courage of young Malays who are able to irritate the Thai authorities in a low-scale guerrilla war. The human rights abuses of the Thai state play into the hands of the insurgents, as they can easily mobilize the hearts and minds of young people agitated by the stories and images of torture and disappearances. The Thai state is also playing into the hands of the military, who can justify large budgets for the defense of sovereignty in the Thai–Malaysian borderland.

HUMAN RIGHTS ABUSES, THE WAR ON DRUGS, ABDUCTIONS, AND DISAPPEARANCES

On January 28, 2003, Prime Minister Thaksin Shinawatra announced that a "war on drugs" would be waged on drug dealers. The use of the term *war* was apt, as more than two thousand people in Thailand were killed as the government effectively declared open season on those accused in the drug trade (Human Rights Watch 2004). Human Rights Watch researchers found that "the crackdown saw rampant human rights violations, including government promotion of violence against drug suspects, extrajudicial executions, blacklisting of drug suspects without due process, intimidation of human rights defenders, and violence and other breaches of due process by the Royal Thai police." Thailand's "war on drugs" began in response to a boom in methaphetamines, locally known as *ya baa* ("crazy pills"). Most ya baa are produced and smuggled from neighboring Burma or, to a lesser extent, Laos, creating large opportunities for enrichment by local mafia and officials alike. In southern Thailand, ya baa are smuggled from

Burma and on to neighboring Malaysia across the Thai–Malaysian border, with huge profits for local warlords who benefit from the special status of the southern Thai border provinces.

Deviating sharply from Thailand's efforts to build the rule of law, then–Prime Minister Thaksin called for his "war on drugs" to be conducted on the basis of an eye for an eye. Prime Minister's Order 29/B. E. 2546, signed on January 28, 2003, called for the absolute suppression of drug trafficking by means "ranging from soft to harsh including the most absolutely severe charges subject to the situation." The document stated, "If a person is charged with a drug offence, that person will be regarded as a dangerous person who is threatening social and national security."

Throughout the drug war, Thaksin and other government leaders repeatedly appeared to give the green light to violence against suspected drug dealers. In the first three-month phase of the crackdown that began on February 1, 2003, the Royal Thai Police reported that some 2,275 alleged drug criminals had been killed. More than 70,000 people allegedly involved in the drug trade were arrested. Murder warrants were carried out in southern Thailand, too. On February 26, a sixteen-month-old-baby, nicknamed "Ice," was in her mother's arms when she and her mother, Raiwan Khwanthongyen, 38, were shot and killed by an unknown gunman in Sa Dao District, Songhkla. Police in Songkla declined an interview with Human Rights Watch and have not found the killer (Human Rights Watch 2004, 10–11). At the beginning of May 2003, Thaksin declared "victory" in the "war on drugs" and announced a second phase that would last until the following December.

The "war on drugs" coincided and overlapped with the new violence in southern Thailand, nourishing and contributing to it with no checks and balances by civil society. Torture, rape, and extrajudicial killings were systematically carried out against Muslim suspects by the Royal Thai Police in the wake of the violence in 2004.

Although the "war on drugs" concentrated on Thailand's northern border with Myanmar, the southern border provinces were among the areas most heavily affected by the antidrug operations. More significantly, the "war on drugs" showed the handwriting of the current prime minister, Thaksin Shinawatra, his rhetorical and practical use of "war," and his determination to counter "criminals, bandits and drug addicts" in the south with an iron fist.

The portrayal of the southern border region as a problem region justified large military budgets. The civil population was squeezed between the claims of military and separatist organizations. According to religious leaders, more than one hundred residents of the four southern provinces were abducted and killed in the four months after January 4, 2004 (International Crisis Group 2005, 36).

In a speech to the Santichon Foundation in Bangkok on February 27, 2004, human rights lawyer Somchai Neelaphaijit publicly accused the police of torturing his clients. On March 11 of that year, he submitted an open letter to five independent bodies calling on them to investigate his allegations. He disappeared the following day and has remained missing since.[4] Many observers believe that Somchai may be dead.

Somchai was defending five Muslims who were released but detained again on murder charges. The five alleged that confessions were extracted under police torture, that police held guns to their heads and in their mouths, that they were urinated on, and that police took them to a beach, blindfolded, and told them if they did not confess they would be thrown into the sea.

Arrests, searches, and seizures increased after the army camp raid of January 4, 2004. On January 14, 2004, a Narathiwat Muslim leader, Matohlafi Maesae, was abducted from his home in Bacho district by ten unidentified armed men (in the style of drug-related death squads). His body was recovered three days later, bearing torture marks.

The extraordinary powers given to the police amounted to a carte blanche to target and eliminate awkward locals for extrajudicial execution. Among those killed were long-standing informers with close ties to the military (McCargo 2005). According to local human rights activists, people are often more scared of the police than of "terrorists" in the south.

In this atmosphere of fear, Amporn Marddent, a native of Phuket, visited women whose sons, husbands, brothers, and other relatives were dead or disappeared during violence in the southern border provinces of Thailand. One informant, Adek (younger sister) Nori, from Pattani, recalled a violent incident she had witnessed in public: "Kak (elder sister) Nori saw a volunteer guard got killed in front of my house. It was terrible! And guess what, the shooter wore a Da'wah cloth [headscarf] but didn't look like a woman and a native from this region. That's what I've heard from talks in the coffee shop. The shooter wore a headscarf, sitting at the back of a motorcycle. Why did they have to do this? It was really scary! You must be careful, Kak. Don't know anything too much" (Amporn 2005, 2).

The story refers to the confusion and horror felt by local Muslims. The killer did not look like a woman or a native yet he displayed a female Islamic cloth. Muslim feminine symbols are being exploited as a tool by militant Muslim organizations. Further, Muslims who have Malay Muslim appearance are targeted by the Thai military, as the following story relates: "When we stopped at a checkpoint, we were afraid even though we had not done anything wrong. If anyone wears a Kapiyoh, he will be thoroughly searched, especially at the checkpoint near the Kampong-Kampong. Women and children were very terrified" (ibid., 3).

In addition to intimidation, sexual abuse of women is used as a tactic to terrify people and to shame them. The story of Yoh, recorded by Amporn (2005), exemplifies the horror and resentment created by the men in uniform who abuse their position: "When we shouted for help, they slapped our faces. My friend spit at them, they punched her stomach—making her choke. They used plaster to shut our mouths. He dragged me further—he did not say anything but immediately raped me. I begged him to stop, but he would not. He slapped and bit me, and it was very painful. He then tied my hands to a wooden log. Then he raped me again for the second time until he was satisfied. After that, he released me and said: 'Thanks, sister, Malays are good.'" The suffering is inflicted not only on a single woman, but on the Malays as a collective, as the insult "Thanks, sister, Malays are good" shows. As Amporn points out, the raping of women by a soldier demonstrates his superior power; as Yoh explains, "And [that man] said further, 'Go back and tell your bandit leader that if you shoot us, we will come back again.'" Rape or torture, besides the suffering inflicted, entails a threat to the local populations, saying, "If you revolt, we will do it again to you."

The mishandling of the Tak Bai demonstration was the most shocking experience for many local Muslims. Early on the morning of October 25, 2004, during the month of Ramadan, around fifteen hundred people congregated outside a small police station in the town of Tak Bai, southern Narathiwat. Plans for a rally had apparently been under way, and security officials had prepared. The Tak Bai protest was apparently planned with the hope of provoking a violent reaction from security forces. Police and soldiers tied the hands of protestors violently behind their backs and loaded them in trucks up to five or six layers deep to be taken to Inkayut army base for questioning. At the end of the journey, 78 protestors were dead, mostly of suffocation (International Crisis Group 2005). According to observers, militant groups used anger over the deaths and the government's insensitive handling of the situation for their own purposes.

THE EMERGENCE OF COMMUNAL VIOLENCE

The local population, Buddhist and Muslim, is squeezed between the almost daily assassinations by snipers, the campaigns of ethnic cleansing, and the human rights abuses, including arbitrary disappearances, torture, and rape, escalating the local conflict to new proportions.

Militant Muslim organizations are certainly interested in embarrassing the government by their ability to stage spectacular operations. In the days following the Tak Bai protest, assassinations increased markedly, in a deliberate attempt to provoke communal violence. In many localities of the southern border provinces, the fragile system of exchange between Malay-speaking Muslims and Thai-speaking Buddhists broke down as a result of the murders.

Several Buddhist civilians were killed, the killers leaving hand-written notes by the bodies, claiming retaliation for the Tak Bai deaths. Militants beheaded a village chief on November 2, 2004, leaving a note reading "For the innocents of Tak Bai." The direction of the random killings remains obscure. Snipers attacked officials, students, and judges. In order to create an atmosphere of fear, militant organizations included civilian Buddhists in their gruesome killings, beheading a local Buddhist farmer and a Buddhist vendor in a tea shop (in front of customers). Militant organizations were making use of Islamist rhetoric, especially the promotion of the concept of jihad and martyrdom.

The directions of the killings became eclectic, including Muslim civilians. A leaflet reads, "Muslims who are not Jihadists do not deserve to live." Leaflets were distributed advising Buddhists to leave the three provinces. The official Web site of the Pattani United Liberation Organization (PULO) gave a stern warning to foreigners not to visit the southern provinces.[5]

The escalating violence has resulted in a widespread collapse of communal relations and mistrust. Buddhist villagers find themselves in a precarious situation as they need to buy from Muslim villagers but do not buy from them anymore, out of fear. Many families, Buddhist and Muslim, escape from the violence by leaving for other provinces in Thailand. This flight from the southern border provinces may well be the intent of militant Muslim organizations who regard the flight of Buddhist population as territorial gain.

Buddhist monks have left about forty-four monasteries, in fear for their lives. The most devastating consequences of violence against Buddhist monks are primarily cultural, because the killings of unarmed monks deeply disturb any sense of community among Buddhist and Muslim communities. As Chaiwat (2004, 3) writes:

> On January 22, 2004, two men on a motorcycle used a long knife to slit the throat of a 64-year-old Buddhist monk [and so killing him]. The monk [had] just returned from his early morning round of alms-begging. Then, on January 24, three monks were attacked, two [died]. A young novice aged only 13 died in hospital after being attacked in the head [by someone] wielding a machete on a motorcycle, while another 65-year-old-monk was killed in the same manner. A third machete attack put another 25-year-old monk in a hospital, with serious injuries.

The lexicon of killings has changed. Two decades ago, there were incidents such as bus robberies in which Thai Buddhist passengers were separated from the Muslims and then shot. In 2003, the main targets of killings were policemen. In the first week of 2004, soldiers became targets. In 2005, everybody, Muslim and Buddhist, could become a target. As Chaiwat (2004) argues in "Facing the Demon Within," the knives did more than kill monks, for the acts cut deep into the cultural ties that bind people of different cultures together.

Conflicts in southern Thailand have mainly been vertical: between state authorities and the local people. Before the recent events, violent inter- and intra-communal conflicts were rare. The facts about the recent violence need to be carefully registered. Monks were killed and injured; the youngest who died was thirteen years old and the oldest was sixty-five years old. They were killed while returning from their daily alms begging. Neither their religious robes nor their ages could offer them cultural protection from acts of brutality. The cold-blooded killings sought to dismantle the fragile balance within the communities. In addition, the equally brutal state response, together with abject poverty and the tide of Islamic resurgence, is prone to create further escalation.

The population is terrified by television images of innocent Buddhists who apparently are slaughtered by "fanatical" Muslims, transforming Muslims into a great "Other" category. On the other hand, the bloodbath in Kru Se mosque and the slaying in Tak Bai make headlines in the Muslim press, creating anger and hate on the part of Muslim audiences in Thailand and the region. Meanwhile, the involvement of ultranationalist forces adds fuel to an already dangerous situation.

Around twenty thousand Village Scouts converged in Bangkok on November 28, 2004, to rally for peace in the south. The Village Scouts, a nationalist militia, was formerly used as one means to suppress student democracy activists (Bowie, 1997). Speakers noted that more nationalist militia would be recruited to "drive out" separatist enemies. According to reports, there are already over seventy thousand Village Scouts in the region. The killing of Buddhist civilians also provokes shrill and violent responses on the Internet; a right-wing militia, "Blood Siam," vowed to send vigilantes to the south to avenge the death of every Buddhist. The twenty thousand Village Scouts who rallied at Sanam Luang from throughout the kingdom, brandishing national flags and yelling a Cold War–era patriotic song, were treated to a promise by one of their leaders that their "separatist" enemies in the south would soon be driven out of the country. "They must be driven off Thai soil within one thousand days!" intoned a clearly emotional octogenarian, Major General Charoenrerk Charas-romrun, chair of the advisory board of Thailand's largest right-wing mass organization. "We shall fight to the death!"

In 1976 alone, about two million Thais became Village Scouts. The movement then fizzled out, in 1981, only to reemerge recently with the aim of protecting the country from the perceived threat of southern separatism.[6]

PEACE PROPOSALS

One reason for the emergence of hatred among Muslim villagers is the perceived neglect of the southern border provinces by the Thai government. Since September 11, Muslims have been criminalized in the name of the war against terrorism. Political defamation and economic deprivation together produce enormous frustration among young people vulnerable to recruitment by radical Islamic organizations.

Peace proposals in the region therefore focus on cultural autonomy, human rights, and issues of equality. Basically, all these peace proposals argue that local people should be allowed to take their fate in their own hands. Local cultural institutions and local expertise, these note, should be strengthened to reestablish a basic intercommunal sensitivity. This can be done by supporting grassroots efforts and civil institutions to solve pressing problems of people in the region. In parallel, the violence against Buddhist civilians must be delegitimized by revitalizing efforts of Muslim villagers and Thai Buddhist nongovernmental organizations to identify common aims and to defend natural resources in the region. With regard to this last aim, it is worth looking into the history of the Federation of Small-Scale Fishermen.

The overexploitation of marine resources by big trawlers was pushing fishing families into poverty. With their large push nets, commercial trawlers devoured the seabed, destroying entire ecosystems and depriving local fishers of their livelihoods. Villagers watched helplessly as the value of their daily catches declined. Intimate relations between the trawler bosses and government officials meant that laws banning large boats were rarely enforced. Without adequate representation, community resources (namely, marine life) were flowing to Bangkok's business interests.

High demand for shrimp, in the tourist industry and for export, produced large-scale shrimp farming in all coastal provinces, leading to environmental degradation from the high proportions of chemicals needed for the farms. The rapid erosion of the subsistence economy in the villages was pushing the villagers either to work in the shrimp farms that polluted their fields, or to work for the big trawlers catching all the fish in competition with the small-scale, mostly Muslim, fishers. The only other alternative was to look for jobs in Malaysia.

Shawn Crispin (2001) introduces us to Sukree Masaning from Tanyongpao, who initiated a coalition of five coastal villages, which became the root of the Federation. The local people received support from local academics who volunteered hard quantitative evidence on the destruction of coastal ecosystems, especially mangrove forests. The December 2004 tsunami showed again how dependent and vulnerable southern Thailand had become on the tourist industry, which often led to booming land prices and turf wars over ownership of primary land.

With the Federation, for the first time, resource management strategies are being designed locally, not centrally (Crispin 2001, 2). Long neglected, coastal conservation is being institutionalized, with local flair. "Coming together during crisis gives movement firepower," says Chaiwat Satha-Anand (2005), a professor of political science at Thammasat University and a member of the National Reconciliation Commission; "remaining free of bureaucracy keeps movements true to their localized needs."

Chaiwat criticizes the top-down approach of the Thaksin government, its ignorance regarding local solutions to pressing community problems. Martial law, a heavy-handed military approach, intimidation, and disappearances destroy the everyday work of nongovernmental organizations (NGOs), civil institutions, and social workers. In southern Narathiwat Province, locally organized volunteer groups are addressing grassroots needs, providing needed education, and helping, day by day, to heal historic wounds, after decades of central repression of religious practices.

The local knowledge of the pressing problems of the villagers may be crucial for finding solutions to the crisis. In Narathiwat, the most contested province in the conflict, over 170 community organizations have joined hands under an umbrella council, pooling resources to rent public radio time slots to disseminate their messages. "We are educating the public," says Hama Mayunu (2001, 3), a core member of Narathiwat's grassroots organizations; "that's something the government never promoted."

STATE OF EXCEPTION IN THE BORDERLAND

Although Thailand's "war on drugs" in no way targeted only people in the border provinces of Pattani, Yala, and Narathiwat—or the Thai–Burmese borderland in northern Thailand, for that matter—the carte blanche by Thaksin and leading government leaders for illegal executions by paramilitary forces presented a new quality of state terrorism in Thailand. The "war on drugs," being backed by a large proportion of the Thai population, effectively eradicated the fundamentals of basic human rights for so-called undesirable elements. The making of blacklists throughout the country meant that human rights abuses of the borderlands were now generalized and imported to the center. Thaksin made it clear that non-Thai elements of the drug trade attacked the integrity and sovereignty of the people in Thailand and should thus be eliminated.

It is very important to consider the continuation of these logics in the so-called war on terror. There was no outcry from the United States on the blunt violation of human rights in Thailand. The "war on terror" was a welcome invitation for Thaksin to review the question of sovereignty in the southern border provinces and to clamp down on "undesirable elements" in the lower south. Thaksin proceeded with the quick arrest of five key members of civic associations, foundations, and movements in Narathiwat who were accused of planning terrorist attacks for Al Qaeda. The evidence for such an allegation was very thin. When the lawyer Somchai went public with his findings of torture in police custody, he himself (as mentioned earlier in this chapter) went missing.

The state of emergency in the three border provinces allowed the state to deliberately arrest any suspect for interrogation. In the wake of the insurgence, four main communal leaders of Pattani Malay society were summoned to an army camp and were never seen again. State killings of selected *ustaz* (Islamic clerics) and Islamic teachers continued. In the Tak Bai case, Malay people demonstrating by peaceful means had their bodies humiliated and crushed in an unnecessary, brutal way. In this manner, the body of the Muslim insurgent is made to have no human identity. Stripped of basic human rights, the body is blindfolded and brought away to one of the secret torture prisons that were established in the worldwide "war on Islamic terror" in the wake of September 11. Meanwhile, documentations of alleged drug dealers continued, with the criminalization of Islamic teachers. Databanks were established to take detailed profiles of Islamic teachers in traditional Islamic schools, including information about their Islamic education and contacts in the Islamic world. Muslim students abroad were suspected of contacts with wider terrorist networks. In 2004, an Islamic teacher from Pattani province disappeared. Since the order of the so-called decree for emergency situations, seventy-six people have been taken into custody without trial.

MILITARIZATION AND THE MEDIA

Even while circulating images of the Iraq war agitated young Muslims in Pattani, images of the slaughter of a Buddhist monk outraged the Buddhist Thai majority in Thailand, leading to portentous statements on widely circulated Web sites—statements like "Kill Them All" or "Drive them out of the country." As with the war on drugs, the Thaksin government seems backed by large sections of the Thai Buddhist population. The media discourse has fueled the discourse on *otherness,* in which the Malay-speaking Muslim majority becomes vulnerable for being un-Thai. This "othering" is a very important issue as the Thai Buddhist population in Bangkok feel extremely uneasy with all issues of Islam and Muslims and do not know how to differentiate between groups and factions within Muslim society in Thailand. Thus, the increasingly militant language of "othering" is paralleled by an actual militarization of the civil population in the south. An example is a new unit of so-called iron ladies, which involves the training, under the protection of the Thai queen, of Thai Buddhist women to shoot. Further, the National Reconciliation Commission, in which Muslim intellectuals, university professors, and Buddhist monks discuss peace proposals, is criticized by the Pattani Buddhist Sangha, which, from vulnerability (as described earlier), has joined in with the ultranationalist tone of the far right. The training of paramilitary forces, the revitalization of the ultranationalist Village Scouts, and the armament of villagers all contribute to the militarization of the borderland and to the emergence of vigilante groups, who, protected by the Thai state, take their destiny into their own hands. This militarization and the distribution of small weapons will have far-reaching consequences on peace proposals, shrinking the possibility of weapon-free peace zones.

REFLECTIONS ON VIOLENCE, SUBVERSION, AND CREATIVITY

While the National Reconciliation Commission was coming up with creative peace proposals that included greater political autonomy for Pattani people, the Thaksin government reverted to fresh attacks on Buddhist citizens with increasing repression, detaining everybody who might look suspect. The decree on emergency situations and the use of the military as the only means of control seems a desperate attempt on the side of the Thai state to reestablish sovereignty in a guerrilla war that nobody is going to win but everybody is going to lose. Ironically, the Thai government, in trying to win the hearts and minds of the majority of Thai people, uses a dominant discourse, including much of the Thai-language media, in a process of "othering" that is countered only by some voices of (Thai Buddhist) NGOs and intellectuals. Meanwhile, the Pattani Malays, especially its diaspora in political exile, seem to reject anything they see as Thai, without reflecting on their long history of coexistence with Theravada Buddhism and Hinduism. Clearly, these confrontational policies support a radicalization of positions on both sides: the nationalist position of the umbrella organization BERSATU makes place for an extremist discourse of Jihad and the utopian pan-Islamist vision of a Melayu Islamic state, while the democratic approach of the Democratic Party is replaced by the chauvinist, ultranationalist discourse of Thaksin's Thai Rak Thai. The escalation of violence ridicules the increasing literature on a borderless world and the decreasing value of territory, while states in Southeast Asia in fact become stronger and the borders become more important. The crisis in the south also reflects processes of *rebordering* taking place simultaneously, processes that do not reflect the actual international border. Rebordering takes the form of ethnic cleansing, with deadly ethnic violence on the rise.

The Thai–Malaysian borderland is characterized by the parallel existence of sex tourism, the presence of karaoke and massage parlors, and the emergence of a radical Islamist discourse in a traditional, tolerant Muslim society, whereby sex tourism gives impetus for the Islamist discourse on piety and purity. It is no surprise that brothels sometimes become symbolic targets of bombs. The failure of the Thai Rak Thai in the elections in February 2005 in southern Thailand, and the folk's nonparticipation in the televised staging of national identity in the stadium of Yala, shows the nonidentification with the state and the increasing polarization as a result of state policies. The growing xenophobia is being fueled by Thaksin's negative attitude toward traditional Malay pondoks and Islamic education. Thus, students with Islamic education gained in Libya, Egypt, or Pakistan are leaving Thailand under fear of arrest. Lately, opposition among Muslims throughout Thailand has grown because of Thaksin's support of the United States' war in Iraq, leading to peaceful demonstrations in which Thai Buddhist students joined. Although Thai support for the U.S. policy in Iraq has hardly been dramatic,

Thaksin's endorsement of U.S. policies in Muslim Southeast Asia has raised frustration and resentment among Muslim communities in Thailand. In addition, Human Rights Watch reported that Thailand is among the eight countries that host secret CIA interrogation centers to which potential Muslim terrorists from anywhere in the world are flown handcuffed and blindfolded.[7] Human rights abuses in the borderland are in such ways generalized and imported to the center.

CONCLUSION

The following are findings and recommendations of the National Commission for Reconciliation's center on the reestablishment of human rights.

On July 15, 2005, Prime Minister Thaksin Shinawatra's government imposed the Emergency Decree on Public Administration in Emergency Situations, B.E. 2548, to replace the existing martial law, effectively legalizing detention without trial. Even though the National Reconciliation Commission proposes the installment of legal defence centers, villagers fear that the government forces will use the emergency decree to detain anybody they want to. The Asian Centre of Human Rights found that 109 human rights defenders and community activists had already been murdered since Thaksin Shinawatra came to power. Most of the cases were never investigated.

Autonomy for the Muslims in the south is not only a matter of administrative control and political participation. Rather, more autonomy encompasses a wide spectrum of human rights, including the right to economic well-being, Islamic education, instruction in Thai and Malay languages, and equality before the law. The recent effort of politicians of the governing Thai Rak Thai Party to close all pondoks (Islamic schools) indicates the growing chauvinism within parts of the government. Only if the government is prepared to change sides is peace to come back to Pattani, Yala, and Narathiwat.

NOTES

The author thanks Amporn Marddent (Bangkok) for sharing her fieldwork on human rights violations in Pattani, Yala, and Narathiwat provinces with the author.

1. For pathbreaking studies of long-time tendencies of Malay national-ism and internal colonialism in Pattani, Yala, and Narathiwat, consult Pitsuwan (1985), Wan Kadir (1990).

2. If not mentioned otherwise, the article is based on ethnographic field-work by the author carried out in southern Thailand in 1995–96, 2001, 2004, and 2005.

3. See Madmarn (2002) for an excellent introduction to the traditional *pondoks* in southern Thailand. See Horstmann (2006) for a critical over-view on the new pressure by reformist forces.

4. Somchai was last seen at the Chaleena Hotel on Ramkhamhaeng Road in Bangkok on March 12, 2004. His car was found abandoned on Kamphaeng Phet Road near Mor Chit 2 bus terminal. After sharp pro-test by civil society institutions, five police officers have been charged with abduction.

5. The Web site of PULO is http://www.pulo.org.

6. The Village Scouts were mobilized as a paramilitary force that was indoctrinated with nationalist propaganda. Volunteers underwent rigorous training, led by officers, in military warfare. See Bowie (1997) for an eth-nography of military training sessions.

7. Human Rights Watch, November 15, 2005.

WORKS CITED

Amporn, Marddent. 2005. "Women in Conflict Situations." Paper pre-sented to the Ninth International Conference of Thai Studies, Northern Illinois University, DeKalb. April 2005.

———. 2006. "From *Adek* to *Mo'ji*: Identities of Southern Thai People and Social Realities." In *States, Peoples, and Borders in Southeast Asia*. Kyoto Review of Southeast Asian Studies 7, ed. Alexander Horst-mann. Kyoto: CSEAS.

Azra, Azyumardi. 2004. *The Origins of Islamic Reformism in Southeast Asia: Networks of Malay-Indonesian and Middle-Eastern "Ulama" in the Seventeenth and Eighteenth Centuries*. Leiden: KITLV Press.

Bowie, Katherine A. 1997. *Rituals of National Loyalty: An Anthropology of the State and the Village Scout Movement in Thailand*. New York: Columbia University Press.

Chaiwat, Satha-Anand. 1987. *Islam and Violence: A Case Study of Violent Events in the Four Southern Provinces, Thailand, 1976–1981*. Tampa: Department of Religious Studies, University of South Florida.

———. 2004. "Facing the Demon Within: Fighting Violence in Southern Thailand with Peace Cultures." In *Seeds of Peace*, February 6, 2006.

————. 2005. "The Silence of the Bullet Monument: Violence and 'Truth' Management, Duson Nyor–1948 and 'Kru-ze'–2004." Keynote presentation to the Ninth International Conference of Thai Studies, Northern Illinois University, DeKalb, April 3, 2005.

Crispin, Shawn W. 2001. "Power to the People." In *Far Eastern Economic Review (FEER)*, November 29, 2001. Source: http://www.feer.com.

Dorairajoo, Saroja. 2004. "Violence in the South of Thailand." *Inter-Asia Cultural Studies* 5, no. 3: 465–71.

Donnan, Hastings, and Thomas Wilson. 1999. *Borders: Frontiers of Identity, Nation, and State*. Oxford and New York: Berg Publishers.

Horstmann, Alexander. 2002. *Class, Culture, and Space: The Construction and Shaping of Communal Space in Southern Thailand*. Bielefeld: Transcript Publishers.

————. 2004. "Ethnohistorical Perspectives on Buddhist–Muslim Relations and Coexistence in Southern Thailand: From Shared Cosmos to the Emergence of Hatred." *SOJOURN, Journal of Social Issues in Southeast Asia* 19, no. 1 (April).

————. 2005. "States, Peoples, and Borders in Southeast Asia." *Kyoto Review of Southeast Asian Studies* 7: 1–9.

Human Rights Watch. 2004. "Not Enough Graves: The War on Drugs, HIV/AIDS, and Violations of Human Rights." *Human Rights Watch Report* 16, no. 8 (June).

International Crisis Group. 2005. "Southern Thailand: Insurgency, Not Jihad." *Asia Report* no. 98 (May 18).

Kraus, Werner. 1982. "Der Islam in Thailand: Über d. Entwicklung u. Problematik d. Islams in Suedthailand, d. islam. Modernismus u. d. seperatist. Bewegung im Süden." In *Der Einfluss des Islams auf die Politik, Wirtschaft und Gesellschaft in Südostasien*, ed. Werner Draguhn. Hamburg: Institut für Asienkunde.

McCargo, Duncan. 2005. "Understanding Conflict in the South through Domestic Politics." Paper presented to the Ninth International Conference of Thai Studies, Northern Illinois University, DeKalb, April 2005.

Pitsuwan, Surin. 1985. *Islam and Malay Nationalism: A Case Study of the Malay-Muslims of Southern Thailand*. Bangkok: Thai Khadi Research Institute, Thammasat University.

Wan Kadir Che Man. 1990. *Muslim Separatism. The Moros of Southern Phillipines and the Malays of Southern Thailand*. Singapore: Oxford University Press.

III

Rethinking Borderscapes:
Mapping Hidden Geographies

7

The Poetry of Boundaries

JAMES D. SIDAWAY

Talking not long ago with one of my friends—there was a time when we used to call them informants—of the changes brought about at his village on the Portuguese-Spanish border by the so-called "Europe Without Frontiers" (or at least without internal frontiers), he thought for a moment and then he replied, carefully and repeating his words: "You may remove the door but the doorframe remains . . . you may remove the door, but the doorframe remains." (Kavanagh 2000, 47)

Here we have the first inkling of a possibility of linking borders away from the ebb and flow, advance and retreat, that are the direct result of battles lost and run, conquests, occupations, and negotiated concessions and withdrawals. Not least of the possibilities is the understanding of lines as active: of flight, of crossings, of the ability to carry us away. (Rogoff 2000, 116)

Reflecting life and journeys along the Portuguese–Spanish borderlands, this is a chapter of detours and departures. Rather like the quotes above, it is a collection of fragments of the border—thoughts and reflections on the boundary stones and rivers, the maps and marks that serve and signify this frontier. My aim (t)here is to think around, through, and against borders. Following Barker (1998, 120), therefore: "To think against is to analyze the level of a surface, not to get closer to or further from the truth or objective reality but to reveal other surfaces and points of contact." The chapter is in part a semiotic analysis, in the broad sense that Elam (1980, 1) describes:

> Semiotics can best be defined as a science dedicated to the study of the production of meaning in society. As such it is equally concerned with processes of *signification* and with those of *communication*, i.e., the means whereby meanings are both generated and exchanged. Its objects are thus at once the different sign-systems and codes at work in society and the actual messages and texts produced thereby. The breadth of the enterprise is such that it cannot be considered simply as a "discipline," while it is too multifacted and heterogeneous to be reduced to a "method."

However, insomuch as this chapter is informed by a "method," the methodology rests not only on the collection and critical scrutiny of published narratives about the Portuguese–Spanish border or about discussions with those who live along it, but on a series of journeys and walks through its landscapes—in particular, across the section of the border, between the Portuguese Alentejo and Spain's Andalucía, that also intersects the rolling hills of the Sierra Morena. In drawing upon these journeys, I have been inspired by Cloke and Jones's rendering (2001, 663) of place and landscape through an exploration of microplaces, ecological and cultural resonances, and networks of social and material relations. Cloke and Jones explore the dialectic between

> the continuing tendency towards framing landscapes as the vista from a fixed point . . . , [which] has the effect of putting the viewer at a fixed point—outside, or on the edge of, the landscape with a single static orientation frozen in time . . . [in contrast to] a perspective which is about being in the landscape, about moving through it, in all the (perhaps) repeating yet various circumstances of everyday life. Being in, and moving through, landscape is different to gazing upon it from a point which always seemingly puts you at the edge of it, or even outside of it. The landscape surrounds you, it will often be unreadable from any one position, and your orientation may be constantly or frequently, even habitually, shifting. It is about fleeting intimate details that your senses can pick up from being in a landscape.

Some inspiration here also comes from Yi-Fu Tuan (1989, 240): "cultural-geographers-cum-storytellers stand only a little above their material and move only a little below the surfaces of reality in the hope of not losing sight of such surfaces, where nearly all human joys and sorrows unfold." However, the starting point of these detours is to reiterate that a border may be read as, among other things, a semiotic system, a system of images and imaginations. As Anssi Paasi (1999, 669) stresses, boundaries are "institutions and symbols that are produced and reproduced in social practices and discourses."

It is in the context of such approaches and understandings that this article begins with descriptions and maps of this border between Spain and Portugal. The chapter then (re)traverses the line between Portugal and Spain to explore border poetics and politics in the contexts of wider European integration. Although I have elsewhere attended more systematically to the geopolitics of Iberian cross-border cooperation projects (Sidaway 2000 and 2001), this chapter unfolds as a series of forays into border poetics. These begin at the border, with maps and boundary stones and the semiotic (and political) system of which these, and the Portuguese and Spanish states, are components and effects. The chapter then considers how the border and the lives around it are reconfigured within the discourses and practices of European integration.

MAPS AND MARKS

Nearly fifty years ago, the Romanian American geographer Ladis Kristof (1959, 272) reminded us that "boundary stones are not the

boundary itself. They are not coeval with it, only its visible symbols." Such classic papers on borders (see note 6 for more examples) bear rereading in the contexts of subsequent geopolitical and theoretical developments. Here, a cartographic and literary detour provides a basis to do this.

Figure 7.1. Border stone, hito. *Photograph by the author.*

Narrating his journeys in Portugal, the British poet and travel writer Paul Hyland (1996, 153) describes a venture into Spain across this border. He explains the culmination of his detour in the borderlands:

> Before crossing the bridge to the old [Portuguese-Spanish] frontier I planned my own pilgrimage, a short one to the Hermitage of Nuestra Señora de los Hilos. Hilos means "threads." I planned to gather up loose ones there. . . . When I found it near dusk, the hermitage comprised a sturdy church. . . . Three men and a dog sat in the porch. It all dated back to the Reconquista from the Moors, but had been rebuilt in 1768 when a new image was installed. The light within was rare, the virgin backlit by a lustrous golden rerdos. Always in Iberia the shock of contrast, disconnection, contradiction between outside and inside, fora/dentro. . . Outside, a sign read "de los Hitos." The name on my Michelin map was a figment of the cartographer's imagination. The men laughed. "Not hilos," one thwacked his thigh with a switch, "but hitos." No loose threads at all, but a line drawn taut for Nuestra Señora de los Hitos, Our Lady of the Boundary Stones.

The toponomy and meaning of the Michelin map (Sheet 444: Central Spain, 16th edition, 1998) is indeed misleading in just the way that Hyland notes. Somehow the hilltop sanctuary known as Nuestra Señora de los Hitos has become, on the map, de los Hilos. The Michelin mapmakers are not alone. Sancha Soria (1995) describes how, with regard to areas further south in the Sierra de Huelva on the Spanish side of the border between Andalucía and the Alentejo, a range of maps, including ones produced by the Spanish Army Geographical Service, the Spanish National Geographic Institute, and the Spanish Institute of the Environment, differ in their naming of hills and notes how all depart from local naming practices. These disparities between the cartographic representations and those attached locally to places are stark reminders that maps are another system of representation, foregrounding certain objects and leaving out others, (mis)naming some, and establishing hierarchies of judgment as to what is represented and by what words and other symbols.[1]

Yet all these maps chose to represent the border between Portugal and Spain. In many of the Spanish maps,[2] beyond the border (marked by a line of crosses) is blank space, marked only by grid lines and bold letters spelling out PORTUGAL. Whether Portugal is blank or not, what makes this border real and worthy of cartographic designation is that the representation on the map *coincides* with other systems of representation in which the border is narrated, cited, and reiterated. First, where the border is not marked by rivers (which it is for about 60 percent of its course), some of which have become reservoirs, it is designated on the ground by boundary markers, known in Spanish as *hitos,* as *marcas* in Portuguese (see Figures 7.1 and 7.2). Either side are different states, *themselves* complex systems of representation.[3] And the border is also demarcated in a series of treaties that (with the one notable exception of a small disputed area outside our focus area here) are recognized and ratified by the two states. Although, throughout its length, there is little or no significant environmental variation to either side of the border, different languages more or less coincide with the border, and with these go different national imaginaries. Some mixing and local dialects complicate this, but the broad coincidence and overlapping of a set of systems of representation (cartographic, legal, linguistic, etc.) make this border seem real and tangible, seemingly worthy of the line of black crosses and the yellow line that winds across the Michelin maps of Iberia.

Moreover, this border, like all others, is variable along its length and through time. It stays in the same places (leaving aside minor variations in the courses of those streams and rivers that, at their deepest points, mark the border), but, since the systems of representation that reproduce it are dynamic, its many meanings and its identity change. As Wilson and Donnan (1998, 12) note, all borders are "complex and multi-dimensional cultural phenomena, variously articulated and interpreted across space and time. This suggests that a priori assumptions about the nature of 'the border' are likely to founder when confronted with empirical data; far from being a self-evident, analytical given which can be applied regardless of context, the 'border' must be interrogated for its subtle and sometimes not so subtle shifts in meaning and form according to setting." In parallel terms, Douglas (1998, 88) demonstrates "that even when borders remain relatively static, as has been the case with the French–Spanish border (arguably the most stable in western Europe), the borderlands themselves are in a constant state of flux. . . . Thus, how, as opposed to where, north meets south is subject to constant negotiation." With this in mind, the Portuguese–Spanish border is read symptomatically (and as an example) here, but in ways that I hope will be suggestive for other students of borderlines. What lends such a project some credibility is that this, the longest of internal EU borders, is usually regarded as the oldest stable frontier (surpassing even the claims made on behalf of the French–Spanish border), not only in Europe, but in the world. Spanish and, particularly, Portuguese historiography trace the origins of the border to treaties of the thirteenth century. In other words, part of its representation is as an ancient demarcation. In particular, the antiquity of this border and its transhistorical presence through the centuries is a key reference point in narratives of Portugal's history (Sidaway 2001). The stable borderline is a totem of Portuguese nationalism, a sign of sovereignty.

In what has become something of a classic account of *Human Territoriality*, Robert Sack (1984) noted how the anchoring of society to place in the nation-state became one of the clearest expressions of mythical-magical consciousness of place in the twentieth century. And among the characteristics of nation-statehood are numerous concentrated sites of mythical-magical performance: monuments and tombs, museums and mountains. Borders are among these. More often seen depicted on maps than actually crossed on the ground, borders have a special place in marking the known and essential limits to the nation-state. Borders are frequently inscribed within narratives of statehood, from maps to history books to popular notions of us and them, self and other. Borders are the very substance of nation-statehood.

In contrast, the European Union as an imagined community lacks such sites. Blue flags on official buildings, and signs alongside infrastructure projects indicating part funding by the EU, are hardly substitutes for war memorials, national monuments, and state borders (Shore 2000).

The European Union is therefore a project of becoming "an ever closer union" (in the terms of the Treaty of Rome) in part via the fostering of connection and flows.[4] This future-orientated project includes the reworking (the *re-placing*, so to speak) of borders between EU members. These borders now enter new texts: those of Brussels. They are scripted as key sites of the integration project. As a special component of what James Scott (2000, 104) terms the "visionary cartography"[5] that emerges from the European Commission,

Figure 7.2.
Border stone,
hito. *Photograph*
by the author.

borders are to be transcended. Borders are thereby subject to special community initiatives that envisage them as *pivotal* spaces of integration. Moreover, places designated as rural, marginal, or underdeveloped (the Portuguese–Spanish border area prominent among these) are subsumed into the European Union's vision of territorial harmonization and development (Richardson 2000).

Since the simultaneous adhesion of Portugal and Spain to the European Union in 1986, the Portuguese–Spanish border has therefore become the subject of EU-funded regional development. Together with the wider EU project of harmonization and integration, this has transformed the identities and meanings of the border. Before returning to this issue in the conclusion, a border (and theoretical) detour is in order.

LIVING ON/BORDER LINES

The study of narratives and discourse is central to an understanding of all types of boundaries, particularly state boundaries. These narratives range from foreign policy discourses, geographical texts and literature (including maps), to the many dimensions of formal and informal socialization which affect the creation of sociospatial identities, especially the notions of "us" and the "Other," exclusive and inclusive spaces and territories. (Newman and Paasi 1998, 201)

What are the consequences of reading boundaries, as do the authors of the above citation, as discourses? It has already been pointed out that what gives the Portuguese–Spanish border, like others, presence is its reproduction through a complex system of representations.

The prevalence of what political geography has called "relict borders"[6] (marked by the many towns that carry the designation *de la frontera*,[7] recalling the mobile frontier of medieval reconquest at the expense of Arab–Berber states) across the lands of the Algarve and Andalucía, might also complicate matters, as would the existence of new and old divisions internal to the Portuguese and Spanish states. The latter in particular has, as part of its transition and reinscription since the end of Franco's dictatorship in 1975, become a quasi-federal entity (in official parlance, an Estado de Las Autonomías; a state of autonomies) divided into a series of autonomous communities, with their own assemblies, executive agencies, flags, and statutes. So, to the north of Andalucía is now an intrastate border with the neighboring autonomous community of Extremadura. Hence, what Romera Valiente (1990) characterized as a "disarticulated and dependent" space of the border between Andalucía and Extremadura has found its relative peripherality reinforced by what Arroyo-Lopez and Machado Santiago (1987, 341) have termed a "frontier effect," whereby boundaries between Spain's *autonomías* produce their own "disfunctions in the management of public services [and] rupture of complimentary economic spaces" (Romero Valiente 1997, 36).

The Portuguese regions lack the formal constitutional powers of those in Spain, but do form a significant part of the sense of Portugal's unity in diversity,[8] and are duly indicated by prominent signs along main roads and on maps. Commenting on the resulting combination of Iberian frontiers, the anthropologist Luis Uriarte (1994, 43) notes:

> One may simultaneously have one foot in Spain and another in Portugal. One can sit down, as I have had the opportunity to do so, on borderpost no. 695 where ecumenically multiple demarcations converge: Spain and Portugal; Cáceres and Badajoz provinces; Portoalegre district; Valencia de Alcántara, San Vicente and La Codosera municipal boundaries. In this border-post, one may dwell and rest your backside simultaneously and respectfully between two nation-states, two provinces, a district and three municipal limits. An excess of frontiers for one backside.

But there is no need to travel with the author all the way to *hito* 695 to become aware of excess in terms of the frontier. A surplus of complexity is present anywhere when the complexities of representation of this (or any other) interstate boundary are fully taken into account. Although the border is easy to identify on a map and not too hard to find on the ground for those who persevere (it would fail, if not), one would find hard to say where the representation of the border is, where it begins and ends and where its limits are. We might start at the thousands of hitos (some duplicated, divided into A, B, C, and so on to mark particularly complex border convolutions) and/or the streams and riverbeds that mark around 40 percent of this boundary. Often encrusted with that strange hybrid plant, lichen (a plant formed out of symbiotic relations between algae and fungi), these hitos form a winding line, but vegetation and topography mean that they are often hidden from one another. And what of the spaces between the hitos, or between them and the rivers? What course does the supposed straight line between these follow, given the complex topography? Indeed, exactly how long is the border? Even allowing for a short undemarcated[9] stretch some two hundred kilometers north of the area of focus of this paper, a stretch that complicates measurement, different accounts provide considerably different lengths. In fact, no two accounts offer exactly the same figure. The lesson from other topographic measurements is that, along the ground, as the scale of measurement becomes finer so does the length of that being measured (Bird 1956). This is what creates such complications to the (at first apparently quite simple) question of the length of such a border.

Fintan O'Toole (1997, 4) therefore comments:

> As it happens modern geometry has given a new sanction to this kind of subjective mapping. In his *Fractal Geometry of Nature*, the mathematician Beniot Mandlebrot asks the apparently simple question "How long is the coast of Britain?" The coast is obviously not smooth and regular. It goes in and out in bays and estuaries and promontories and capes. If you measure it at one hundred miles to an inch, all these irregularities appear. But if you measure it at twenty miles to an inch, new bays open up on the coastlines of promontories and new promontories jut out from the sides of bays. When you measure these as well, the coastline gets longer. At a mile to an inch it is even longer . . . and so on, until you crawl around on your hands and knees measuring the bumps on the side of each rock that makes up the coast. The more accurately you measure it, the more uncertain it becomes. What matters, in the end, is your point of view. Mandlebrot compares the length of the border between Spain

and Portugal in a Portuguese and a Spanish atlas. In the former it is 20 per cent longer than in the latter, not because the territory is disputed, but because Spanish surveyors used a larger scale, and thus measured fewer squiggles.

In a related sense, we may note, the border exists not only as a system of signs (rivers, hitos, road signs, and so on) inscribed on the text(ure) of the landscape, but simultaneously in the texts, signed, sealed, and ratified, that declare and demarcate it. Before the border could be marked on the ground, it had to be agreed on between Lisbon and Madrid. Joint teams of military surveyors were dispatched, but it took years to mark and agree on the border. The northern half was agreed on in 1864 and the southern section in 1926, the contracts for the hitos were drawn up, and they were put in place. But well into the twentieth century, a few parts of the border remained undemarcated. These were either disputed or of shared (condominium or common) status, with access regulated by feudal *usfruct* rules. Today, with the exception of the short disputed section mentioned earlier, the border is signed and sealed in legal treaties. But (t)here too, in the state archives, things become very complex. These texts have no fixed end. They speak and act in the name of other representations, notably those of the state, the government, or the king. Or they cite earlier texts (of which there is no end) and then refer back to that which they define (the border). Which comes first—where exactly is the border? The answer is that one cannot limit the demarcation of the border either to the text of an agreement or to the marks and features that supplement or are appropriated by it on the ground. Amid all these signs and wonders, it is impossible to decide exactly what comes first (in a manner analogous to the relationship between the "main" text and the footnotes/references and images of hitos here in this chapter). The border derives a significant part of its identity precisely from such *undecidability*: from the combinations and cross-references of authority, texts, and symbols in different places. Consider a case from another European border, where this combination is intensely evident:

> The most standard nationalist/republican response to the Border [between the Republic of Ireland and Northern Ireland] is that it has cut the city off from its natural hinterland of Donegal and has thereby damaged both places. But it is even more important to recognise that the Border reproduces itself in every area within the North. It is and

always has been a sectarian border; it embraces a fertile progeny of internal borders, all of which enhance the unnatural, defensive atmosphere of the State. These are not flexible or porous borders; they are not indicators of a community's autonomy. They are prison walls. Their function is to immure communities and, with that, to fossilise the political situation in its original form. Since the Troubles [the political and sectarian violence of the 1970s through 1990s] began, the existence of these borders has been signalled in every conceivable way by flags, murals, graffiti, painted kerbstones. . . . Derry is a border town with internal borders that make themselves manifest even in the slash mark between the names Derry/Londonderry; in the sectarian housing estates; in the old walled architecture of the town and in the competing histories of its development. . . . There is nothing "natural" about borders; they are all created to assert power and control. The most unfortunate aspect of the Border's history is that, to survive, it has forever to insist on its presence. It never therefore became naturalised. If it could have been forgotten about, it would have been more secure. But it can never be ignored. So it will remain assertive, creating division within the territory it was designed to consolidate. (Campbell 1999, 29)

The Irish border is comparatively recent, among those carved out of decaying European empires in the aftermath of the 1914–18 world war.[10] Its contested status has given it a special character, and the project of a boundary treaty (and, with this, a formal and full demarcation) fell foul of the deteriorating relations and mutual antipathy between the two states in Ireland (Kennedy 2000). However, though Ireland is an extreme case, the endless and uncontainable replication of "the Border" is symptomatic of how modern borders operate. Consider another case, rendered from the same decaying empire:

Partition. For a long time, and certainly all the time that we were children, it was a word we heard every now and again said by some adult in conversation, sometimes in anger, sometimes bitterly, but mostly with sorrow, voice trailing off, a resigned shake of the head, a despairing flutter of the hands. All recollections were punctuated with "before Partition" or "after Partition," marking the chronology of our family history. . . . How do we know Partition except through the many ways in which it is transmitted to us, in its many representations: political, social, historical, testimonial, literary, documentary, even communal. We know it though national and family mythologies, through collective and individual memory. Partition, almost uniquely, is one event in our recent history in which familial

recall and its encoding are a significant factor in any general reconstruction of it. In a sense, it is the collective memory of thousands
of displaced families on both sides of the border that have imbued
a rather innocuous word—partition—with its dreadful meaning: a
people violently displaced, a country divided. Partition: a metaphor
for irreparable loss. (Menon and Bhasin 1998)[11]

There were Cold War partitions, too: Korea, Germany, Europe, and
Vietnam. But those that predate the Cold War, India and Ireland
among them (and some partitions since, such as Cyprus), reproduce
logics of difference that are tied up with colonialism (Kumar 1997).
Yet, despite their apparent abnormality, such special cases also be-
tray the ways that borders are reproduced elsewhere and are thereby
never simply a line on a map or on the ground.

All this complexity and undecidability calls for another line of
analysis, even with regard to the relatively peaceful Portuguese border, to the trace of which we now return. What has long been proper
to border identity is that it is simultaneously *liminal* and *regulated*.
One the one hand, the border is an edge, a set of peripheral places.
These have something of the characteristic of what Robert Shields
(1991) has described as *Places on the Margins*. Indeed, most of the
Portuguese–Spanish border is sparsely populated and relatively
underdeveloped in material terms. Low-intensity agriculture, min-
eral extraction, energy (hydroelectricity), and water extraction all
reinforce the sense of an area of socioeconomic marginality. The
border provinces and districts have the lowest average per capita
income of any in Spain or Portugal, and among the lowest in the entire European Union. Historically, neither country has attached priority to transport connections with its neighbor. Indeed, at the start
of the 1990s, the entire section of the border between Andalucía
and Portugal had only one official road border-crossing point, at
Rosal de la Frontera, and a few years earlier this had been closed
at night. No railway[12] linked the two countries along this stretch
of the border, and the river Guadiana that formed the course of the
Andaluz–Portugal border for its last 30 kilometers or so was without any bridge until 1991. The opening paragraph to the first joint
Portuguese–Spanish study on *Transfrontier Territorial Articulation*
(conducted with EU funding by the territorial planning authorities
holding jurisdictions in Andalucía, Algarve, and Alentejo) notes:

The frontier space between Andalucia and the contiguous Portuguese regions (Algarve and Alentejo) has historically represented a barrier between two neighbouring territories, which besides being a frontier in the European sense, are separated by a veritable wall. The existence of a natural barrier (the Guadiana and Chanza rivers) in part of the frontier line, is not sufficient explanation for this fact, given that valleys in Europe have represented more a means of communication and integration than a separation between its two banks. (Junta de Andalucía 1995, 11)

Yet this border has not only been a liminal or marginal space. It has also been highly regulated, not only through its demarcations on the ground and in treaties, but as a place that has been subject to patrol by police and customs. The border is characterized by a series of border posts, observation posts, and patrol tracks and roads. In common with many other European boundaries, these border markers were reinforced through the twentieth century. Anxiety about the integrity of the borders and accompanying efforts to demarcate, survey, patrol, and police—in short, to regulate—the borders forms part of the moment when, as Löfgren (1999, 2) explains, "modern (and centralizing) nation-making shifted the energy [of inscribing statehood] to the periphery where the state, its power, its cultural capital, its routines, rules and ideas were materialized and challenged." However, the identity of the border changes as the balance and forms of liminality/regulation shift. It is in this context that, for a contemporary west European border like the one under discussion here, the project of the European Union intervenes. Today, therefore, the frontier posts are abandoned; the patrols are gone. Moreover, the border becomes caught in a wider space of networks and possibility (referred to earlier) that is the EU. Indeed, the European project, as envisioned by the commission, seeks to overcome the peripherality of border (and other marginal areas) through a series of cross-border interventions.

A key means of this is EU regional policy, offering capital investments, usually in terms of transport infrastructure for marginal regions. In the terms of the EU, such strategies are about *cohesion*. At the same time, the EU seeks, in its terms, *harmonization* of European space through the removal of barriers to mobility, including borders. A new European scale[13] is envisaged, with new effects. In the 1990s, this combined cohesion and harmonization was expressed in the concept of a "single market" and the removal of customs and passport control at the borders between most EU members. However, such a process is not without contradictions, since it means the disintegration of the space hitherto configured by the identity of the border. A detailed anthropological study of *Frontiers, Territories, and Collective Identities* by Valcuende del Rio (1998) provides a densely narrated account of the history of the Portuguese–Spanish frontier as a *resource* for local inhabitants. Although focused on a particular locale on this border, the study's findings about the border as a resource are echoed in studies from elsewhere along the same border (Garcia 1997; Hernández León and Castaño Madroñal 1996). This history is shown to have had two aspects. First, until the European Union's 1992 "single market" eliminated significant tariff differences, smuggling and contraband provided income and a way of life for many along the border. Particularly in the rural areas, smuggling became a *genre de vie* in which liminality and regulation were reinscribed as a local resource. However, this capacity has been eroded in the context of the so-called integrating and harmonizing European space into which Portugal and Spain were formally incorporated beginning in the mid-1980s.

The second aspect of the frontier as a resource was legal commerce, notably the sales of goods and services to cross-border travelers, including those who came to the border (in view of the same exchange rate and tariff differentials that drove smuggling) to shop, to purchase, or to consume something that was cheaper on the other side of the border. This significant aspect of the frontier genre de vie has also undergone recent transformation and relative decline, as border controls and currency differentials disappeared and, at the same time, new (frequently EU-funded) bridges and roads speeded flow and increasingly removed the border as an essential stopping point.

CONCLUSION

This chapter has sought to develop theoretical reflections on how borders operate within complex systems of meaning. In the more specific terms of the Portuguese–Spanish border, the chapter has also noted how the *harmonization* associated with the EU is predicated on a certain local *disintegration* of ways of life based on the Portuguese–Spanish border as a space of liminality and regulation. The production of a new European scale is, in this case, also the decomposition of another scale. But how does this relate to the issue of borders as semiotic systems? Among the detours of this chapter has been a concern to trace how, in Rogoff's words (2000, 114), the "minute gesture of the border stone alerts us to the imaginary power of borders as a concept." In the European Union, we have come to hear much about the project of a *Europe without frontiers*. But the borders endure, incorporated and cited now in European discourses of integration. As this chapter has argued, a certain visualization (or, if you prefer, a perspective) of borders as that which must be *overcome* is at the heart of the European project. This is not so much the end of borders as their radical reinscription as something to be transcended, as spaces of European integration. Hence, to adapt the argument of Jens Bartelson (1998, 322): "Only when this perspective [of transcending borders] itself has long been forgotten, will we be totally entitled but not the least tempted to speak of the end of the state [boundary]." In other words, the reference to, and the meanings of, the border are reinscribed inside another project, another set of discourses and powers, another vision. It is (t)here, in such inscriptions and actions, that Europe is remade, and the old marks of its borders are rearranged into new networks and systems of meaning.

Or, in the more poetic terms of the border villager cited by William Kavanagh (2000, 47) in the extract with which this chapter opened, "but the doorframe remains."

NOTES

This chapter is dedicated to the memory of Jasmin Leila Sidaway (1997–2007), who walked with me.

1. A point made in Harley (1988 and 1989), and more widely in "critical geopolitics" analyzing the writing of words (Ó Tuathail 1996). For a review of cartographies of power that works with Harley, but also points out his narrow reading of *deconstruction,* see Carib (2000).

2. For example, the 1:25´000 topographic maps produced by the Spanish National Geographical Institute.

3. On the state as a system of representation, see Bartelson (1998); Constantinou (1996); Dillon and Everard (1992); Mitchell (1991); and Weber (1995). On statehood/sovereignty as recited in social sciences, see Agnew (1994).

4. For critical reflections on the EU as a project of connections and mobility, see Barry (1993 and 1996) and Sparke (1998). For a reading of the poetics and politics of connection within an EU-funded cross-border (INTERREG) project, see Hebbert (2000) and Jönsson, Tägil, and Törnqvist (2000).

5. Sparke (2000) signifies a similar concept with the term *anticipatory geographies*.

6. The classic account is Hartshorne (1936). His account of relict boundaries is one of a series of interpretations of frontiers that appeared in the *Annals* between the 1930s and the 1960s, a series including works by, e.g., Kristof (1959), Jones (1959), and Minghi (1963). All drew on prior British, French, and German traditions of political geography, geopolitics, and international law. These papers are briefly reviewed in Newman and Paasi (1998).

7. The literature on the reconquest and medieval Iberian frontiers is utterly vast, since it forms a centerpiece of both Portuguese and Spanish historiography. Rather than enter into this subject here, it seems more appropriate to cite from a short article that appeared in the Andalucia edition of the Spanish daily *El Pais,* whose author expressed his indignation at the proposal by the authorities of his native town of Jerez de la Frontera (famed for sherry production) to drop "de la Frontera" from the official name of the town: "I can't make sense of it *[No me lo explico]*. How could they cross out *[corregir]* a tradition by the process of annulling it in the stroke of a pen. One recalls that the towns of lower Andalucia *[Andalucia la Baja]* that call themselves 'of the Frontier'—Jerez, Arcos, Castellar, Vejer, Conil, Chiclana . . .—have been protagonists through centuries of living together *[convivencia]*—or the collision—of Muslim and Christian civilisations. All these localities are also found in the memorable scenes of the so called frontier romances, those fascinating epic poems of love and war. . . . Although I admit that it is a matter of a symbol or of a poetic emblem lodged in the regional memory, it is not acceptable for Jerez to erase from its archives that which history has documented, that is to say, doing without the evocation of a frontier fixed in medieval maps and conserved in the collective imagination. . . . What is going on is that I am not resigned to lose my native frontier, including amongst other things because I feel more a person of the frontier than of Jerez *[sentirme más fronterizo que jerezano]*" (Caballero Bonald 1999, 2).

8. This unity and singularity of Portugal is a prominent theme in Portuguese geography, notably the works of Orlando de Ribiero inspired by a similar genre (evident in Vidal de la Blache) in French geography. Proposals to introduce a regional system of government in Portugal were put to a referendum in November 1998. They failed to attract the necessary majority

and were shelved. Portugal therefore remains among the most centralized of European states. The backdrop to the proposals, their failure and the way that interregional rivalries appeared stronger than demands for devolution is analyzed in Gallagher (1999). For a historical perspective, see Nogueira da Silva (1998).

9. This undemarcated sector, near to the town known in Spain as Olivenza, consitutes a gap through which many confusions and dissonances have passed, notably an ongoing Portuguese irredentist movement (Sidaway 2001). Such places, along with enclaves and exclaves, are places where an excess of border signification/representation overflows. Europe has several of these places; indeed, there are even enclaves within exclaves, for example, at the Belgium–Netherlands border around Baarle. See Robinson (1959).

10. For a contextual analysis of the range of political debates about the Irish border, see Howe (2000).

11. For other critical works on the aftermaths of Partition, see Chaturvedi (2000); Tan and Kudaisya (2000).

12. An 1877 plan for the construction of a Lisbon-Huelva-Seville railway still gathers dust in the archives of the Portuguese Ministry of Public Works.

13. On the EU as the production of a new scale of reference and actions, see Swyngedouw (1994); see also the material cited in note 4. There is also a fast-growing literature on the EU as an articulation of different scales (multiple levels) of governance. For an overview, see Jordan (2000).

WORKS CITED

Agnew, J. 1994. "The Territorial Trap: The Geographical Assumptions of International Relations Theory." *Review of International Political Economy* 1: 53–80.

Arroyo-Lopez, E., and Machado-Santiago, R. 1987. "Areas deprimidas y límites administrivos." *Tenth Congresso de la Associación de Geografos Españoles* (A.G.E.), Zaragoza, 341–56.

Barker, P. 1998. *Michel Foucault: An Introduction.* Edinburgh: Edinburgh University Press.

Barry, A. 1993. "The European Community and European Government: Harmonization, Mobility, and Space." *Economy and Society* 22, no. 3: 314–26.

———. 1996. "The European Network." *New Formations* 29: 26–37.

Bartelson, J. 1998. "Second Natures: Is the State Identical with Itself?" *European Journal of International Relations* 4, no. 3: 295–326.

Bird, J. 1956. "Scale in Regional Study Illustrated by Brief Comparisons between the Western Peninsulas of England and France." *Geography* 41: 25–38.

Caballero Bonald, J. M. 1999. "Fronteras." *El País* (Andalucía), January 19, 2.

Campbell G. 1999. "Personal Reflections." In *The Border: Personal Reflections from Ireland, North and South,* ed. P. Logue. Dublin: Oak Tree Press, 27–29.

Carib, R. B. 2000. "Cartography and Power in the Conquest and Creation of New Spain." *Latin American Research Review* 35, no. 1: 7–36.

Chaturvedi, S. 2000. "Representing Post-colonial India: Inclusive/Exclusive Geopolitical Imaginations." In *Geopolitical Imaginations,* ed. K. Dodds and D. Atkinson. London and New York: Routledge, 211–36.

Cloke, P., and O. Jones. 2001. "Dwelling, Place, and Landscape: An Orchard in Somerset." *Environment and Planning A* 33: 649–66.

Constantinou, C. M. 1996. *On the Way to Diplomacy.* Minneapolis: University of Minnesota Press.

Dillon, G. M., and J. Everard. 1992. "Stat(e)ing Australia: Squid Jigging and the Masque of State." *Alternatives* 17: 281–312.

Douglas, W. A. 1998. "A Western Perspective on an Eastern Interpretation of Where North Meets South: Pyrenean Borderland Cultures." In *Border Identities: Nation and State at International Frontiers,* ed. T. M. Wilson and H. Donnan. Cambridge: Cambridge University Press, 62–95.

Gallagher, T. 1999. "Unconvinced by Europe of the Regions: The 1998 Regionalization Referendum in Portugal." *South European Society and Politics* 4, no. 1: 132–48.

García, E. M. 1997. *Estudio sobre el contrabando de posguerra en olivenza y su área de influencia.* Mérida: Gabinete de Iniciativas Transfronterizas.

García Crespo, J., and S. Serra González. 1995. *Los Pueblos de Huelva: Sanlúcar de Guadiana.* Madrid: Editorial Medeterraneo.

Harley, J. B. 1988. "Maps, Knowledge, and Power." In *The Iconography of Landscape,* ed. D. Cosgrove and S. Daniels. Cambridge: Cambridge University Press, 277–312.

———. 1989. "Deconstructing the Map." *Cartographica* 26, no. 2: 1–20.

Hartshorne, R. 1936. "Suggestions on the Terminology of Political Boundaries." *Annals of the Association of American Geographers* 26: 56–57.

Hebbert, M. 2000. "Transpennine: Imaginative Geographies of an Interregional Corridor." *Transactions of the Institute of British Geographers NS* 25: 379–92.

Hernández León, E., and A. Castaño Madroñal. 1996. "Una frontera, un espacio social cambiante: La raya de Portugal." *Demófilo, Revista de cultura tradicional de Andalucía* 20: 139–53.

Howe, S. 2000. *Ireland and Empire: Colonial Legacies in Irish History and Culture.* Oxford and New York: Oxford University Press.

Hyland, P. 1996. *Backwards Out of the Big World: A Voyage into Portugal.* London: Harper Collins.

Jones, S. B. 1959. "Boundary Concepts in the Setting of Place and Time." *Annals of the Association of American Geographers* 49, no. 3: 241–55.

Jönsson, C., S. Tägil, and G. Törnqvist. 2000. *Organizing European Space.* London: Sage and New Delhi: Thousand Oaks.

Jordan, A. 2000. "The European Union: An Evolving System of Multi-Level Governance . . . or Government?" *Policy and Politics* 29, no. 2: 193–208.

Junta de Andalucía. 1995. *Articulación Territorial Transfronteriza, Algarve–Alentejo–Andalucía: Diagnóstico y Estrategia para la Articulación/ Articulação Territorial Transfronteiriça, Algarve–Alentejo–Andaluzia: Diagnostico e Estratégia de Articulação.* Junta de Andalucía, Consejería de Obras Públicas e Transportes, Dirección General de Ordenación del Territorio e Urbanismo, with the collaboration of the Comissão de Coordenação da Região do Algarve and the Comissão de Coordenação da Região do Alentejo, Sevilla.

Kavanagh, W. 2000. "The Past on the Line: The Use of Oral History in the Construction of Present-Day Changing Identities on the Portuguese–Spanish Border." *Ethnologia Europaea* 30, no. 2: 47–56.

Kennedy, M. 2000. *Division and Consensus: The Politics of Cross-Border Relations in Ireland, 1925–1969.* Dublin: Institute of Public Administration.

Kristof, L. K. D. 1959. "The Nature of Frontiers and Boundaries." *Annals of the Association of American Geographers* 49, no. 3: 269–82.

Kumar, R. 1997. *Divide and Fall? Bosnia in the Annals of Partition.* London and New York: Verso.

Löfgren, O. 1999. "Crossing Borders: The Nationalization of Anxiety." *Ethnologia Scandinavica* 29: 5–26.

Menon, R., and K. Bhasin. 1998. *Borders and Boundaries: Women in India's Partition.* New Brunswick, N.J.: Rutgers University Press.

Minghi, J. V. 1963. "Boundary Studies in Political Geography." *Annals of the Association of American Geographers* 53: 407–28.

Mitchell, T. 1991. "The Limits of the State: Beyond Statist Approaches and Their Critics." *American Political Science Review* 85: 77–96.

Newman, D., and A. Paasi. 1998. "Fences and Neighbours in the Postmodern World: Boundary Narratives in Political Geography." *Progress in Human Geography* 22, no. 2: 186–207.

Nogueira da Silva, A. C. 1998. *O modelo espacial do estado moderno.* Lisbon: Editorial Estampa.

O'Toole, F. 1997. *The Lie of the Land.* London and New York: Verso.

Ó Tuathail, G. 1996. *Critical Geopolitics: The Politics of Writing Global Space.* London: Routledge.

Paasi, A. 1999. "Boundaries as Social Practice and Discourse: The Finnish–Russian Border." *Regional Studies* 33, no. 7: 669–80.

Richardson, T. 2000. "Discourses of Rurality in EU Spatial Policy: The European Spatial Development Perspective." *Sociologia Ruralis* 40, no. 1: 53–71.

Robinson, G. W. S. 1959. "Exclaves." *Annals of the Association of American Geographers* 49, no. 3: 283–95.

Rogoff, I. 2000. *Terra Infirma: Geography's Visual Culture.* London and New York: Routledge.

Romero Valiente, J. M. 1990. "La frontera interautonomica de Andalucía: Un espacio periferico, deprimido y desarticulado." *Revista de Estudios Andaluces* 15: 29–43.

———. 1997. "El espacio limítrote entre Andalucía y extremadura: Medio físico y estructura socio-territorial." *Demófilo: Revista de Cultura Tradicional de Andalucía* 21: 29–43.

Sack, R. 1984. *Human Territoriality: Its Theory and History.* Cambridge: Cambridge University Press.

Sancha Soria, F. 1995. "Estudio geohistorico de la Comarca de la Sierra." *Jornadas del Patrimonio de la Sierra de Huelva* 10: 41–71.

Scott, J. 2000. "Euroregions, Governance, and Transborder Cooperation within the EU." In *Border Regions and People,* ed. M. van der Velde and H. van Houtum. London: Pion, 104–15.

Shields, R. 1991. *Places on the Margin: Alternative Geographies of Modernity.* London and New York: Routledge.

Shore, C. 2000. *Building Europe: The Cultural Politics of European Integration.* London and New York: Routledge.

Sidaway, J. D. 2000. "Iberian Geopolitics." In *Geopolitical Traditions,* ed. K. Dodds and D. Atkinson. London: Routledge, 118–49.

———. 2001. "Rebuilding Bridges: A Critical Geopolitics of Iberian Transfrontier Cooperation in a European Context." *Environment and Planning D: Society and Space* 16, no. 6: 743–78.

Sparke, M. 1998. "From Geopolitics to Geoeconomics: Transnational State Effects in the Borderlands." *Geopolitics* 3, no. 2: 62–98.

———. 2000. "'Chunnel Visions': Unpacking the Anticipatory Geographies of an Anglo-European Borderland." *Journal of Borderlands Studies* 15, no. 1: 187–219.

Swyngedouw, E. 1994. "The Mammon Quest: 'Glocalisation,' Interspatial

Competition and the Monetary Order: The Construction of New Spatial Scales." In *Cities and Regions in the New Europe: The Global-Local Interplay and Spatial Development Strategies,* ed. M. Dunford and G. Kafkalas. London: Belhaven, 39–67.

Tan, T. Y., and G. Kudaisya. 2000. *The Aftermath of Partition in South Asia.* London and New York: Routledge.

Tuan, Yi-Fu. 1989. "Surface Phenomena and Aesthetic Experience." *Annals of the Association of American Geographers* 79: 233–41.

Uriarte, L. 1994. *La Codosera: Cultura de frontera y fronteras culturales.* Mérida: Editorial Asamblea de Extremadura.

Valcuende del Rio, J. M. 1998. *Fronteras, territorios, e identificaciones colectivas: Interacción social, discursos políticos, y procesos identitarios en la frontera sur Hispano–Portuguesa.* Sevilla: Fundación Blas Infante.

Weber, C. 1995. *Simulating Sovereignty: Interventions, the State, and Symbolic Exchange.* Cambridge: Cambridge University Press.

Wilson, T. M., and Donnan, H. 1998. "Nation, State, and Identity at International Borders." In *Border Identities: Nation and State at International Frontiers,* ed. T. M. Wilson and H. Donnan. Cambridge: Cambridge University Press, 1–30.

8

The Sites of the Sino–Burmese and Thai–Burmese Boundaries: Transpositions between the Conceptual and Life Worlds

KARIN DEAN

The concept of a boundary cutting through naturally connected space has forged the division of world space into fixed, sovereign units—both on the tangible political maps and in the less palpable but pervasive practices of international relations. The borders separating territories and people are subject to complex dynamics stemming from countless and imminently contradicting state, global, and local factors. Thus the spaces at the sites of borders are diverse, multiple, and overlapping at the same time, and may be confusing, chaotic, and contested. Borderland interaction can be alienated, co-existent, interdependent, or integrated, according to Oscar Martinez (1991), and thus can establish unique "borderland milieus." The spaces on the borders often operate as "one unit on two sides of the border" (Baud and Van Schendel 1997), incorporating economies and dynamics on both sides of the border into a single unit that is distinctive or even disregardful of the respective states' enforced conceptions.

Some moments of space are not so obvious, as these are submerged by the conceptual, helped by those who have secured "legitimate" power and tools in the modern world system of states. Although de facto Earth space has no visible territorial lines, spaces are not restricted to the two ends of a binary: the conceptual de jure border and the "real" de facto. The de jure border, often only recently imposed onto the lived realities, may easily dismiss many long-established practices as illegitimate. Since several such realities may coexist, no objective reality can be designated. If such designation happens, as in the binary objective-subjective designations, this is so done from the acquired power positions. However, the "realness" in people's consciousness has been shifting away from both the "real" (with no visible territorial lines) and from the defined and imposed "conceptual"; local people may *know* that the particular river, mountain range, or indeed invisible line through an apparently homogenous terrain acts as a border, and they have started to utilize its effects (stemming from the differences between the two sides) in a variety of ways.

The taken-for-grantedness of the powerful conceptual enforcement of the border by state rhetoric and representations in the practices of international relations, everyday news reports, and colorful political maps that represent world space visually as distinct and unique units of space has offered theoretical challenges. Anssi Paasi (1996) has emphasized the need to study spatial socialization, stressing that, apart from being preoccupied with territorial lines, we should focus on the process of becoming that constitutes special constructions in state-spearheaded sociospatial consciousness. Carl Grundy-Warr (1998) has invited scholars to look backward, into history, and inward/outward by turning the political maps inside out to reveal the hidden geographies that can disclose the sociopolitical relations buried from view by those brightly colored units, the states.

This chapter introduces three interdependent and interacting moments of space—lived, perceived, and conceived—that Lefebvre (1974), Soja (1999), and Allen (1999) theorized in the framework of general social sciences. The chapter holds that a view through this tricameral lens helps to comprehend the complexities and contradictions surrounding the spaces at boundaries much more finely than would antagonizing the de jure and de facto. The discussion of the transformation of the overlapping sovereignties in nineteenth-century Southeast Asia into modern states with exclusive sovereignties and boundaries will additionally provide a context to state practices of territoriality and to its constructions of space. The main part of the chapter extends the inquiry of the boundary phenomena both in depth and scope, by examining the conceived, lived, and perceived moments of space of the Thai–Burmese and Sino–Burmese boundaries.[1] The effort of viewing the border through this multicameral lens requires the scholar to leave the imaginary global village, where our professional life worlds do tend to make most of us remote from the daily life worlds of many of our subjects. This chapter attempts to bridge this gap. Through descriptions, vignettes, and quotes from the Thai– and Sino–Burmese borders, I seek to run the life world of the ordinary people living on the two boundaries over the academic pages of this book.

A TRICAMERAL VIEW OF SPACE AND BORDERS

In Western thought, the subject/object and idealist/materialist binaries have prevailed, having "colonised spatial thought in modernity" (Allen 1999, 254). Modernist spatialities have been viewed as either real or imagined. The illusion of having a choice between two antagonistic ends of a binary simplifies and underestimates the objects of any spatial inquiry. In political geography, for example, it has led to endless debates about the porosity of boundaries, with more scholars declaring the boundaries porous and leaking, supported by the acknowledged amount of activities and practices defying the modern boundaries in the life world. A scholar daring to propose that an international boundary might be impermeable risks consignment to Cold War–era spokespersons of political geography. However, even in our postmodern world of global flows, information highways, networks, popular global culture, and free trade zones, a boundary can still be a line that separates realities of life and death and tangible extremes of livelihood conditions for people living on either side.

The choice should not be limited to the two ends of a binary, whether of permeable/impermeable, de facto/de jure, black/white, or objective/subjective. Dualism has become the primary target of critical postmodern spatial theory, and Lefebvre (1974) was among the first to attack it. He criticized the "double illusion" consisting of the "realistic illusion," which fetishes the real, and the "illusion of transparency," which fetishes the imaginary. Both simultaneously reference and critique each other, thus creating a rationale "that masks alternative spatialities vis-à-vis a delimiting binocular vision" (Allen 1999, 254). Stretching beyond the limits and confinements of such antagonisms and confrontation would help to achieve a richer spatial inquiry (see Lefebvre 1974, 292). Different forms of space do not have different existences. Lefebvre (1974) and Soja (1999) view space as a complex whole constituting simultaneously three moments interrelated to and interdependent on the real-and-imagined. Each of these three moments (which Lefebvre refers to as the conceived, perceived, and lived moments of space) must be defined through the other, as none can exist or be understood separately. This is an intricate way of viewing space (and boundaries) that helps to explain the different ways of seeing.

Conceived space is the imagined representation of space as found in normative forms of spatial knowledge. It is powerful; it dominates and codes that which is perceived, while submerging lived space. Perceived space is the mundane space of everyday life and of its spatial practices easily recognized and discussed. Although *close to* the real space, it is labeled *perceived* to insist that there is no objective reality: that what we consider reality is actually perceived as such. *The* reality is the world as each individual perceives and senses it. Lived space is produced and obscured by conceived space, and it is also different from the culturally normative perceived space. It is re/produced in direct contradistinction to the homogenizing influences of conceived space, while being the creative source of the latter. Whereas conceived space is convergent, lived space is divergent, resistant, and marginal (Allen 1999, 259–60). These three moments of space are interactive and interdependent, and cannot exist independent of one another.

At the site of the border—in the life world—there is no boundary line. Daily practices may actually undermine or transcend the boundary in numerous ways. There may be state manifestations of territoriality such as gates, checkpoints, or flags helping the observer to see (that is, to *perceive*) that the boundary "really exists." If such signs and symbols are missing in the life world, locating the boundary (so sharp, clear, and cutting on the political map) requires deep conviction and planted knowledge of its reality. The states are the main agents in constructing such conceptions, through strategies ranging from military force to more subtle modes such as careful deliberation, mapping, school geography, various administrative-governmental policies.

"AMBIGUOUS SPACE" OF BORDERS

Space in Southeast Asia before the end of the nineteenth century was organized through reciprocal tributary relationships involving ritual submission to the more powerful overlord, who in turn provided protection. In practical terms, the tributaries were obliged to provide workforce and other material supplies to the overlord if required. In practice, the overlords hardly involved themselves in the vassals' political and internal affairs. On many occasions, local chiefs paid tribute simultaneously to several more powerful overlords and to the supreme overlord of Siam, Burma, China, or Vietnam, and upon the arrival of the French and British offered to pay similar tribute to them as well. Thus sovereignties were multiple, hierarchical, and overlapping. For example, when the British turned to Bangkok to settle its colonial border, they were confused that the Bangkok court let the matter be decided by the local ruler (Thongchai 1994).

The differing approaches to viewing space and sovereignty at the end of the nineteenth century were the source of several misinterpretations and confusions among the Siamese, the British, and the French. The Siamese court at first did not take seriously the British desire to establish a boundary line through its buffer zone with Burma. However, it quickly learned to speak the language of modern political geography for self-motivated gains. Siam "voluntarily entered the contest for the ambiguously sovereign space" and tried to "extract its own share of territories to be allocated" in the scramble (Thongchai 1994, 97–101). Where, earlier, the economic value of the territory was important, now every bit of soil became relevant to the meanings of sovereignty, royal dignity, and nationhood. The 2401 km Thai–Burma border was settled for final delimitation peacefully, through inquiry and negotiation, in 1890–93, during which the Siamese and the British colonial authorities interrogated the local people on where their loyalties belonged (Thongchai 1994, 107–9). Vis-à-vis the French, however, the Siamese adopted methods of modern geography, utilizing both mapping and military to forge the new "geo-body" of Siam. New centralized administration and maps became the tools in the construction of the geo-body of Thailand, and persist as the recognized maintenance tools, as they do everywhere else in the modern organization of world space.

The Sino–Burmese border was a colonial heritage imposed by the British on China from its then-power position, although the British claims were based on similarly overlapping and ambiguous sovereignties and tributary systems. After prolonged negotiations between China and Burma (some of which took place in the Burmese media and reminded the public of the arbitrariness of settling boundaries) the 2185 km border reached the final conceptual settlement, with compromises from both states, on January 28, 1960. Nothing was settled on that day in the lived space of some Kachin villagers who suddenly found themselves in China, or surprisingly, in Burma.

As modern states, both Thailand and China have since made efforts to enforce, signify, manifest, and institutionalize their borders. One of the powerful nationalist themes in Thai history and in the making of an us and another has been the Thai rop phama ("the Thai fought Burma") (Thongchai 1994, 163). This historic enemy image is today aggravated further by the modern perceived menace from Burma: the guerilla warfare between the ethnic armies and the Burmese government, the war refugees, the flows of illegal immigrants bringing lawlessness, poverty, and diseases to Thailand, the drugs that the government declared the greatest threat to Thai national security. Thus policing and establishing the presence of state power at the border through intensive policing seeks to maintain the line of separation between us and the other, while simultaneously maligning that other.

China, on the contrary, has built its strategy toward its borderlands on a different principle: it views the disintegration of the state as the most serious security threat. The Chinese government has very carefully deliberated on the concentration, on its borders, of minorities whose tribal kin spread across several countries. In Yunnan province, the government's objective has been appeasing its transborder minorities through positive discrimination and generous provision of economic and social incentives to stay in the state of China. Autonomous governments are granted flexibility in administering central policies and in passing regulations related to specific local affairs. Areas where such flexibility can be practiced include local economic construction and planning; arrangements for foreign trade; management of local funds, subsidies, and budgets; retaining local revenue; education; and the development of local culture. There is wide space to pursue local cross-border interests and arrangements, as long as these do not jeopardize China's national security, that is, encourage separatist tendencies. The state has not been policing its borders as Thailand. However, the inevitable instability of the illegitimate military regime in Burma that has persisted through coercion, and of its by-product—particularly drugs, HIV/AIDS, and cross-border gambling that drains "prohibitively huge amounts of Chinese money" (Xu Er 2003)—is emerging, and China's relaxed attitude may be changing toward more intensive policing, as in Thailand.

The Burmese military regime that calls itself the State Peace and Development Council (SPDC), although recognized by the international system of states as a sovereign government, has yet to establish internal sovereignty. The SPDC continues the nation-making efforts of the illegitimate State Law and Order Restoration Council (SLORC) that in the 1990 elections lost overwhelmingly to the National League for Democracy (NLD) but never let it rule. Peter Taylor (1995, 6) uses the term "internal sovereignty" to mean "effective control of a territory"; he has remarked that "external sovereignty" (recognition by the international community) is the basis on which a state is considered sovereign. The unpopular military regime in Burma has failed to capture the mechanism of a modern state agency, and the territorialities within its borders remain fiercely contested and continue to generate unique dynamics in lived space that spill over its boundaries. On the Sino—Burmese boundary, the Burmese army is attempting to occupy the territory delimited by state boundaries that are the product of historico-political contingencies, and competes with the Kachin Independence Army (KIA), which sees itself as the legitimate protector of the historic Kachinland. The KIA still controls and administers relatively large areas on the border and maintains truly *inter-national* relationships with the Chinese and Burmese authorities.[2] Thus, being a de facto political agency without recognized international sovereignty does not necessarily prevent that agency from engaging in *inter-national* relations across political boundaries.

MOMENTS OF LIVED SPACE

A man in shorts and a T-shirt is sitting at the riverbank; he is most likely from the village nearby. A few houses and a temple can be seen on the other side of the narrow river. A few boats are docked at the shore. Some people are sitting in one and waiting—because only after a certain number of passengers embark will the boat operator steer it across the river. A monk in an orange robe approaches the man sitting on the bank and gives him a pack of Thai baht. He receives Burmese kyat in return, and then heads to the boat. The number of people is now sufficient for the boat operator to pull the boat off the shore—and in four to five minutes it drops the passengers on the other side.

> The self-assumed money changer continues his job of waiting for new customers. Up on the river bank, the Thai border rangers in black uniforms, equipped with loaded AK-47s and turned-on walkie-talkies, are sitting under the hoisted national flag. They keep an eye on the developments on both sides of the river. (Observation, Baan Tha Song Yang, Thailand, May 9, 2003)

> The road curves and the Chinese border checkpoint appears out of nowhere. The border guards ask me where I am going. I say that my friend's house is still three hundred meters down the road, and, after some negotiation and confirmation from my friend, they let me go. I spend the night beyond the last manifestation of the Chinese state territoriality. "It's good to have a border. The border identifies what happens where," my friend, a local Kachin pastor, notes. He lives "on the China side" of the border but services churches "on both sides," thus crossing the border several times a day. (Observations and communication, Man Hai, Dai-Jingpo AP, People's Republic of China, October 14, 2000)

In the life world no territorial line separates the two predominantly Karen villages on the opposite banks of the river that is the Thai–Burma border. A river is not a natural border, as that demarcation has been determined by people, the authorities with power. "Simply because a line is marked by nature does not necessarily imply that it is a 'natural' thing to utilize it for boundary purposes or that it may constitute a desirable or 'natural' line of separation between neighboring peoples," Boggs (1940, 23) argues, demonstrating a long tradition in geography of discarding any boundaries as natural. The local residents at Baan Tha Song Yang continue their daily or weekly activities in their life worlds regardless of the conceived border. So does the pastor on the Sino–Burmese border, by servicing the churches daily on both sides. By commuting across the river, primarily for buying or selling goods, for social or religious visits, or by commuting between the churches, the people cross the invisible line that on the maps is clear and sharp. Do they really defy and ignore the international boundaries, or do the international boundaries, on the contrary, disturb and interrupt—*challenge*—the local residents' everyday lives? This question only recognizes the binary options.

Theoretical discussions in political geography engage in analyses of the phenomena that defy and ignore, *challenge,* the boundaries in the (post)-modern world, and predict yet more "unbundlings" to come (Anderson 1996; Marden 2000).

Thus we can talk about the legitimization of the dominating power position and the power of such legitimatization. In any case, such legitimization has firmly established the conceived boundary in the everyday lives of the "simple" (or clever) border people, exemplified best by the changing of money at a particular riverbank or by approving of the existence of the border that separates the Kachin congregation into two countries. The people at the two adjacent villages are likely not to contest the act of money-changing or their belonging to two different states. This perception is enhanced and continuously reinforced by the hoisted Thai flag high up on the riverbank, and by the Karen Buddhist flag that in fact is hoisted on the opposite bank to mark the territory controlled by the Burmese ceasefire group, the Democratic Karen Buddhist Army (DKBA). The nearby Thai border rangers in black uniform, with loaded machine guns, who under the law can shoot suspicious "intruders" if warning shots are ignored, magnify the seriousness attached to the attempts of making the boundary "real" in lived space. These spatial practices constitute the perceived moment of space, the acknowledged and normalized "real." Perhaps it was the strong perception of the conceptual border that prevented me from getting on the boat and crossing the river despite the encouragement by the boat operator who emphasized the low cost of the trip. "Come—it's only 10 baht (US$ 0.25)—there are some shops and a market to see across the river" (personal communication, Baan Tha Song Yang, May 9, 2003).

The above example neatly demonstrates how all three moments of space coexist at the sites of the boundary. The perceived moment of space constitutes the practices generated by (the power of) the conceptual border. A boundary, if strongly enforced by the state, becomes "real"—that is, perceived. The lived moment of space, on the contrary, constitutes the practices that take place regardless of the interstate boundary. Lived space is resistant to what the conceptual tries to obscure and to what has been normalized as perceived space.

The above example of spatial practices also demonstrates the limitations of the de jure/de facto binary. For in this binary, the phenomena that are not de jure must be lumped together as de facto. In the above example, all de facto actions (money-changing, daily crossing of the river, and the armed Thai border rangers) send very different messages. The policing of the border and exchanging money are generated by the powerful and dominating conceived space, while the people's uninterrupted daily activities take place in the resilient and resistant lived space. The former code what is perceived; by changing money, recognizing and referencing to the boundary, the perceived moment of space is deeply rooted in the local consciousness. It coexists with the lived and conceived moments of space.

> What is Burma like? I have only been to Tachilek [the largest official border crossing in North Thailand, opposite Mae Sai]. It is amazing how different it is. The differences start right on the [Friendship] bridge [between the two countries]. (Communication with an educated professional in Bangkok, November 10, 2003)

This is another illustration of the power of the conceived moment of space in coding perceptions. A widespread and strong conviction among common Thais is that a different world starts on the Thai–Burmese border, a world that is predominantly dangerous and lawless, where the police do not help and the impoverished people, who look very different, use any opportunity to steal or take advantage of a visitor. These perceptions are also the consequence of the state constructions of Bamar (Burma) as the historic enemy.

"Both lived and perceived space in modernity are monitored and coerced by [conceived space]" (Allen 1999, 264).

RESILIENT, RESISTANT, AND SUBMERGED LIVED SPACE

Some of the wildest, roughest, and remotest landmass in Asia—predominantly uninhabited mountainous jungle areas distant from larger regional centers—constitutes the life worlds of the Thai– and Sino–Burmese boundaries.

Thus mostly the towns, the villages or refugee camps, or those locations where armies (political or ethnic opposition groups, proxy armies, or criminal cartels) are based near, or control, the border are the sites of spatial practices. The Burma side is literally maintained by the goods and services from, and relationships with, Thailand or China that resist the meanings of a conceived international boundary.

Four types of resistant activities that undermine the conception of an international boundary can be distinguished at the Sino−Burmese boundary.[3] First, local border residents, mostly the Kachin, the Chinese, and the Shan, continue their centuries-old five-day market system, where a market moves around from village to village for the convenience of buyers and sellers. The imposition of the border that now divides the participatory villages into Burma and China has not interfered with the tradition; consequently, the villagers cross the international boundary every day to buy or sell vegetables, meat, eggs, household items, etc. The second type of activity is the commuting and communication via China by the Kachin from Burma. When someone needs to go from northern parts of Kachin State to the south, the common practice is to travel along Chinese roads by bus and then cross back into Burma. This practice results from the location of the better roads and infrastructure on the "China side." Third, the Kachin family lines extending on both sides of the boundary, continuously fortified by new cross-boundary marriages, ensure frequent visits, social connections, and feelings of belonging, with the respective states' ideologies and normative loyalties peripheral. Fourth, the maintenance of the enclaves controlled by the Kachin Independence Organization (KIO), and the latter's cross-border activities and communications defy most de jure meanings of an international boundary and state sovereignty. According to the 1994 ceasefire between the Burmese military government and the KIO, the latter retains administration of the territories it controlled at the time of signing. Most of these territories are near the boundary with China, thus putting the KIO in control of large tracts of the Sino−Burmese boundary, where it has established several official border-crossings with China, complete with the KIO flag and uniformed armed guards at checkpoints.

Many submerged cross-border practices on the Thai−Burma border similarly resist the conceptual meanings of an international boundary and sovereignties, but these practices are mostly of one kind. Local trade in food and supplies caters predominantly to the armed resistance groups, proxy armies, and criminal cartels. Such trade has established some mutually symbiotic relationships and odd marriages between local agents of power or money and such groups, and is not based on long-established relationships on the ground, as on the Sino–Burmese border.

The KIO ceasefire on the Sino–Burmese border has secured temporary peace and business activities (though no political solution to the contested territorialities), while the dynamics on the Thai–Burmese border are dictated by the ongoing guerilla war. In this resistant lived space, the border is controlled by the Burmese army, the DKBA, the United Wa State Army (UWSA), the Karen National Liberation Army (KNLA), the All Burma Students Democratic Front (ABSDF), Shan State Army (SSA), or the Karenni National Progressive Party (KNPP). The KNLA, the ABSDF, the SSA, and the KNPP are ethnic organizations contesting the Burmese state territorialities. The DKBA and the UWSA are notorious drug-producing cartels and allies of the Burmese army. Since the takeover of the Karen National Union (KNU)/KNLA headquarters, the Burmese army and the DKBA have gained control of most of the Thai–Burmese boundary and planted their bases in the immediate vicinity of the border. The KNU/KNLA and ABSDF pockets of territory near the Burmese army and the DKBA camps can be accessed from the Salween River north of Mae Sam Laep, and serve as a route for the military, humanitarian, and political support for the democratic guerilla effort. The bewildering complexity of territorialities (in, furthermore, a continuous flux) rules the lived world on the Thai–Burmese boundary.

War-inflicted inhumanities such as raids, looting, burning, kidnapping, killing, torture, rape, and resultant flights and displacement also constitute a large share of the lived space, where civilians are the most vulnerable. The already-described village opposite Baan Tha Song Yang—with a temple, market, and a few shops—belongs to one of the most murderous units of the DKBA, known to loot both the Karen villages in KNU-held territories and occasionally the Karen refugee camps on Thai soil. The 140,000 official Karen refugees now in refugee camps in Thailand knew nothing of political geography when escaping the Burmese army through landmine-infested jungle, but they knew that crossing the Thai–Burmese boundary would make a huge difference. It would save their lives.

PERCEIVED MOMENTS OF SPACE

Both borders also boast several large official border crossings that powerfully manifest the state. The biggest exit port on China's Yunnan border is at Je Gau/Muse, and is correspondingly signified by its sheer size and the government's attention and investment. Similarly, displays of state regalia and manifestations of authority easily make the border visible, and loom large in people's daily lives on the Thai–Burmese border at the Mae Sot/Myawaddy and Mae Sai/Tachilek crossings. Enforcement and enhancement of a border crossing and of the border through each other in the state rhetoric obscures the distinctions between these two. The crossing (the fence, gate, and other functional constructions) is "real," touchable; thus the territorial line on the ground is readily perceived. Smaller and locally important official crossings, too, try to establish a boundary in local perceptions through such means: the presence of authority, gate, flags.

In the predominantly rugged, mostly uninhabited terrains covered by jungle, mountain ranges, hillocks, and rivers, the territorialities are marked and boundary made visible by army camps and respective flags on distant hilltops—if one knows where to look. Thus even amid seemingly unpopulated, rugged spaces, the Thai–Burma border can be perceived rather easily because of the intensive manifestations of authority. Authority on the Burma side may be in flux, but the act of manifestation is constant: the flag on the hillock simply changes depending on military success. The armies in Burma that control the border are particularly keen on manifesting their authority, induced by the contestedness of territorialities. The issue is control over territories, and the various armies imitate the state in their practices and manifestations of territoriality.

The Thai–Burmese boundary thus displays, due to the prevailing diverse pattern of territorial control on the "Burma side," an abnormal variety of flags on the riverbanks and hillocks, eliciting a matching response from Thai authorities. The Thai response has included intensive policing since the fall of the KNU headquarters in 1995 and the assuming of control of the border by its historic enemy, the Burmese army, and its proxy DKBA (which also increased the drugs flowing into the Thai kingdom). Goodden, who traveled the border in 1995 and again in 2001, commented that, though the border between the Karen state and Thailand in the past used to be "relatively quiescent, even benign" (when the KNU/KNLA were in control), "now it is 'hot,' troubled and even dangerous again" (Goodden 2002, 18). Similarly, the ABSDF leaders point out that during the KNU/KNLA and the ABSDF control of the border, only a few Thai checkpoints were present, while currently there are many, "because the Burmese army control the opposite side of the Salween" (personal communication, leaders of the ABSDF, Mae Sariang, May 9, 2003). The Burmese army, not only better equipped but operating under the internationally recognized legitimacy of Burmese state sovereignty, is taking control of the areas mentioned, thus stepping closer to enforcing internal sovereignty to match the external.

In addition to manifestations of multiple territorialities, other phenomena such as abrupt and stark differences in living standards and levels of economies, and an enforcement of the border as a time line, facilitate perceptions of an international boundary. A borderline can be "seen," since one side (in China and Thailand) boasts cars, four-wheel drives, sleeper buses, Internet shops, international telephone booths, and ATM machines, in contrast to the buffalo carts and WWII–period trucks and dirt roads, the nonexistence of Internet and other forms of modern technology, of the Burma side. Electricity, phone lines, and most roads generally stop at the border on the China and Thai side; this includes even the *information highway*, which disregards most other international boundaries in the world. Only mobile phones, both Thai and Chinese, can be used on the other side of the border, and even this depends on the location of a respective network station near the border. On the Thai side, all roads leading to Burma are further blocked by checkpoints run by a Thai army unit, border police, rangers, or paramilitary volunteers (observations on the roads between Mae Hong Son and Ranong, May 2003).

Local terminologies exemplify the strengthened perception of the border, as most border residents use references such as "China side," "Thai side," and "Kachin/Burma side," while the ethnic opposition based on the Thai–Burmese border refer to Burma as "inside the country" and Thailand as "outside the country."

All these factors have easily created the perception of a boundary line. The ethnic villagers might not have seen the borders when these were initially established between the states, but they have learned to see, and utilize, these today.

A very interesting arrangement has been reached by the Thai and Burmese governments in Pilok, near the well-known crossing-point of Three Pagodas Pass, where a territorial dispute between the states remains unresolved. Both disputed hillocks have the flags of both countries. One of the hillocks is open for visit from the Thai side, and tourists can photograph one another with the two state flags. The opposite hillock, however, hosts the Burmese army camp—under the two flags. Is this site a true in-between space, belonging to no one state and at the same time belonging to both?

CONCLUSION

Discussions in borderland studies predict that we are moving toward a borderless world again. Perhaps the most convincing argument for this "future prospect" is the expansion of information highways creating a "global village." The latter rhetoric plus the multiplying instances transcending states and sovereignties, such as common markets, international fairs, transnational functional regimes, and political communities not delimited primarily in territorial terms, not to speak of transnational corporations and information flows (Marden 2000), convince us that the world is becoming borderless. However, borders are unlikely to disappear from political maps in any near future.

The boundary set in conceived space sooner or later "strengthens" in perceived space and starts creating differences in people's everyday lives. The perceptions of the boundary can vary and depend mostly on the power of conceived space, including the deliberations of those holding power in establishing the physical presence of readily recognized manifestations such as border institutions (checkpoints, border guards, gates, customs, immigration), symbols and signs like flags, and other state regalia, boards, or announcements that help to inscribe the boundary in people's consciousness and everyday lives. While the states' rhetoric and the power of conceived space work toward enforcing stronger perceptions of the boundary, lived moments work to diverge these. Thus the space between the borderless lived space and the conceptual is undergoing continuous shifting. The borders were drawn (and in the distant future, maybe will be erased) only on maps. The life-world, lived space is always resiliently borderless. "Spatial practice regulates life—it does not create it. Space has no power 'in itself,' nor does space as such determine spatial contradictions. These are contradictions of society" (Lefebvre 1974, 358).

The mundane spaces of everyday life, and the spatial practices that try to make sense of the imposed conceptual and of the life world are in a continuous flux. These spaces are between the conceptual and the lived, and their location depends on the presence, or absence, of power and hegemonic practices that determine the acceptance, modifications, understanding, or rejection of the boundary.

The Sino–Burmese and Thai–Burmese boundaries are the sites of diverse unique phenomena, in the lived space. Some of these practices and activities continue uninterrupted regardless of the international boundaries; others have been generated solely by the boundary and unresolved territorialities in Burma. The former type of lived space dominates on the Sino–Burmese boundary; the activities generated by the boundary reign on the Thai–Burmese boundary. The perceptions of the Thai–Burmese boundary have been firmly established; policing, monitoring, and intensive manifestations of authority and territoriality by the state and local powers contribute to consolidating the border in people's perceptions. These perceptions are in a space of flux, in between the lived and conceived moments of space, revealing the role of states, geopolitics, and historical contingencies: "our understanding of the present must thus be based on their 'becoming' rather than on their 'being'" (Paasi 1996, 31). On the Sino–Burmese border, the trend clearly indicates that China is moving toward more intensive control of its border. The dynamics and relations at the site of the present borders are different between China and Thailand, and related to constructed, imagined, and perceived security threats, and to historic linkages, dynamics, and relationships.

NOTES

1. The name of Burma was changed to Myanmar (and of Rangoon to Yangon) in 1989 by the State Law and Order Restoration Council (SLORC), the government of the Union of Burma; the SLORC in 1997 was reconsti-

tuted into the State Peace and Development Council (SPDC). In this chapter, the name Burma is used in accord with the choice of most respondents.

2. For the in-depth analysis of the activities on the Sino–Burmese border that, from the normative state point-of-view, and in the modernist real-vs.-imagined spatial binary, are "unbundling" the Burmese state territoriality, see Dean (2005).

3. These are based on observations and conversations, January–February, October–December 2000, April–May 2001, and January 2002 in Kachin State, Burma, and Yunnan Province, China.

WORKS CITED

Allen, Ricky Lee. 1999. "The Socio-Spatial Making and Marking of 'Us:' Toward a Critical Postmodern Spatial Theory of Differences and Community." *Social Identities* 5, no. 3: 249–77.

Anderson, James. 1996. "The Shifting Stage of Politics: New Medieval and Postmodern Territorialities?" *Environment and Planning D: Society and Space* 14, no. 2: 133–53.

Ball, Desmond. 2003. "Security Developments in the Thailand–Burma Borderlands." A paper prepared for the Mekong Discussion Group, Sydney, September 26, 2003.

Baud, M., and W. Van Schnebel. 1997. "Toward a Comparative History of Borderlands." *Journal of World History* 8, no. 2: 211–42.

Boggs, Whittemoore S. 1940. *International Boundary: A Study of Boundary Functions and Problems.* New York: Columbia University Press.

Dean, Karin. 2005. "Spaces and Territorialities on the Sino–Burmese Boundary: China, Burma, and the Kachin." *Political Geography* 24: 808–30.

Goodden, Christian. 2002. *Three Pagodas: A Journey down the Thai–Burmese Border.* Bangkok: Jungle Books.

Grundy-Warr, Carl. 1998. "Turning the Political Map Inside Out: A View of Mainland Southeast Asia." In *The Naga Awakens: Growth and Change in Southeast Asia,* ed. Victor R. Savage, Lily Kong and Warwick Neville. Singapore: Times Academic Press, 22–86.

Lefebvre, Henri. 1974. *The Production of Space,* trans. D. Nicholson-Smith. 1999 edition. Oxford, U.K., and Cambridge, Mass.: Blackwell.

Marden, Peter. 2000. "Mapping Territoriality: The Geopolitics of Sovereignty, Governance, and the Citizen." In *Migration, Globalization, and Human Security,* ed. David T. Graham and Nana K. Poku. London and New York: Routledge, 47–70.

Martinez, Oscar J. 1991. "The Dynamics of Border Interaction: New

Approaches to Border Analysis." A presentation at the Conference on International Boundaries, IBRU, University of Durham, England, July 18–21.

Paasi, Anssi. 1996. *Territories, Boundaries, and Consciousness: The Changing Geographies of the Finnish–Russian Border.* Chichester, U.K.: John Wiley.

Soja, Edward W. 1999. "Thirdspace: Expanding the Scope of the Geographical Imagination." In *Human Geography Today,* ed. Doreen Massey, John Allen, Philip Sarre. Malden, Mass.: Polity Press, 260–78.

Taylor, Peter, J. 1995. "Beyond Containers: Internationality, Interstateness, Interterritoriality." *Progress in Human Geography* 19, no. 1: 1–15.

Thongchai Winichakul. 1994. *Siam Mapped: A History of the Geo-body of a Nation.* Chiang Mai, Thailand: Silkworm Books.

Xu Er. 2003. "China Moves on Myanmar: PLA Masses on the Border." *Asia Times,* November 22.

9

A Pacific Zone? (In)Security, Sovereignty, and Stories of the Pacific Borderscape

SUVENDRINI PERERA

> There was no book of the forest
> no book of the sea, but these
> are the places people died.
>
> —MICHAEL ONDAATJE,
> "THE DISTANCE OF A SHOUT"
>
> What the map cuts up, the story cuts across.
>
> —MICHEL DE CERTEAU,
> THE PRACTICE OF EVERYDAY LIFE

First, three boat stories.

On November 4, 2003, the day of the Melbourne Cup, the most significant sporting event in Australia ("the race that stops a nation"), a fishing boat, the *Minasa Bone,* landed on Melville Island, about twenty kilometers off the northern capital of Darwin. The Islanders, Indigenous Tiwi people, were surprised to come across obviously foreign men on the beach who asked them, "Is this Australia?" Perhaps the arrivals were confused by the large number of black faces and the general third world look of the place. The Islanders' answer marked a subtle distinction: You are on Melville Island. Yes, it is Australia. *In but not of.* Did the arrivals register any qualification?

There were fourteen of them plus the Indonesian crew of four. They requested water, indicated they were from Turkey, and asked for asylum. Only a few weeks earlier, the Islanders had been instructed by visiting officials what to do in such an eventuality. The men were provisioned, quickly dispatched back to their boat, and the authorities notified. These Melville Islanders were the first and last Australians, apart from the navy and immigration bureaucrats, that the new arrivals would set eyes on.

Within hours, three armed navy ships were headed to Melville Island. An exclusion zone was established around the small boat. A second injunction prohibited planes from flying over it. And, just to make certain, the navy towed the small vessel out to sea and put a guard on it. Thousands of miles away in Canberra, ministers met in urgent session and determined to excise Melville Island from the migration zone. They threw in another four thousand or so north Australian islands for good measure. This ensured that any "suspected asylum seekers" making landfall in these places would not have access to the domestic legal system—they had been effectively deterritorialized. The governor-general was summoned from his race-day festivities to approve the proceedings. The ministers made the legislation retrospective, then went to bed satisfied. They had performed miracles that day. By this exercise of their excising imaginations, they had turned back time and commanded the waves to flow outward. That boat never arrived in Australia. *Boat? What boat?*

Next day no one could find the *Minasa Bone*. The administrator of Christmas Island, severed from the mainland in a previous excising operation, disclosed that he had been asked to turn the community sports center into a detention camp for the men. He refused: there was already an expensive, new, custom-built detention center on Christmas Island. But the recent arrivals couldn't be allowed to talk to other would-be refugees. Although the full implications of this cordon sanitaire were yet to emerge, from the beginning the *Minasa Bone* was encircled in silence; the asylum seekers would later tell how each request and plea they made was met with "keep quiet" or "shut up" (see chapter 11, this volume).

A day later, it was disclosed that the navy had towed the damaged boat to the edge of international waters, then pointed it toward Indonesia. The Indonesian government was holding the men, now identified as Kurds from Turkey. Unlike Australia, Indonesia is not a signatory to the 1951 United Nations Convention on Refugees protection. The country contains large numbers of internally displaced people. Unsurprisingly, it was expected to return the fourteen Kurdish men to Turkey. Only a few months earlier, in preparation for joining the U.S. and the U.K. invasion of Iraq, Australia had had a lot to say about the plight of oppressed Kurds in the region. But, challenged about its responsibility for the *refoulement* of these Kurdish asylum seekers to a country that oppresses them, official responses fell into a tawdry, predictable sequence: (1) the men did not claim asylum in Australia; (2) if they did claim asylum, these officials are not aware of it; (3) yes, they did claim asylum, but they couldn't have, because they never entered Australia in the first place.[1]

This is the marvelous, brutal, incontrovertible logic of excision.

The technology of excision, by which certain parts of a state's territory are decreed by law *not* to be accountable to law, is one of a repertoire of technologies for producing hybrid spaces: what may be described as spaces of exception, in Giorgio Agamben's terms, spaces that are both inside and outside the law. These spaces, designed to isolate, contain, and punish asylum seekers, include onshore detention centers, deterritorialized zones, and the offshore arrangements known as the "Pacific Solution." In their turn, these new organizations of space alter geopolitical and cultural alignments and produce new border relations and spatial reconfigurations within a region. Fixed boundaries are displaced by flows and diversions (Agnew and Coleridge 1995, 214) that undermine as well as reinforce existing spatial divides. At the junction of the Pacific Ocean with the Arafura and Timor seas, Australia's coastline, its outlying islands and territories, its varied kinds of tenure over place, form a mobile, unstable, racialized border traversed by the tortuous itineraries of bodies seeking asylum.

WALLS IN THE WATER, LINES IN THE SEA

In Australia, spring is the season of boats from the north.

In 2001, the most significant boat in our recent memory, the Norwegian container vessel *MV Tampa,* arrived on the horizon a week before September 11, carrying 450 or so asylum seekers, most of them Afghan, Iraqi, or Sri Lankan, who had been rescued from their sinking vessel. The *Tampa* made for Christmas Island but was denied permission to enter Australian waters. The port was closed to all, including the island's inhabitants, who were not even allowed to go fishing. As days passed and the sick asylum seekers on board called for help, the *Tampa*'s captain, later to receive humanitarian awards for his actions, defied the ban and sailed into Australian territory, invoking the authority of international law. After a long stand-off, as conditions on the overcrowded ship deteriorated, the *Tampa* was intercepted, forcibly boarded, and taken over by the military. With the government vowing these asylum seekers would never set foot on Australian soil, the 450 or so men, women, and children were divided up between New Zealand and a hastily established camp on an exhausted mine site on Nauru. Some of the voyagers were later transferred to an old army base on Manus Island in Papua New Guinea. Other sites, from a disused leper colony in Fiji, to newly independent East Timor, to tiny Tuvalu, Palau, and Kiribati, were canvassed to ensure they would also make themselves available. So began the exercise known as the Pacific Solution. The term is a misnomer in every sense. Beginning with an armed action by the Special Armed Services (SAS) and imposing the full weight of Australia's economic and diplomatic authority over the region, the plan is neither peaceable nor a genuinely regional Pacific move. And, as events prove increasingly ominously, even as I write in late 2003, it is no solution.

In the months after the *Tampa*, asylum seekers became the supreme national preoccupation. In a special sitting as the Western world was still reeling from the September 11 attacks, Parliament agreed to excise Christmas Island, Ashmore Reef, and Cartier and Cocos Islands, outlying territories targeted by asylum seekers' boats, from Australia's migration zone to prevent any new arrivals. Boats carrying asylum seekers, termed SIEVS (Suspected Illegal Entry Vessels), were shadowed, intercepted, fired on, and instructed to turn around, with those that persisted being boarded by force and the occupants hauled off to deterritorialized camps in the Pacific.[2] On the boat known as SIEV 10, two women, Nurjan Husseini and Fatima Husseini, died in the chaos when their vessel caught fire and sank during one of these operations after it was fired on and then boarded by the navy.

SIEV X, the unknown SIEV, was a wretched, unsafe vessel packed with over 450 people, which sailed for Australia a few weeks after the *Tampa*. It broke into pieces in a site yet to be definitively identified, somewhere in the indeterminate space between Indonesian and Australian waters (Kevin 2004). Wherever SIEV X went down, apparently out of the range of Australia's elaborate "saturation surveillance" and monitoring operations, 353 of those on board drowned, the largest recorded death toll ever of asylum seekers at sea.[3] Of these dead, a little under half, 146, were children; 142 were women. Many of their names are known from family members who survived that night, but almost a hundred remain unidentified (Perera 2006).

These are neither the first nor the last deaths in the waters around Australia. The lament of the group Algeciras Welcomes, spoken over the body of yet another anonymous asylum seeker, a young man washed up by the waters of the Mediterranean in 2003, calls for repetition:

> There are those . . . who died in the sea and who died forgotten because here no one knew who they were. . . . They suffocated in the water and drowned in anonymity. We do not know whom to weep over when we have gathered here. Their relatives, on the other side, do not know we are weeping for them. They died in the sea, but politics, outlined in dispatches sent from the West, murdered them. They built walls in the water, they demanded that visas should appear out of thin air; politics assured that people would be moving on from one place to another. (quoted in Fekete 2003)[4]

This passage graphically evokes all the malignant perversity of recent regimes of border control in the West: walls rise up out of oceans; queues are expected to form in the desert; visas lose their materiality, floating free of the state and bureaucratic regimes that give them life, and demand to be plucked out of thin air. Yet, as this lament also powerfully demonstrates, the moving on of peoples from one place to another, the passages of uncertain transit, the unmarked places where people die, make cracks in the walls rising out of the water, allow for new spaces across the lines drawn in the sea. And, as Nevzat Soguk points out, the movements themselves are, "for those who are part of them, acts of resistance by which the 'moving' people are able to shape their own experiences in ways hitherto unprecedented" (Soguk 1995, 293). This border space of moving people and bodies is a site where new relations, practices, possibilities, and forms of connection may emerge.

THE BORDERSCAPE

The endless voyaging back and forth of the people on the *Tampa*, the unknown terminus of SIEV X, the strange voyage of the *Minasa Bone*, with its crossing and recrossing of territorial and temporal boundaries, are all possible starting points for conceptualizing a complex border zone. How do the moving bodies of asylum seekers reconfigure this multiethnic, transnational, transborder space of islands and archipelagos, coastlines and oceans, constituted by a mesh of discourses and practices? There are multiple actors in this geo-politico-cultural space, shaped by embedded colonial and neo-colonial histories and continuing conflicts over sovereignty, ownership, and identity. The bodies of asylum seekers, living and dead, and the practices that attempt to organize, control, and terminate their movements bring new dynamics, new dangers and possibilities, into this zone. Allegiances and loyalties are remade, identities consolidated and challenged, as border spaces are reconfigured by discourses and technologies of securitization and the assertion of heterogeneous sovereignties.

My essay attempts to outline the multilayered spaces of this unstable border zone. The relations between Australia and its outside(s)—those places that are, in one way or another, not-Australia—defy representation by a linear divide: the border. Under the Pacific Solution, Australia's border both contracts, as it excises its outlying territories for specific purposes, and expands, as it annexes the sovereign space of other states for its own uses. Both movements are assertions of territorial authority and suzerainty, acts that bring space under differential forms of control, making new borders that in turn give rise to multiple resistances, challenges, and counterclaims. This making and remaking of different forms of border space in the Pacific is what I describe as a *borderscape*.

Against the flat and static representation or "tableau" of the modern map as described by Michel de Certeau, the notion of a borderscape is multidimensional and mobile, drawing on de Certeau's discussion of "spatializing practices" that encompass "geographies of actions," histories of place, and the itineraries of moving bodies:

> Space is composed of intersections of mobile elements. It is in a sense actuated by the ensemble of movements deployed within it. Space occurs as the effect produced by the operations that orient it, situate it, temporalize it, and make it function in a polyvalent unity of

conflictual programs or contractual proximities. On this view, in relation to place, space is like the word when it is spoken. . . . In contradistinction to the place, it has thus none of the univocity or stability
of a "proper." (1988, 117)

My conceptualization of the borderscape allows for the inclusion
of different temporalities and overlapping emplacements as well as
emergent spatial organizations. A shifting and conflictual space, this
Pacific borderscape is currently being reconstituted through technologies and discourses of securitization as well as through forms
of new and ongoing spatial relations and practices that defy the categorizations of the border, and unsettle the univocity and stability of
the "proper."

MARITIME HIGHWAYS

Like the detainees of the war in Afghanistan held on Guantanamo
Bay by the United States, the asylum seekers held in Australia's offshore detention camps under the Pacific Solution occupy a legal
limbo, a deterritorialized space of indeterminate sovereignty. Their
position recalls that of Haitian asylum seekers held at Guantanamo
in the 1980s, outside U.S. jurisdiction, until international protest
forced a change in policy (Simon 1998; Perera 2002b). Under the
Pacific Solution, asylum seekers are considered out of reach of both
Australian law (and the international obligations it entails) and the
domestic law of the states where the camps are located. As deterritorialized spaces, these Pacific camps infringe on the sovereignty
of the small states that act as their official "hosts" and testify to
the economic and diplomatic power wielded by Australia over its
neighbors.

In his essay on the Pacific Solution, Prem Kumar Rajaram discusses the politics of Australia's "emplacement" in the region
through metaphors of presence, visibility, and materiality. Studying
a map of the region, he comments that it "shows a disparate group
of indistinct landmasses, floating in the wide Pacific, with the exception of Australia, large and whole; concrete in a way that Nauru
and even PNG [Papua New Guinea] cannot be. Australia dominates
the region; the idea of Australia gives focus and orientation to the
'the region.' Australia is not entirely of the region, but the region is
Australia's; it is Australia's dependent backyard" (2003, 290). This
passage beautifully evokes the solidity and stability of Australia's

official self-representations in the region, as against the "pathological degeneration" (Rajaram 2003, 291) that characterizes its representations of the diminutive neighbors dotted to its north and west. Other representations and understandings of this space, however, make it possible to trace different cartographies and patterns of sovereignty in the region.

The massivity and *thereness* of Australia, "large and whole," is very much a matter of *how* we read the space on the map. The seeming largeness and wholeness of Australia on the map are effects of historical processes that remain incomplete, and that continue to be troubled and challenged by contradictions, most obviously by racialized tensions and anomalies between the imagined large and whole(some) body of the white nation and its unassimilable black or colored bits, often envisaged as dysfunctional, grotesque, or diseased. Nor are these contradictions only internal; I want to propose a reading of this space on the map that, instead, confounds the distinction between internal and external, between Australia and its apparent "outside(s)."

As a form of representation, the map privileges national boundaries as it overwrites alternative geographies and contested spatialities. Not representable on the static surface of the map are, precisely, the complexities of a mobile and multilayered borderscape. The map as an artifact of Western modernity, de Certeau argues, must be understood as a totalizing representation that "colonizes space" (1988, 121). Whereas the earliest medieval maps recorded *itineraries* and were forms of "a memorandum prescribing actions," in modernity the map "slowly disengaged itself from the itineraries that were the conditions of its possibility" (1988, 122). This "erasure of the itineraries" wipes from the map other configurations, "geographies of actions" and living histories of border practices. It ignores the complex and contested nature of national or natural boundaries.

The processes that put Australia on the map as a large and whole nation-state actively obscure processes that suggest other, more complex and layered, geopolitical and spatiotemporal configurations. Intricate, long-established maritime ties reach across and between national borders linking the northern coastline of Australia and the islands of the Torres Strait with Papua New Guinea and Indonesia. As Anna Shnukal, Guy Ramsay, and Yuriko Nagata write, "For the Islanders of the Torres Strait, South-East Asia and the Pacific . . .

the notion of the shifting sea as a boundary is alien, the inverse of the European terrestrially focused perception. . . . Their surrounding territorial seas are not boundaries . . . which serve to separate them from their . . . neighbors, but rather maritime highways which connect them with others for reasons of trade and ritual" (2004, 2). This mesh of traffic along the maritime highways of the Indian and Pacific Oceans created communities looking outward across the ocean at least as much as inland to the center of the continent. Henry Reynolds points out that "For many Aboriginal groups right around the north coast their relationship with the Macassan (Indonesian) fishermen was much more significant than their experience with Europeans passing through their country or sailing by in pearling luggers" (2003, 12).

Centuries-old border practices and affiliations with Macassan fishing communities are "woven into legend, kinship networks and the coastal economies" of the Aboriginal peoples of the north and west (Reynolds 2003, 13). Reynolds reports that the Macassans arrived in fleets of up to a thousand with the northwest trade winds each year and returned three months later with the southeasterlies, sometimes accompanied by Aboriginal passengers who would live with them until the next year's voyage. The coercive power of White Australia attempted first to prohibit and then to erase these networks of connection and exchange across borders. In the early years of the twentieth century, Reynolds records, Yolngu people of the north were incredulous at the idea that Balanda [whites] could interfere with these well-established ceremonial links and the cyclical itinerary of the Macassan arrival and return: "There were stories that some Macassan captains had said . . . they might not be able to come in future years because the Balanda out of Port Darwin would not let them land (some Yolngu elders today remember their fathers in tears of disbelief when the Macassan captains told them this news). But many Yolngu dismissed these stories. They said 'Who are these Balanda? They have no say in the legal agreements between our clans and the Macassans'" (2003, 14).

But if colonial power could suspend the border practices and arrangements that structured relations among the coastal communities of the Arafura Sea and Torres Strait, it could not succeed in erasing the memories and traces of those practices. Since the 1980s, various forms of affiliation among the coastal and island communities of the Pacific have begun to be reasserted and tested (Balint 2005). Indigenous sea rights and claims of title to water and intertidal zones are part of this shifting, multidimensioned borderscape, following logically on the land rights activism of previous decades.

Similarly, the excised territories of Christmas and Cocos Islands testify to other regional identities and itineraries overwritten by the map. These islands, populated mostly by descendents of imported colonial labor, once "belonged," though perhaps equally tenuously, on the Indonesian side of the border. Many inhabitants of Christmas Island display a markedly different attitude to asylum seekers than do the majority on the Australian continent. On the night before the navy shipped the *Tampa* asylum seekers out of Australian waters, the Christmas Islanders farewelled them with an outburst of fireworks (Jameson 2003, 13). This wordless display of defiance and support between ship and shore is one of a series of gestures and signals improvised by communities to assert alternative itineraries and affiliations, to break out of their imperializing location on the map and reach across the proliferating borders within borders.

"NON-AUSTRALIANS"

The seemingly self-evident nature of Australia as a unitary, sovereign geo-body with boundaries that naturally coincide with its landmass is undone, or at least brought into question, in several ways by its colonial history. As Indigenous claims to land and ocean make clear, here sovereignty over *country* was never ceded (unlike in other settler societies such as New Zealand, Canada, or the United States) by the signing of a treaty between colonizer and colonized. Contestations of the colonizers' title have continued for over two hundred years and in the 1990s resulted in two significant gains, the Mabo and Wik legal judgments. The Mabo judgment is a landmark ruling establishing that ownership of land did exist prior to colonization, and reversing the ingenious fiction of *terra nullius,* nobody's land, that legitimizes colonial occupation. The Mabo and Wik judgments established that, in certain (highly restricted) circumstances, native title might not have been extinguished by colonization. Following these rulings, calls for recognition of Indigenous rights to land and for a treaty between the state and the Indigenous owners resurfaced strongly, reaching a crescendo in 2001, the centenary year of the federated Australian state.

In response, fears of Aboriginal people making claims on private property were aroused, leading to fevered anxiety in parts of Anglo-Australia over the security of suburban backyards. The intensity of the fears stirred up over ownership of family backyards (never remotely under threat by the rulings) suggests the elemental level at which many reacted to the notion of native title to land. The "backyard" thus remains a loaded term in Australian discussions of both domestic and regional concerns, suggesting not safe tenure and ownership but a territory riven by the tensions between public and private claims (Perera and Pugliese 1997, 1998).

Renewed anxieties over the legitimacy of the Australian state, unresolved issues of native title, and the sense of Anglo-Australia as an anomalous racial/ethnic presence in the region are all factors to consider in the disproportionate and hysterical response to the arrival of a few hundred asylum seekers in the country's surrounding waters or "backyard." The presence of these moving bodies, retracing old routes of connection and creating fresh links across borders, introduces an unpredictable dynamic to ongoing conflicts over internal sovereignty, legitimacy, and title, and, in Reynolds' words, reawakens an "ancestral unease about an empty and vulnerable north [that] continues to reside just below the surface of the [white] Australian psyche" (Reynolds 2003, 193). Spectacles such as the repulse of the *Tampa* and the acts of excision are, as Rajaram points out, *performative* assertions of sovereignty in the face of these perceived threats (2003, 292). As such, however, they also have the effect of giving rise to counterclaims and assertions, and provoking debate about competing styles or modes of exercising sovereignty.

One of the most potent of these discussions concerns the ethical and moral responsibilities of the host in enacting ownership of place. Although the state has assumed the stance and rhetoric of a besieged householder turning away uninvited guests and queue jumpers intruding on private property, other models of ownership emphasize hospitality and generosity to strangers in need as part of the inexorable obligations of ownership (Derrida 2000; Levinas 1969). Indigenous Australians draw on their own traditions of inviolable duty toward others to assert a stake in the debate over the treatment of asylum seekers (Perera 2002a). Tony Birch has written that Indigenous people must assert and claim their ownership of the land through the exercise of their ethical responsibilities of

hospitality and care toward people seeking protection as much
as through legal claims to sovereignty: "As Aboriginal people . . .
we must also assert moral authority and *ownership* of this country. Our legitimacy does not lie within the legal system and is not
dependent on state recognition. . . . We need to claim our rights,
beyond being stuck in an argument about the dominant culture's
view of land rights or identity. And we need to claim and legitimate
our authority by speaking out and for, and protecting the rights
of others, who live in, or visit *our* country" (2001, 21). To assume
the role of host is to claim and enact ownership of the land. But
Indigenous people, while retaining moral authority over the land,
also share with asylum seekers experiences of being physically dislocated and dispossessed. As Birch says elsewhere in the same essay,
Indigenous people have been for too long positioned as strangers in
their own land, and as dispossessed refugees in their own nation.
Dispossession has taken the form of Indigenous peoples' expulsion
from their traditional country, their social exclusion from the body
of the nation, and their positioning as other to its cultural norms
and way of life. These aspects of their exclusion from the realm of
the nation informed the responses of some Melville Islanders when
they heard of their island's excision from the body of Australia: "We
watch the news and read the paper. We're not stupid people, we're
educated. We know what it means to be non-Australians. If that
boat comes back, we'll welcome them and give them food and water.
You know why? Because we're all one group—non-Australians"
(Hodson 2003). Responding to a history of exclusion, the Islanders
invert the logic of deterritorialization. They assume the position of
being alien to Australia, but also take on the counterrole of host
and of legitimate owners of the country by vowing to provide food
and water for the asylum seekers. The assertion "We know what it
means to be non-Australians" exposes the forms of disenfranchisement and dispossession that fissure the idea of a unified and whole
nation. The deterritorialization of Melville Island only reinforces for
its inhabitants the provisional and incomplete nature of their own
belonging in Australia. Now formally relegated to an ambiguous
and expendable status by the excision of their territory from the
state, the Islanders embrace the identity of the non-Australian. Their
use of this term also refers back to the frequent use of the descriptor
"un-Australian" as a barely coded reference for racial and cultural

otherness. The Islanders mark the fractures that already exist within the body of the nation and claim solidarity with the deported asylum seekers, voluntarily excising themselves from the nation in the process: "We're all one group—non-Australians."

The responses of the Islanders also turn the logic of deterritorialization back on the state. They declare: "If they want to talk about this, they have to come out here and see this place. Next time an election comes around, Labor Party, Liberal Party, they'll be out here asking us to vote for them. We'll say, 'Sorry, we're asylum seekers, we can't vote'" (Hodson 2003). Alienated from the confines of the national, the Islanders strategically adopt a position that allows them to reject citizenship, a status that they never fully owned in the first place, and disavow the state's authority over them. Although statist practices of excision and deterritorialization are acts that assert and perform sovereignty, these acts also expose the random and contingent nature of a power that is exerted or withdrawn at the stroke of a pen. For Indigenous people who, over two hundred years, withstood and contested through a range of means the sovereignty claims of the state, the ease with which more than four thousand islands were struck off in the course of a single afternoon can only confirm the arbitrary, transient, and discontinuous nature of colonial sovereignty. The presence of asylum seekers in the waters around northern Australia has, in this sense, the potential to act as a lightning rod for existing contradictions surrounding the Islanders' membership in the nation, and to bring to a crisis their ongoing resistance to the sovereignty of the Australian state.

These insights underlie a mock newspaper report that appeared in the *Sydney Morning Herald* a few days after the excision of Melville Island, under the headline "Proclamation of a 'Great Southern Rainbow Republic of Antipodea' by the spirit of Eddie Mabo [the author of the legal challenge resulting in the Mabo judgment that erased *terra nullius*]." The report imagines that the excision of Melville and other islands triggers an immediate declaration of independence:

> The unexpected but proud declaration by Mabo of the world's newest sovereign territory . . . came after the Australian representative of the British Caretaker Government, Mr. Howard, finally surrendered his long struggle to claim ownership of the various islands to the "Australian" mainland's north on behalf of his Monarch, Her Majesty Queen Elizabeth the Second.

Mr. Howard . . . is understood to have finally conceded the point-lessness of fighting to retain British sovereignty over the islands . . . by simply "chopping the bastards off the Aussie map" and declaring them "terra nullius" again.

This in turn allowed Mr. Mabo's spirit to take possession of the islands and re-assert a newly-independent sovereignty. (Robertson 2003)

The excision of Melville and the other islands from the migration zone in effect creates a border inside the border, confirming unspoken truths about the deep racial faultlines of the nation, the parts of it that never really belonged. This in-between zone, a not-quite-Australia, opens up new spaces for understanding the limits of membership in the nation and for the possibility of new sovereignties, itineraries, and affiliations.

"NOT-QUITE-AUSTRALIA"

The notion of a "Great Southern Rainbow Republic of Antipodea" arcing triumphantly over the Pacific is of course the nightmare scenario for a state concerned with asserting its internal wholeness and sovereignty, and consolidating regional hegemony. The Pacific Solution and "Great Southern Rainbow Republic" are in this sense alternative cartographies through which the region is being reimagined as borders are remade and the moving bodies of asylum seekers engender new spatial practices and configurations.

Where the idea of a "Great Southern Rainbow Republic" allows us to imagine a final severing of links with the colonial motherland and a remaking of geography by cutting through existing borders and divides, the Pacific Solution must be understood as a neocolonial act of geographical violence:

In one fell swoop Australia created a new international "practice": the export of a refugee problem from one area to another, thereby creating, in a callous display of neo-colonialist guile, a refugee problem in an area where there was previously none. In the process Australia . . . artificially set the scene for the description of the problem by the . . . government as a "regional" one requiring, according to the rather shameless . . . argument now being used, a "regional" solution based on regional "burdensharing." (Fonteyne 2002, 19)

The creation of a refugee problem in a location where none previously existed once again remaps the relations between Australia and its outsides, creating deterritorialized spaces of ambiguous sovereignty. It immediately introduces a new element into the internal dynamics of the affected states, and into the relations of the states to one another and to Australia. The introduction of the Pacific Solution also prepares the way for a number of other regional interventions under the overarching theme of ensuring secure borders. To quote from the Department of Foreign Affairs and Trade (DFAT) policy white paper for 2003, bluntly titled "Advancing the National Interest":

The South Pacific matters to Australia

Instability in the South Pacific affects our ability to protect large and significant approaches to Australia. The Government also has a duty to protect the safety of the 13 000 Australians resident in the countries of the region—some 7 000 in Papua New Guinea alone. And transnational crime in and through the region—terrorism, drug trafficking, people smuggling, illegal immigration and money laundering—is a growing threat to Australia and the South Pacific countries themselves. Cooperation with the South Pacific on such issues, particularly people smuggling, has delivered real benefits to Australia. The establishment of Australian-funded processing centres in Nauru and PNG showed how regional countries can cooperate to deal with an issue of concern to the region as a whole. (DFAT 2003)

This passage reveals the continual slippage and sleight of hand by which Australian interests are represented as regional interests, and the creation of a rationale for increasing Australia's reach into domestic as well as external affairs of South Pacific states (Perera 2007). The Pacific Solution is cynically represented as a collective response to a regional threat instead of what it is, the "export of a refugee problem from one region to another," to extricate Australia from an international impasse of its own making.

In its operation, the Pacific Solution weakens the sovereignty of the participating states, fosters or heightens internal tensions in some of these states, and increases Australia's reach into other aspects of their government.[5] In this sense, the Pacific Solution cannot be separated from a range of other moves recently put in place to reshape the region, forging new geopolitical entities and creating new border zones through the mobilization of discourses of security and its more acceptable counterpart, "human security."

OUR PATCH?

In an essay written in the weeks after the September 11 attacks, Agamben discusses the rise of security to the status of a preeminent principle of governance. Drawing on an unpublished lecture given by Foucault in 1978, Agamben suggests that "in the course of a gradual neutralization of politics and the progressive surrender of traditional tasks of the state, security becomes the basic principle of state activity." He argues: "While disciplinary power isolates and closes off territories, measures of security lead to an opening and to globalization; while the law wants to prevent and regulate, security intervenes in ongoing processes to direct them. In short discipline wants to produce order, security wants to regulate disorder" (Agamben 2001, 1). Security may thus be understood to override or collapse existing borders as it produces new entities to protect and control. It operates through multiple forms, military and extra-military, promoting new identities, itineraries, and regionalities.

The collusion between security and globalization is clear in recent Australian moves in the Pacific. The two entwine in the 2000 military operation in East Timor and the 2003 policing expedition aimed at averting a "failed state" in the Solomon Islands. Both expeditions are represented as acts of good neighborliness, peacemaking, and regional nation-building. The Solomons expedition mirrors the U.S.-led intervention in Somalia in the 1990s, complete with its own approximation of a "rebel warlord" figure, Harold Keke. The earlier entry of the Australian military into the conflict in East Timor was cast as a down-to-earth goodwill operation: in the words of Adrian Vickers, "The diggers were presented as a sporting team with guns" (2003, 108). Yet this benign image of the local champion stepping in to sort out neighborhood bullies has its more sinister aspects. The enabling role of Australia in the liberation of East Timor is undermined by competing claims over the oil deposits in the Timor Gulf, disputes in which Australia exercised the full weight of its power to its own advantage (Kehi 2005).

These Australian peace and security interventions in the Pacific can be read against the "mythologies" of peacemaking that Sherene Razack examines in detail in the Canadian context. Razack argues that "modern peacekeeping is constructed as a colour line with civilized white nations standing on one side and uncivilized Third World countries on the other . . . [T]he peacekeeper . . . is entrusted with

the task of sorting out the tribalisms and the warlords that have mysteriously sprung up in regions of the world where great evil dwells"
(Razack 2004, 10). The question Razack poses for Canadians is:
"When a nation announces itself as peacekeeper to the world, and
when its national subjects derive from this and related mythologies a
sense of self, history, and place . . . what racial hierarchies underpin
and are supported by such apparently innocent beliefs?" (Razack
2004, 9). Australia's engagements in East Timor, Solomon Islands,
and Papua New Guinea, underpinned by these same racial hierarchies, are played out in a more complex arena because of continuing
anxieties over the legitimacy of its presence in this "backyard," a
term that signifies not secure ownership but, on the contrary, the
domestic baggage of uncertainty over title.

Extending Razack's thesis, I would suggest that, in addition to
the practice of peacekeeping, the color line between Australia and
its neighbors is enacted through the rhetoric of "good governance,"
as the peace builders are confronted with the waste, inefficiency,
and corruption that dwell in the dark places of the region. Like policing and peacekeeping, the rhetoric of good governance partakes
in the mythologies of bringing civilization to benighted locales, and
is inextricable from the narratives of colonization.

The Pacific, the Australian prime minister has taken to declaring since 2001, is "our patch" to cultivate, beautify, and protect.
Here, regionalism becomes a surrogate for another kind of empire,
an empire with advisors, experts, and consultants at its vanguard,
all speaking the language of stability, good governance, and security. The rise of security as the preeminent business of the state is
accompanied by measures to promote nation-building, new fiscal
practices, and "human security" through the installation of a class of
Australian bureaucrats in key areas of government across the Pacific.
In addition to sending civil forces to Solomon Islands in "Operation
Helpem Fren," to avert its expected collapse into a "failed state," the
Department of Foreign Affairs and Trade also, in late 2003, signed an
agreement to send close to four hundred Australian officials to work
in areas such as justice, policing, immigration, taxation, transport,
and customs in Papua New Guinea (DFAT 2003b). The Australian
government dismissed PNG protests that this was the beginning of
a new neocolonial phase in the region, although the demand that
Australian officials be granted blanket immunity in advance from

local law bears out fears that the operation is something more than a friendly hand extended across the backyard fence (Perera 2007).

Australia also recently called for "pooled regional governance" in the neighborhood, on the grounds that some states were "too small to be viable" as independent entities (Maclellan 2003–4, 22), again setting off fears of incursion on the sovereignty of smaller neighbors. Together all these moves suggest a clearly expansionist itinerary, premised on an implicit color line and promoted in the name of security, border control, and protection of the region from external and internal threats. Australia's defensive response to the entry of asylum seekers into the spatial economies of the region, then, has enabled a series of other movements that, coupled with the internal coercion or collaboration of smaller states, infringes on their sovereignty and attempts to make over the Pacific borderscape through practices and discourses of security and nation-building. The resistances and countermoves these attempts provoke are, necessarily, more difficult to outline. They operate in subterranean ways at local and micro levels, often through gendered and communal collectivities that promote alternative modes of territoriality and identity (Maclellan 2003–4). Counterhegemonic and resistant practices also often occur at the level of the nonverbal, gestural, and performative. The arrival of *Tampa* asylum seekers at Nauru is a good example of the latter.

When they finally disembarked in Nauru, the people from the *Tampa* had been at sea for close to four weeks, on three separate boats, crossing and recrossing the ocean, or simply waiting. They had almost drowned and had lost most of their possessions on the damaged *KM Palapa*, and had endured long days and nights in the open aboard the *Tampa* while lawyers, agencies, and governments haggled over their fate. Transferred by the military onto the *HMAS Manoora*, they were held in limbo while a high court appeal to return them to Australian waters was turned down. During their weeks on the sea, worlds changed around them: they now faced the baggage of the September 11 attacks, the anticipated war in Afghanistan, and a climate of heightened hostility and suspicion toward anyone who might fit the category of *Muslim* or *Middle Eastern*.

The first asylum seekers to disembark into this new world (some would later have to be forced) came ashore with two men holding up a banner thanking the Nauru government for protection and shelter. They were met by rows of slow-moving Nauruan dancers, arms widespread in welcome. The dancers offered a small handful of flowers to each asylum seeker stepping onto the shore (Marr and Wilkinson 2003, 164). In this wordless exchange, relations of host and guest were enacted and established, momentarily holding in suspension the sordid transaction in which both parties had become enmeshed. The asylum seekers, briefly, were not "human cargo" or the international rejects they had become in their wanderings through the Pacific. Simple motions of welcome acknowledged their human dignity and inscribed them in a long tradition of wayfarers coming ashore. In turn, the Nauruans, who had insisted that no one should be sent to their country unwillingly, claimed their role as hosts, not jailers, performing sovereignty where it could not be spoken.

I want to read this beach scene on Nauru as a site where new modalities of connection and alternative spaces of relationality are constructed between peoples *in the shadow of the border.* The beach, a meeting place between land and ocean, acts as both boundary and border zone. Even as the color line is reinforced and extended, and borders encircled by other borders, border spaces engender their own defiances, practices of connection, and potential for contestatory actions.

IN THE SHADOW OF THE BORDER

Another beach is the final scene in my attempt to introduce a many-dimensioned map of an emerging Pacific borderscape.

In the days and weeks after the nightclub bombings on Kuta Beach on the Indonesian island of Bali, grief-stricken and incredulous Australians could be frequently heard protesting "Bali is our backyard!" Australians had suffered the largest number of casualties in the bombings, with eighty-eight dead. In "A Paradise Bombed," his lament for "our island of domesticated exotica," the historian Vickers develops the connection, locating Australia as a space suspended from the region, with Bali as its gorgeous extension:

> Before the bombings, paradise island . . . stood apart from the rest of Indonesia at the idyllic end of a spectrum in the Australian imagination, an extension of [the local resorts of] Byron and Noosa. The bombings targeted our sense of place, a paradoxical reminder that we are still in Asia, and Indonesia is our nearest neighbor.
>
>   At the other end of the spectrum of Australian views about the region is the Asia of danger where Indonesia proper is found—summed up in the cliché of "living dangerously." (2003, 107)

Vickers describes his essay as "a eulogy for 'Bali' before it was subsumed by the Asia of 'Living Dangerously.'"

Bali is easily located on this map as part of Australia, detached from the rest of Asia: paradise island versus third world hellhole. But on the other side of the Aratura Sea a more differentiated geography necessarily prevails. Whereas, for Australian observers like Vickers, Bali signifies simply as its main tourist strip, Kuta Beach, the Indonesian commentator Ida Ayu Agung Mas observes that "[p]lacing Kuta on the Balinese conceptual map would be very difficult. . . . The famous and glittering Kuta, now in ruins, is actually a 'faraway' place for the majority of local people" (Anggraeni 2003, 94). From this Balinese perspective, Kuta is a site distinct from Bali, although connected to it. The bombings on Kuta Beach, in Dewi Anggraeni's words, hurt the Balinese greatly, "but in an abstract manner. They were devastated to hear that their spiritually protected soil had been so cruelly destroyed, and so many of Bali's guests murdered" (Anggraeni 2003, 94).

Kuta Beach is both remote from Bali and an inextricable part of it. In an interview immediately after the bombings, Luh Ketut Suryani, professor of psychiatry and a public figure in Bali, suggested that all Balinese were implicated in the terrible events because they shared responsibility for what Kuta had come to represent: "We should be able to look back at what we have done. Many years ago, when we developed tourism, we wanted it to be cultural tourism. We introduced our culture to those who came to visit our land. Now, we have the tourism of iniquity, where we are no longer in control, where we are chasing the dollar. So we have been punished" (Anggraeni 2003, 79). Instead of displacing the bombings onto outsiders ("terrorists"), Suryani articulates a sense of collective implication in and responsibility for the violence, involvement from which neither hosts nor guests could be exempted.

A fraught and complex site, Kuta Beach is a border zone where difference is displayed, transacted, and negotiated, and where exchange among peoples—unequal, asymmetrical, even iniquitous—happens. After their capture, the bombers, who are non-Balinese, revealed that their motive was to purify or cleanse Bali by driving the aliens out of Kuta. This is a more loaded desire than first appears. The bombers' own claims on Bali were mediated by differences in religion, culture, and geography. The Bali they aimed to produce was informed as much by a nativist vision of Indonesia as by global anti-Western Islamism. Anggraeni suggests that the prelude to the bombings on Kuta Beach extends further than the West's preoccupation with September 11 to local Indonesian histories, specifically the 1998 campaign of rape and murder against ethnic Chinese in Jakarta. For the victims of the Jakarta violence, Bali was a favored refuge because "it was known as a peaceful place, where the population showed tolerance toward others and otherness" (Anggraeni 2003, xix–xx). What was targeted in the Kuta bombings, along with the Western presence, was the idea of Bali *within Indonesia* as a multiracial and pluralist society.

In opposition to the inclusive form of ownership signified by Bali (despite the excesses of the Kuta tourist trade that perverted traditional host-guest relations), the bombers sought to impose on the island a sovereignty based on exclusion. Their act can be read as an assertion of the ultimate logic of border protection in response to the perceived threat of difference from within and without. This response, authorized by the ambition to exert absolute sovereignity over space, creates an unmentionable link between the violence on Kuta Beach and events in the waters between Australia and Indonesia ("We decide who comes into this country").[6] Both are in-between places where the bodies of moving people become entrapped in the violent logic of the border.

The victims of the Bali bombings (October 12, 2002) died just one week short of a year after the sinking of SIEV X (October 19, 2001) somewhere in the waters between Indonesia and Australia. To read the deaths on Kuta Beach and the drownings of people on SIEV X as in some way mirroring each other is not to compare or balance these events. In more ways than I can specify, they are impossible to compare. But the public *meanings* of these deaths, and the spaces assigned to them in our official memories, do bear comparison. Except in the memorials of families, survivors, and a few dedicated activists, the victims of SIEV X have vanished from public view with barely a trace (see http://www.SIEVXmemorial. org). They have been disappeared even from the name of the senate investigation into their deaths, the obscurely titled "Inquiry into a Certain Maritime Incident" (Kevin 2004; Perera 2006). The victims of Kuta, on the other hand, will never cease to be remembered in the official memorials. Already, their deaths have achieved the monumentalized resonance of other national historical events, from ANZAC Day to September 11.

But "What the map cuts up, the story cuts across" (de Certeau 1988, 129). Alongside the role that the deaths at Kuta perform in legitimizing the Australian state's assertions of sovereignty and in buttressing nationalist sentiment, their insertion in other public or private acts of remembrance testifies to the resilience of border spaces and the flows of moving peoples against the dividing logic of the border. These are living memorials that enact border practices, connected by the itineraries of the dead and living across oceans.

My final story begins with two bodies linked by the casual carnage on Kuta Beach (they are also, of course, part of other stories). Josh Deegan, a teenager holidaying at Kuta, was one of the Australian dead. After Josh died, his father, Brian, devoted his energies to lobbying for the young children (Safdar Sammaki, aged seven, and Sara Sammaki, aged three) of another victim of the bombings, an Indonesian woman, Endang. Endang was in Kuta that day to seek legal advice about the case of her husband, Ibrahim, an asylum seeker from Iran. At the time, Ibrahim, intercepted by the navy on his crossing to Australia, had been held in a detention camp in South Australia for over fifteen months. For a year, Brian Deegan used whatever opportunities he could to remind Australian authorities about Safdar and Sara, now virtual orphans in Bali. He asked the Australian government to let the children be reunited with their remaining parent. His work was supported by a handful of Australian activists and politicians, but met with refusal after refusal.

The unexpected ending to Sara and Safdar's story tells of a happy conjunction of place, politics, persistence, and the power of images in circumventing the border. After a year of rejections, at the first anniversary commemoration of the bombings, the children mysteriously sneaked into a photograph where they appeared holding the hand of the Australian prime minister on Kuta Beach. The photograph appeared on the front pages of newspapers and on television. It refused to go away. The prime minister protested that he hadn't had any idea who the children were. But now, suddenly, things changed. Walls in the water came down. Ibrahim Sammaki was released from the Baxter detention camp. The immigration department plucked visas for Sara and Safdar out of thin air. Within weeks, the three remaining members of the family were together again, authorized to become legal residents of Australia. Brian Deegan had a single comment on the remarkable ending to this story: "I'll treat this as a gift from my son to this family" (ABC Online).

Through Josh's gift, the stories of the two families become inextricably entwined, living testimony to forces and flows that cut across the map, weaving new routes of connection even as others are violently terminated.

DIFFERENT EDGES OF THE NATION?

In Thongchai Winichakul's terms, my retelling of the story of Josh's gift, cutting across lines on the map to configure an Australasian and Pacific borderscape, can read as an attempt to contour "different edges of the nation" (Winichakul 2003, 13), where new spatial identities are enacted and formulated:

> A spatial identity usually has its story, probably many stories, without which such a place or identity would be meaningless. . . . On the one hand, stories can change the ways people think about a place, and therefore redefine it, or give birth to new spatial identity. Stories become resources for the new spatial identities, for formulating the narratives of its birth, development, characteristics, and so on. On the other, as a place changes or a different spatial identity emerges, its story usually changes accordingly. A potentially new spatial identity may inspire, and project, stories that help its emergence. (2003, 9)

Winichakul directs these remarks at the weakening of nationalist narratives in Southeast Asia as "the nation is . . . perhaps losing its predominance as the primary historical Subject and the privileged site of history" (2003, 9). Appeals to anticolonial sentiment are becoming less effective here, he argues, as it becomes apparent that "The real dynamism that propels changes in the spatial identity of a nation is increasing diversity and complexity within a society. The old national story has served its purpose. . . . Now other narratives of nonnational subjects perhaps, begin to emerge and blossom. We are at this transition, when national history is on its way out and narratives of new spatial identities are emerging" (2003, 9).

My stories of the edges of Australia, bordering on Winichakul's Southeast Asia, map a space where the authorized national story (that is, its spatial identity) is both violently (re)asserted and potentially weakened. In tracing the itineraries, irregular, intermittent, interrupted, of three boats as they traverse and constitute the space

I name the *borderscape,* different stories are heard and other relations and practices enacted; beaches and backyards, hosts and
guests, inside and outside recombine across multiple state borders
and regional and racial divides. Itineraries of bodies moving in time
and space confound the intelligibility and stability of place-names
on the map. They weave interrelations, engender border practices,
reanimate contested sovereignties, and give rise to new geographies,
spatial identities, and territorial claims and counterclaims.

In naming the space of these practices as constituting a distinct
zone, the borderscape, I am suggesting the need for an alternative
conceptual and spatial frame for analyzing what are usually read as
disparate elements—negotiations of Indigenous sovereignty, regional
governance initiatives, security and border control operations—
within it. Cutting across the conventional classifications into separate "domestic" and "foreign" policy concerns, the notion of a
borderscape allows for differentiated understandings of space, territoriality, sovereignty, and identity across this zone. It also opens the
way for theorizing emergent formations and practices that are mobilizing in response to the exclusionary assertions and new territorial
violences attempting to overrun the region. To repeat Winichakul's
words, "A potentially new spatial identity may inspire, and project,
stories that help its emergence" (9). I hope the stories I outline here
perform this double function, at once anticipating and projecting
the possibility of new spatial identities and affiliations across the
cut-up pieces and divided fragments of the map.

NOTES

1. For a further discussion of the *Minasa Bone* and the reversal of time,
see chapter 11, this volume.

2. See Fry and Fonteyne on the dubious legality of these naval
interceptions.

3. See http://www.SIEVX.com and http://www.SIEVXmemorial.org for
comprehensive information about SIEV X, its implications and aftermath.

4. *El Pais,* August 20, 2003. Translated by Virginia McFadden and
quoted by Liz Fekete.

5. For example, the dispute over use of force in disembarking the *Tampa*
asylum seekers in Nauru, and arguments about responsibility for the care
of hunger strikers in the camps.

6. If the parallels I am suggesting here seem exaggerated, evidence is readily available that exhortations to "sink the *Tampa*" or "dump them all in the sea" were not infrequent responses by Australians on talk radio and media Web sites to the presence of asylum seekers on the horizon (Burke 2002).

WORKS CITED

ABC Online. 2003. "Magistrate Brian Degan Celebrates Sammaki Release." ABC Online, November 6. http://www.abc.net.au/pm/content/2003/s984023.htm (accessed December 16, 2003).

Agamben, Giorgio. 1997. "The Camp as *Nomos* of the Modern," trans. Daniel Heller-Roazen. In *Violence, Identity, and Self-Determination*, ed. Hent de Vries and Samuel Weber. Stanford, Calif.: Stanford University Press, 106–18.

———. 1998. *Homo Sacer, Sovereign Power, and Bare Life*, trans. Daniel Heller-Roazen. Stanford, Calif.: Stanford University Press.

———. 2001. "On Security and Terror," trans. Soenke Zehle. *Frankfurter Allgemeine Zeitung*, September 20. http://www.egs.edu/faculty/agamben/agamben-on-security-and-terror.html (accessed September 6, 2003).

Agnew, James, and Stuart Coleridge. 1995. *Mastering Space*. London: Routledge.

Anggraeni, Dewi. 2003. *Who Did This to Our Bali?* Melbourne: Indra Publishers.

Balint, Ruth. 2005. *Troubled Waters*. Crows Nest, Australia: Allen and Unwin.

Birch, Tony. 2001. "The Last Refuge of the 'Un-Australian,'" *UTS Review* 7, no. 1: 17–22.

Burke, Anthony. 2002. *In Fear of Security*. Sydney: Pluto Press.

Certeau, Michel de. 1988. *The Practice of Everyday Life*, trans. Steven Rendall. Berkeley and Los Angeles: University of California Press.

———, Luce Giard, and Pierre Mayol. 1998. *The Practice of Everyday Life*, vol. 2, trans. Timothy J. Tomasik. Minneapolis: University of Minnesota Press.

Department of Foreign Affairs and Trade (DFAT). 2003a. *Advancing the National Interest: Australia's Foreign and Trade Policy White Paper*. http://www.dfat.gov.au.

———. 2003b. "Australia and Papua New Guinea to Work Together against Terrorism." Media release, December 11. http://www.dfat.gov.a.

Derrida, Jacques. 2000. *Of Hospitality*, trans. Rachel Bowlby. Stanford, Calif.: Stanford University Press.

————. 2001. *On Cosmopolitanism and Forgiveness,* trans. Mark Dooley and Michael Hughes. London: Routledge.

Fekete, Liz. 2003. "Xeno-Racism and the Demonisation of Refugees: A Gendered Perspective." Workshop on Women, Crime, and Globalization, Onati, Spain, September.

Fonteyne, John-Pierre. 2002. "'Illegal Refugees' or Illegal Policy?" In *Refugees and the Myth of the Borderless World,* ed. William Mayley et al. Canberra: Department of International Relations, RSPAS, 16–31.

Fry, Greg. 2002. "The 'Pacific Solution'? Australia's Refugee and Humanitarian Policies." In *Refugees and the Myth of the Borderless World,* ed. William Mayley et al. Canberra: Department of International Relations, RSPAS, 23–31.

Hodson, Michael. 2003. "Government Lies Again—Tiwi Islanders: 'We're All Non-Australians.'" *Green Left Weekly,* November 19.

Jameson, Julietta. 2003. *Christmas Island, Indian Ocean.* Sydney: ABC Books.

Kehi, Balthasar. 2004. "Australia's Relations with East Timor: People's Loyalty, Government's Betrayal." *Borderlands* 3, no. 3. http://www.borderlandsejournal.adelaide.edu.au/vol3no3_2004/kehi_timor.htm.

Kevin, Tony. 2004. *A Certain Maritime Incident: The Sinking of SIEV X.* Melbourne: Scribe.

Levinas, Emanuel. 1969. *Totality and Infinity: An Essay on Exteriority,* trans. Alphonso Lingis. Pittsburgh: Duquesne Press.

Maclellan, Nic. 2003–4. "Helping Friends or Helping Yourself?" *Spinach* 7, no. 2: 19–23.

Marr, David, and Marion Wilkinson. 2003. *Dark Victory.* Crows Nest, Australia: Allen and Unwin.

Ondaatje, Michael. 1998. *Handwriting.* London: Bloomsbury.

Perera, Suvendrini. 2002a. "A Line in the Sea: The *Tampa,* Boat Stories, and the Border." *Cultural Studies Review* 8, no. 1: 11–27.

————. 2002b. "What Is a Camp?" *Borderlands* 1, no. 1. http://www.borderlandsejournal.adelaide.edu.au/issues/volno1.html.

————. 2006. "They Give Evidence: Bodies, Borders, and the Disappeared." *Social Identities* 12, no. 6: 637–56.

————. 2007. *Our Patch: Enacting Australian Sovereignty Post-2001.* Perth: Australian Public Intellectual Network, 119–46.

Perera, Suvendrini, and Joseph Pugliese. 1997. "'Racial Suicide': The Relicensing of Racism in Australia." *Race and Class* 39, no. 2: 1–20.

————. 1998. "Wogface, Anglo-Drag, Contested Aboriginalities: Making and Unmaking Identities in Australia." *Social Identities* 4, no. 1: 39–72.

Rajaram, Prem Kumar. 2003. "Making Place: The 'Pacific Solution' and Australian Emplacement in the Pacific and on Refugee Bodies." *Singapore Journal of Tropical Geography* 24, no. 3: 290–306.

Razack, Sherene. 2004. *Dark Threats and White Knights: The Somalia Affair, Peacekeeping, and the New Imperialism.* Toronto: University of Toronto Press.

Reynolds, Henry. 2003. *North of Capricorn: The Untold Story of Australia's North.* Crows Nest, Australia: Allen and Unwin.

Robertson, Jack. 2003. "Eddie Mabo Proclaims Great Southern Rainbow Republic." *Sydney Morning Herald,* November 14.

Shnukal, Anna, Guy Ramsay, and Yuriko Nagata. 2004. *Navigating Boundaries: The Asian Diaspora in the Torres Strait.* Canberra: Pandanus Books.

Simon, Jonathan. 1998. "Refugees in a Carceral Age: The Rebirth of Immigration Prisons in the United States." *Public Culture* 10, no. 3: 577–607.

Soguk, Nevzat. 1995. "Transnational/Transborder Bodies: Resistance, Accommodation, and Exile in Refugee and Migration Movements on the U.S.–Mexican Border." In *Challenging Boundaries: Global Flows, Territorial Identities,* ed. Michael Shapiro and Hayward R. Alker. Minneapolis: University of Minnesota Press.

Vickers, Adrian. 2003. "A Paradise Bombed." *Griffith Review* (Spring): 105–13.

Winichakul, Thongchai. 2003. "Writing at the Interstices: Southeast Asian Historians and Postnational Histories in Southeaest Asia." In *New Terrains in Southeast Asian History,* ed. Abu Talib Ahmad and Tan Liok Ee. Singapore: Singapore University Press, 3–29.

IV

Rethinking Borderscapes:
The New Political

10

"Temporary Shelter Areas" and the Paradox of Perceptibility: Imperceptible Naked-Karens in the Thai–Burmese Border Zones

DECHA TANGSEEFA

"BEFORE THE LAW" AND NAKED LIFE: ARRIVING AT THAILAND'S "DOOR"

The fate and struggles of forcibly displaced peoples[1] from the Burmese nation-state along Thailand's "door" can be articulated in the spirit of Franz Kafka's "Before the Law":[2]

> Before the Law stands a doorkeeper. To this doorkeeper there comes a teenage girl from Burma, who prays for admittance to the Law. But the doorkeeper says that he cannot grant admittance at the moment. The girl thinks it over and then asks if she will be allowed in later. "It is possible," says the doorkeeper, "but not at the moment." Since the door leading into the Law stands open, as usual, and the doorkeeper steps to one side, the girl stoops to peer through the door. Observing that, the doorkeeper laughs and says: "If you are so drawn to it, just try to go through it despite my veto. But take note: I am powerful. And I am only the least of the doorkeepers. From door to door, there is one doorkeeper after another, each more powerful than the last. The third doorkeeper is already so terrible that even I cannot bear to look at him."
>
> These are difficulties the girl from Burma has not expected; the Law, she thinks, should surely be accessible at all times and to everyone,

but as she now takes a closer look at the doorkeeper in his trench coat, with his big sharp nose, and hugely vicious eyes, she decides that it is better to wait until she gets permission to enter. The doorkeeper allows the girl to build a shed to stay by the door. There she lives for days and years. With her perseverance, she makes countless attempts to be admitted. The doorkeeper frequently has little interviews with her, asking her questions about her home and other matters, but the questions are put indifferently, as bureaucrats put them, and always finish with the statement that she cannot be let in yet.

The woman, who has furnished herself with things for her journey, sacrifices all she has, however valuable, in the hope of bribing the doorkeeper. The doorkeeper accepts everything, but always with the remark: "I am only taking it to keep you from thinking you have done nothing." During these long years the woman fixes her attention constantly on the doorkeeper. She forgets the other doorkeepers, and this first one seems to her the sole obstacle between herself and the Law. She curses her bad luck, and since in her attentive observation of the doorkeeper she has come to know even the ants on his trench coat, she begs the ants as well to help her and to change the doorkeeper's mind. Later, as she gets very weak and very ill, she only grumbles to herself and to her tiny daughter.

Soon her health deteriorates and her eyesight begins to fail, and she does not know whether the world is really darker or whether her eyes are only deceiving her. Yet in her darkness she is now aware of a radiance that streams inextinguishably from the door of the Law. She does not have very long to live and her tiny daughter is very sick, too. Before she dies, all her experiences in these long years gather themselves in her head to one point, a question she has not yet asked the doorkeeper. She waves him nearer, since she can no longer raise her ailing and stiffening body, hugging her sick, tiny daughter. The doorkeeper has to bend low toward them, for the difference in height between them has altered much to the woman's disadvantage. "What do you want to know now?" asks the doorkeeper; "you are insatiable." "Everyone strives to reach the Law," says the woman, "so how does it happen that for all these years no one but myself and my daughter have ever begged to enter the door?" The doorkeeper recognizes that the woman has reached her end, and, to let her failing senses catch the words, roars in her ears: "No one else could ever be admitted here, since this door was made only for you. I am now going to shut it."[3]

Following Giorgio Agamben's reading of Kafka's "Before the Law" and Gerschom Scholem's formula for the status of law in Kafka's novel *The Trial* (of which "Before the Law" is a part), one recognizes the sovereign power over human life *"being in force without significance (Geltung ohne Bedeutung)"*(Agamben 1998, 51, 49–58, and 1999, 169–70). In effect, the girl/woman from Burma becomes a biopolitical body: a primary object of a sovereign power.

The production of a biopolitical body is the original activity of sovereign power; it is the "originary inclusion of the living in the sphere of law," which in turn results from the sovereign's decision of the exception (Agamben 1998, 25–26). That is, the structure of law locates its force in the possibility of the suspension of the rule and order such that a state of exception emerges.[4] The decision of the state of exception does not decide whether the girl/woman or her act is licit or illicit. Rather it inscribes life from "outside" in the sphere of law so as to animate the law and suspend it (Agamben 1998, 28). Law is, therefore, "made of nothing but what it manages to capture inside itself through the inclusive exclusion of the *exceptio:* it nourishes itself on this exception and is a dead letter without it. In this sense, the law truly 'has no existence in itself, but rather has its being in the very life of men'" (Agamben 1998, 27). Without (human) life, law is dead, becoming nothing at all. Law needs (human) life to breathe life into itself. A zone of indistinction between law and life thus emerges through the logic of the paradox of sovereignty (Agamben 1998, 9, 27). In effect, biopolitics is at least as old as the sovereign exception.[5]

The girl/woman from Burma is a *naked life,*[6] who encounters her nakedness and extreme vulnerability. My deployment of the notion of *nakedness* follows Agamben. In a nutshell, this notion has two angles. The first refers to the sheer fact of living *(zoe),* as opposed to form-of-life *(bios);* the second refers to the life quintessentially abandoned through sovereign exception. Whenever the sovereign threat is materialized, the first angle of naked life emerges: we as forms-of-life (bios) are stripped, and we as sheer facts of living (zoe) are revealed. As humans are always protected by a certain sovereign power, it is thus the fact of life that we are always potentially threatened. Political power always founds itself, in the last instance, on the separation of a sphere of naked life from the context of the form-of-life. Naked life thus constitutes the originary cell of the sovereign power; it is the "hidden foundation of sovereignty" (Agamben 2000a, 5).

Many of the forcibly displaced peoples from Burma arriving at Thailand's "door" find their quotidian lives have been placed under a state of exception; various parts of the Thai–Burmese border zones have regularly been transformed into *spaces of emergency* by the two sovereignties. When one follows, in the Agambenian sense, the forcibly displaced peoples' flight from Burma to Thailand, one recognizes that admission to asylum is a strategy of inscribing lives from "outside" into the sphere of Thai laws and the agreements Thailand has with the United Nations High Commissioner for Refugees (UNHCR).

These forcibly displaced lives animate the relevant laws; the Royal Thai Government and the UNHCR endorsed closer cooperation on displaced persons from Burma in May 1988 (Department of International Organizations 1988). The agreements signify intended cooperative relations between two sovereign powers, one national, the other international, to act upon forcibly displaced peoples from Burma. As an initial step, it was also intended that the agreements would contribute toward the resolution of the issue of the displaced peoples. The two entities agreed to coordinate with each other on the following issues: admission to asylum, registration, UNHCR access, repatriation, relocation of temporary shelter areas, UNHCR assistance parameters, and long-term strategies.

The first issue, admission to asylum, centered on the question of how displaced persons should be perceived and how they should be recognized. But the two sovereign powers' inscriptive strategies have not coincided. The Thai government wants to grant temporary shelter only to *peoples fleeing fighting* whereas the UNHCR has been trying to push the criteria to also include *peoples fleeing effects of civil war.*[7] For those who would be granted temporary shelter, the result would be, in a Kafkaesque sense, to be given stools to sit at Thailand's door.[8] A genealogy of the Thai nation-state's inscription of lives from Burma will be instructive.

INSCRIBING THE DISPLACED PEOPLES FROM BURMA

At the international level, Thailand's stance toward the issue of displaced persons from Burma has been articulated consistently in statements by heads of the Thai delegation at the annual sessions of the Executive Committee of the Programme of UNHCR (ExCom), held in Geneva in October of each year. The sessions are venues where

committee members present conditions of protection in their countries, with particular regard to situations that will impact upon the
work of the UNHCR. It has been reiterated at these venues that
Thailand continues to regard repatriation and prevention as the
best solutions for those displaced peoples classified as ethnic nationalities: "repatriation represents the durable and viable solution for
these displaced persons; . . . the best solution to the refugee problem
is the prevention or termination of the root causes which force the
people to flee and become refugees."[9] And the best solution for the
group that the Thai government categorized as "Burmese students"
was resettlement to a third country (Kachadpai 2000, 5).

In an Agambenian sense, Thailand's Regulations Concerning Displaced Persons from Neighboring Countries, issued by the Ministry
of Interior on April 8, 1954, and Thailand's 1979 Immigration
Act[10] are the two juridical fabrics that an immigrant encounters
when traversing through the Thai–Burmese in-between spaces into
Thailand's territory. Many parts of the in-between spaces are mountainous, rugged, isolated, and densely forested; they were demarcated along two main divisions, north and south of the confluence
of the Salween and Thaungyin (Moei, in Thai)[11] Rivers. In some
areas, the two rivers are the Thai–Burmese state boundary. There
are many "doorways" where people can traverse across the two
countries' state boundary, especially during the dry season when
the Moei River is very shallow in areas. Arriving along Thailand's
doorways, or "gates," the forcibly displaced peoples from Burma
can choose to be "before the law" or to proceed through these un-
policed gates. The difference is that the life of those who choose to
be before the law would be inscribed into Thai laws, whereas those
who pass through the doorways are beyond the law—at least until
they are caught.

The Thai government, however, has preferred to deal with the
refugee issue on the basis of discretionary policy decisions, rather
than to be bound to international law or specific national law. The
Thai 1979 Immigration Act contains no reference to refugees, and
no permanent legal mechanism is in place for making a determination whether an individual qualifies for protection as a refugee.
Hence, the Thai state apparatuses have consistently avoided using
the terms *refugee* or *asylum seeker* (Alexander 1999, 40; Vitit
2005) except in international venues like the annual sessions of the

Executive Committee of the Programme of the UNHCR. The generic term for Indochinese and others arriving in Thailand is *displaced persons* (Lang 2002, 92) and all are prima facie *illegal immigrants* unless they arrived before March 9, 1976.[12] The official designation in law of displaced persons relates back to clause 3 of the 1954 regulation, which defines a displaced person as someone "who escapes from dangers due to an uprising, fighting or war, and enters in breach of the Immigration Act" (Vitit [n.d.], 7; Lang 2002, 92; Alexander 1999, 40).[13] At first glance, this definition clearly fits with the 1951/1967 United Nations refugee juridical fabrics, but the Thai kingdom has its own reasons for not acceding to the refugee instruments.[14] Genealogically, the displaced peoples have been classified by the kingdom into three major groups: *Burmese-national displaced peoples, peoples fleeing fighting*, and *"illegal" economic immigrants*.[15]

Burmese-national displaced peoples. Officially designated as such, these peoples fled to Thailand *before* March 9, 1976, and have lived along the bordering provinces of Thailand; some had even arrived before 1957. These peoples are, for example, the Shans, the Mons, the Karens, the Laotians, the Tais, and the Nepalis. According to the record of Thailand's Department of Local Administration, Ministry of Interior, as of 1986 there were about 47,000 people under this category (Kachadpai 1997, 67; cf. Lang 2002, 83n5). The Thai government allowed these peoples to temporarily stay in—and wait to be pushed out of—the country. In reality, however, the temporary stay has become an unlimited stay (Kachadpai 1997, 80), and these peoples have been allowed to work in some occupations. They nonetheless have not been granted citizenship. In contrast, those who arrived in Thailand *after* March 9, 1976, are illegal immigrants and must be "decisively blocked and pushed out [of the country]" and officials would "capture and strictly conduct legal proceedings to every single one [of them]."[16]

Peoples fleeing fighting.[17] Even though the Thai government had issued the Ministry of Interior's announcement on March 9, 1976, stemming the flow of the forcibly displaced peoples from Burma, in reality there had always been dry-season annual attacks that drove small numbers of ethnic peoples across the Thai–Burmese boundary. These peoples, however, returned to their villages when the rainy season began and the fighting ceased. The Thai government usually allowed these peoples to set up temporary shelters while the danger persisted (Alexander 1999, 39; Lang 2002, 82). It was in the dry season of 1983–84 (Lang 2002, 83) that the Tatmadaw was able to mobilize troops to crush the ethnic nationalities' armies along the Thai–Burmese border zones and start to set up permanent bases there. Hence, many forcibly displaced peoples were not able to return when the rainy season came. The year 1984 therefore marked the birth of a string of semipermanent "camps" being set up in Thailand's Tak Province first by the Karens and subsequently for the Karennis and Mons (in 1989 and 1990 respectively) (Lang 2002, 81–99). Many of the fleeing ethnic peoples have since then been living on the Thai side of the Thai–Burmese state boundary.

The 1988 massacres of pro-democracy demonstrators in Burma's urban areas also resulted in large numbers of peoples fleeing to the Thai–Burmese border zones, many of whom were dominant ethnic Burmans (e.g., Alexander 1999, 40; Lang 2002, 101n1). The Thai government allowed these new waves of forcibly displaced peoples to live in small camps, usually close to the groups that had come before 1988. Simplicity, self-sufficiency, and self-management were the keys: these peoples planned, built, and administered the camp communities themselves (Alexander 1999, 39; Lang 2002, 84, 91).

In contrast to its reaction when displaced peoples from Indochina arrived in the 1970s, however, the Thai government did not want to "put the world spotlight on Burmese refugees": a low profile was the norm (Alexander 1999, 39). It was the armies of ethnic nationalities who provided security to these communities of displaced peoples, and a number of international NGOs working in Thailand, composing the Burma Border Consortium, supported the displaced peoples materially, educationally, and medically (Alexander 1999; Lang 2002, 84). The UNHCR had no role in these. Not until fourteen years later was the UNHCR allowed by the Thai government to have any presence along the border zones. Hence, since 1998, despite its nonaccession to the Refugee Convention of 1951, Thailand has allowed the UNHCR roles in five aspects: witnessing the process of admission, assisting the Thai authorities in registration, assisting the Thai authorities on the relocation of temporary shelter areas, providing complementary assistance in existing temporary shelter areas, and assisting the displaced peoples from Burma for their safe return (Department of International Organizations 1998,

1–2; UNHCR 1998, 1–2). As of May 2005, 146,058 people in the category of persons displaced from Burma have officially been living in "camps" along the Thai–Burmese border zones (TBBC 2005). From the perspective of the international protection regime, by defining as forcibly displaced only those peoples from Burma meeting the condition of "fleeing from fighting," Thailand has sometimes refused to provide first asylum to new arrivals: those who have not fled directly from fighting but have taken flight from the *effects* of war inside Burma.

"Illegal" economic immigrants. This third group comprises those who have escaped Burma to be laborers in Thailand. They are the Burmans and other ethnic nationalities who have escaped from economic hardship or who have been smuggled into Thailand for better employment opportunities. There is no official figure for this group, but the former secretary general of Thailand's National Security Council (NSC), Kachadpai Burusapatana, estimated that there were at least 500,000 in Thailand in 1997 (Kachadpai 1997, 9). Amnesty International, which estimated in 2002 that there were 1,000,000 such people (2002, 36), stated in 2005 that "it is almost impossible to estimate the number of unregistered migrant workers" due to their intended imperceptibility to the Thai authorities (2005, 1n3). In August 2001, the Thai government established a new registration system for migrant workers from neighboring countries, and 560,000 workers subsequently registered, some of whom renewed their registration again in March 2002, the majority being from Burma (Amnesty International 2002, 36). As of June 2004, altogether there were 111,189 migrant workers registered (Kritaya and Pramote 2005). By registering, migrant workers are, in principle, exempt from arrest and deportation by the Thai authorities unless they are found without a registration card.[18] These peoples have either been working in low-paying jobs or looking for work all over Thailand.

All three categories of displaced peoples lack adequate juridical protections; this is especially true for those who have traversed through Thailand's door instead of deciding to be inscribed "before the law." Even as the inscribed displaced peoples are trapped at the door, the others have more freedom to roam Thailand, yet little protection when they choose to become registered. It is argued here, however, that the three categories (especially the peoples fleeing fighting and the "illegal" economic immigrants, as the Thai nation-state classifies them) need not be absolutely separated. In reality, many displaced peoples from Burma can be fleeing fighting and be "illegal" economic immigrants at the same time. From the statist perspective, all three categories signify both security and humanitarian problems. Behind these problems lie a set of entanglements produced by the confrontations between the international protection regime and the Thai nation-state's sovereign power. These are the entanglements that shape the fate of the forcibly displaced peoples.

"WHY IT HAPPENED?"

As of May 2005, there were nine registered temporary shelter areas, or "camps," along the Thai–Burmese border zones, under the jurisdiction of Thailand's Ministry of the Interior, with relief support from a network of international relief agencies called the Committee for Coordination of Services to Displaced Persons in Thailand (CCSDPT). There were two shelter areas where the Karennis (red Karens) were the majority, and seven where the Karens were the majority. Moreover, there were three resettlement sites for the Mons, who were all located on the Burmese side and not registered.[19] Altogether, there were 158,348 people (up by 2,563 from the figure of December 2004). The figure for the forcibly displaced Karens, as of December 2004, according to the Karen Refugee Committee, was 104,002 (TBBC 2005, 59).[20]

During my fieldwork in a temporary shelter area, a Karen shelter member told me of his belief that there was an insidious collaboration between the Thai armed forces and the Burmese junta's army (the Tatmadaw) together with its ally, the DKBA (Democratic Karen Buddhist Army). Such collaboration resulted in an attack on his shelter area in early 1997—which would not have been possible without permission from the Thai army, as the shelter area is located about seven kilometers from the Thai–Burmese state-boundary.[21] When I interviewed a Thai government official taking care of the shelter, I inquired about any possibility of the shelter being raided by either the Tatmadaw or the DKBA troops from the other side of the river. It was confirmed to me that there was no way that those troops would be able to attack the shelter without the knowledge of the Thai armed forces patrolling the Thai–Burmese state-boundary.

The following short essay by a student of mine in the shelter where I was teaching English (the student was responding to my question

"When was the hardest time in your life? Please explain in detail") provides a glimpse of a shelter raid that occurred on January 29, 1997. I copy it here verbatim except to omit some information for the safety of the student.

> From 1995 we always heard the news that DKBA has threatened to attack us all the time. I couldn't bear with this news because it's a bad news and made me tired. From then, we hardly had good opportunity to celebrate Christmas, Karen New Year, and New Year.
>
> On 28th January 1997 night we slept peacefully. That night was a quiet night and I slept very well. In the morning at 6:00, the shooting started like a raining which made all of us got shock and get up to see what was happening. Mortars shelling came from every direction and I pray while my heart is shattering with fears. I tried to find a refuge but no way and everyone seems to find his or her own refuge as well. Within few minutes a man came to our house and said, "Seventeen houses are burnt down by the DKBA." It makes me more hopeless and fearful. From then we move to the other side of the camp. When I came close to my friend's house, she looked at me and sadly said, "My grandma was hid [hit] by a mortar shell that fall near my house, and died." I was shocked and I couldn't imagine myself being in that kind of situation. After hearing the news, "Why?" came to my mind while moving my foot step to the other side of the camp. When I sat in a quite [quiet] place, certain questions came to my mind such as, "Why my people kill my people?"[22] and "Why life is very uncertain?" "The grandma whom I saw yesterday, watching T.V. together with me is now gone." "Why it happened?"
>
> People gave me food to eat and I said, "No." Then, it was a time for me to study, cook, eat, go to church and pray . . . play and so on. But with this kind of situation, I couldn't do what I wish to do.
>
> In the evening, we came back to our house and were not allowed to open the generator, and even a candle. In a very dark night, I thought, "What should I need to do for my people and the grandmother who died recently?" For sure, I can't do big things, but . . . one little task for me is to share to the people who do not know about this cruel event happing in . . . Refugee Camp on 29th January, 1997.

While I was conducting fieldwork in the temporary shelter area, I heard other similar stories and witnessed other shocking events, but I do not feel able to record these here because of concerns for the safety of those involved. Many of these borderline peoples' struggles have not been perceptible because their subjectivities fell outside the authoritative territorial mapping of what constitutes political subjectivity. As others to the Burman imagined community, their maps of allegiance (Appadurai 1996) do not coincide with the Burman nation's map. As *strangers* to the Thai nation-state, these peoples do not belong to Thailand's juridical map.

Not only does the geopolitical map of states represent the structure of approved sovereignties, it is also one of the primary forces determining recognized political subjectivity (Shapiro 1994, 482). In effect, only those who exist in *and* belong to the nation-state's juridical map are qualified political subjects. The univocity of the statist discourses render so-called unqualified political subjects both inaudible and invisible (Guha 1996, 11). Moreover, certain events and actions are assigned to history by specific values and criteria, which Guha calls the "ideology of statism": "the life of the state is all there is to history." This ideology "authorizes the dominant values of the state to determine the criteria of the historic" (Guha 1996, 1) and becomes the *common sense* of our understanding of history, which has invariably disregarded many "unqualified political subjects" as outside history. Consequently, those who have been living and struggling along the nation-states' border zones have more often than not fallen between the cracks of human awareness. Many of the struggles by these so-called unqualified political subjects, no matter how prolonged and bloody, never receive media attention because they do not exist in, nor do they belong to, the nation-state's juridical map. The states of exception under which these peoples have lives, and that do not comport with the homogenous continuum of the nation-state's history, are simply invisible and inaudible. Hence, when one considers those labeled by the state as internally displaced peoples, refugees, or stateless peoples, one finds that these parts of humanity have more often than not been left out of the community. These peoples' plights have been inadequately (ac)counted for.

STORIES THAT MUST NOT BE PERCEIVED

It is in such light that this chapter is a Rancièrian practice of politics. That is, for Jacques Rancière, politics is to make perceptible that which has been rendered imperceptible (here, combining Rancière and Agamben, I use the term *imperceptible naked lives*). Rancièrian politics is contained in a specific mode of relation called "parttaking" (*avoir-part,* which in French means both a *partaking* and a *partition*). It is a mode of relation that disturbs, deviates from, or supplements the normal order of things (Rancière 2001, ¶18). This political relationship allows one to think the possibility of a political subject(ivity) *(le sujet politique)*, and not the other way around (ibid., ¶1).[23] Politics disappears the moment one undoes this knot of a subject and a relation in which a political subject is defined by "its participation in contrarieties" (ibid., ¶4–5). This anomalous relation is thus expressed in the nature of the political subjects who "are not social groups but rather forms of inscription of 'the (ac)count of the unaccounted-for' . . . 'the part of those who have no-part'"; whether or not this part exists is *the* political issue (ibid., ¶18–19). Rancière's idea of the "unaccounted for" *(l'hors-compte)* refers to those who have no qualifications to part-take in the *arche*, i.e., no qualification for being taken into account by the logic of beginning/ruling.[24] From the *arche*'s point of view, there-fore, not everything is supposed to be seen or heard. The presup-positions that designate what is perceptible become customs in a community (in every sense of the term). These customs or general laws define the form of part-taking. Rancière terms these general laws "the partition of the sensible" *(le partage du sensible)*: the laws that operate by first defining the modes of perception in which forms of part-taking are inscribed (ibid., ¶20). The partition of the sensible always concerns the things that a community regards as "'to be looked into,' and the appropriate subjects to look into them, to judge and decide about them" (Rancière et al. 2000, 11, 12).[25] I am interested here in connecting Rancière's politics of im-perceptibility with Agamben's classification of *naked lives* by coin-ing the term *imperceptible naked lives* as an apparatus of recogni-tion to account for the experiences of the forcibly displaced Karens. Agamben's concept of naked life illuminates the interlocking rela-tions of sovereign power and human life, and hence enables me to discern the state terror inflicted upon the existence and bodies of the displaced Karens. Rancière's conception of the political allows one to understand that the struggles of forcibly displaced peoples— their practices of enunciating and/or demonstrating themselves as qualified political subjects—*are* the process of constructing po-litical spaces, even if labeled as illegitimate, and often as illegal, by the state. Moreover, naming the forcibly displaced peoples *im-perceptible* naked lives produces a reversal effect (it is an ironic strategy): rendering them *perceptible* outside the police logic of the state (see Decha 2003a and the shorter version in 2003b).[26]

Before I attempted to enter one of the so-called temporary shelter areas, I did not understand that unauthorized people are officially not allowed in these security spaces—let alone an academic wishing to do research there. In other words, these security spaces have been off-limits areas for unauthorized activities, including the production of knowledge. During my visits to the Operations Center for Displaced Persons in Bangkok, where I interviewed the director and later picked up a permission letter, I was told by an official that, without a letter from the Office of the National Security Council, it would take at least one month to process a letter asking for permission.[27] And not everyone who had asked for permission had been approved; even a member of the Thai army working on a master's thesis was refused entry. At least one foreign university lecturer's request to enter a shelter was rejected. A person asking for permission would typically have to prepare a package summarizing his or her research purposes and rationalizing the necessity to be in the intended shelter, together with references from the person's educational institutions and sample questionnaires, if applicable. I was informed by the official, furthermore, that mine would be the first doctoral dissertation research to be allowed to occur in shelter areas since they had become closed areas.

The whole procedure therefore reinforced a statist partition of the sensible, with its specific general laws: how the temporary shelter areas are usually not the place to look into; and how only certain *appropriate* subjects are qualified to look, to judge, and to decide about them. Even so, the police logic that governed the perceptibility of the temporary shelter areas, and of whatever was inside, in this case, also ordered an approved researcher's body: defining the allocation of his ways of doing, ways of being, and ways of saying. For instance, the official in Bangkok told me that I would not be allowed to stay overnight in the temporary shelter areas during my research, which meant that I would have to commute back and forth from a town with a hotel, about thirty-five miles each way, unless I could find a place to stay with local people near the temporary shelter area. I, too, was supposed to be there temporarily.

It is the figure of sovereign power who decides whether or not to learn more about the forcibly displaced peoples from Burma. The figure of sovereign power controls the production of knowledge about these peoples; it is through this control over what constitutes knowledge of these peoples that the Thai state-centered partition of the sensible is maintained. From the Thai nation-state's perspective, there is nothing else to know about these peoples' pasts. One could imagine that the figure of Thai sovereign power says: "We have *already* and adequately known them and we are not interested to know more." And *this* is understandable. For the production of knowledge about the histories of the forcibly displaced peoples and their ways of locating themselves in time threatens to tear apart the self-vindicating constructs of the Thai sovereign power (hence potentially destabilizing its sovereignty), as well as to empower the forcibly displaced in the process.

I contend that these peoples' situatedness and their pasts must be known not only for their sake, but also for our sake, because *tragedies* have been occurring and, at times, exploding, as in the Ratchaburi hospital siege at the dawn of the new millennium, along the Thai–Burmese border zones. The siege exemplified a group of forcibly displaced (Karen and Burman) peoples' making themselves perceptible and transfiguring a field of experience dominated by statist practices, but at great and horrifying costs. Although the statist dismissal of nonstatist subjectivities succeeded, the hospital siege and the execution of the hostage-takers within twenty-four hours shook the Thai nation-state to its core as a peaceful country, according to its collective self-perception.[28] We must, therefore, be vigilant, or, better yet, we must wake up.[29]

As a statist discourse, the languages and practices surrounding temporary shelter areas have inflicted security forces on any kind of subjectivities that come to be involved with the discourse, whether politically qualified or not, and whether living "naked" or "covered" lives. The permission letter opened a "door" for me to enter a temporary shelter area and to conduct my field research. I was another form of life that was inscribed and managed by the sovereign power, as part of the discourse of the temporary shelter areas. Although revealing myself to the sovereign power as a researcher, as someone who conducted research on this very sensitive issue, was my choice, my being inscribed was by no means an option. It will be possible, therefore, for the sovereign power, if it deems necessary, to abandon me, ripping off my form of life and exposing my nakedness.

After picking up my letter, I started to ask myself certain questions. Is there a link between a nation-state's security, the security of the forcibly displaced peoples, and the production of knowledge? Can sovereign power and knowledge not conjoin for the betterment of the forcibly displaced? It became obvious to me later that they can. As a result, the rest of this chapter will evince that, as far as the forcibly displaced peoples are concerned, the sovereign power that has attempted to silence their plight is ironically the pathway for their perceptibility and recognition. In other words, the forcibly displaced peoples' sufferings and struggles—their histories and temporalities—cannot be perceived without sovereign power: they are what I term *imperceptible naked lives*. What, then, are the characteristics of the entanglements embedded in perceiving their plight and paths? More specifically, what are the conditions for an *apparatus of recognition* to account for the plight of the forcibly displaced Karens living in these shelters and to enunciate their agonies?

FORMS OF LIFE OR SHEER FACTS OF LIVING?

So-called temporary shelter areas exemplify statist security discourses. They are security spaces that must be placed out of sight; and as little noise from inside as possible should be heard—the littler, the merrier. These are spaces of exception, spaces of pollution (cf. Malkki 1995), heterotopias (Foucault 1986) to the attempted utopia of the Thai nation-state. These spaces contain elements of chaos from another failed attempted utopia, Burma. When these suffering elements arrived at the door of the neighboring countries, they therefore had to be contained, disciplined, and ordered. The discourse of humanitarianism has been the sole impetus that has obliged the kingdom to receive the forcibly displaced peoples from Burma. The kingdom has invariably hoped that it could exert as little effort as possible. For Thailand to open the temporary shelter areas to the outside world would be to sow unruly seeds, questioning the territorial demarcation of political subjectivity, and hence from the kingdom's perspective, to instigate chaotic elements embedded within these suffering bodies.

In attempting to discern the forcibly displaced Karens' situatedness, it is critical to problematize a conception of forcibly displaced peoples as universal *victims* (cf. Allen and Turton 1996, 9; also cf., for instance, Rajaram 2002). Such is a view held by the international community, particularly by many international relief organizations. But when one focuses on the displaced Karens' signs of impoverishment and injury, more often than not one ignores these peoples as unqualified political subjects. Worse, many forcibly displaced peoples have not even been recognized as political subjects. Through this latter view, the forcibly displaced Karens become entities without histories, a view that both strips them of their pasts and silences their articulation of the present. Even though many Karens have been displaced from their "homeland" and many have nevertheless lost their lives, still their identities *as a people* have not been lost: they have been reconfigured. Although the displaced Karens today are not in the strong positions that their ancestors were when, for instance, in the nineteenth-century the latter demanded to be perceived in the Burmese public sphere under the semirule of the British Empire (see Decha 2003, 89–105), many of today's displaced Karens have been struggling relentlessly to survive and to be perceived as a nation again.

Everyone with whom I talked and who was not a Thai official used the term "camp" instead of the official term "temporary shelter areas" *(Phunthi phakphing chuakhao).*[30] I hence was corrected and reminded during an interview with a Thai official supervising the temporary shelter area, where I conducted my fieldwork and teaching, that this was indeed a "temporary shelter area," never a "camp." He wanted to make certain that I *saw* where I was, where he had been working, and where the forcibly displaced peoples had been inhabiting as temporary shelter areas. The official reasserted a statist discourse of "temporary shelter areas," a discourse that inscribes the camp in terms sensible before, and assertive of, the articulation of the authority and sovereign power of the Thai nation-state. It was, therefore, supposed to be clear to me that only such statist conceptions of the space were thinkable and enunciable.

The temporary shelther area where I visited is extremely hot in summer, very cold in Thailand's winter, and very muddy after rains during the rainy season. The shelter has always been very crowded. As of May 2005, there were 46,855 people in the shelter, with a density of about 105 persons per acre (TBBC 2005).

At first glance, an outsider could get an impression that the shelter residents, most of whom were the Karens, were docile bodies living in bamboo houses. Initially, it seemed to me that many people in the shelter had little sense of self-reliance, future, or certainty, compared to those of humans who are not forcibly displaced. The shelter residents did not know the duration of their stay: when that would be terminated by repatriation. Many youngsters got married not long after graduating from high school, some even before then; there was not much to do in the shelter. There was a saying among the shelter members that many just "eat sleep, eat sleep" *(au mee, au mee)*. Some told me that some Thai villagers living nearby scolded their dogs, "Don't be lazy like the Karens in the camps. They only sleep and eat." Coming from the Thais, the statement not only reinforced perceptions about life in the shelters, but, equally important, it also degraded the Karens to the level of animals. Such a statement pushes the shelter members back to the state of sheer facts of living *(zoe)*.

Being physically confined and provided with monthly food rations, with nutrition intake carefully monitored, the shelter residents are disciplined by the containment strategy of the statist discourses and practices.[31] As temporary shelter areas are located within the Thai state's jurisdiction, and as lives within are supported by an international protection regime, these forcibly displaced peoples appear as biopolitical bodies. Nonetheless, unlike the biopolitical bodies of citizens, the shelter members have been perceived by the Thai nation-state as politically unqualified beings, and thus have no juridical protection from the state. As a result, their voices by and large express merely their state of being, and their speeches are not recognizable before the law, unlike statist-territorialized fellow members of the human community.

Most shelter residents had no income or land to farm, and they have therefore become more dependent on aid.[32] This state of dependency becomes more drastic when one considers the rate of births each month and the amount of food that can be produced without income and land. For example, there were about one hundred births monthly—i.e., an approximately 3.18 percent population increase annually (Thai Ministry of Interior 2001, 7); and, except for some small plots for gardening, available only for earlier shelter members, the Thai government did not allow the residents to cultivate any crops.

From the Thai state's perspective, allowing cultivation could engender a sense of permanency, both for the shelter residents and to the international community. The shelter members must not be allowed such a perception. Time and again, the Thai government officials reiterate that theirs is a *temporary* hosting of forcibly displaced peoples. Hence, the sense of dependency in the shelter reinforces the statist discourse of temporariness. As aberrants to the Thai nation-state—epidemically, socially, culturally, juridically, politically, and environmentally[33]—they have therefore been abandoned and left outside Thai society as imperceptible naked lives *before the law.* One wonders, however, to what extent the kingdom has succeeded in its attempt to maintain them as sheer facts of living. One effective way to respond to this question is to explore the education being provided to the young in the temporary shelter areas, for it is through education that the Karen nationhood has been reenacted and their forms of life reinvented.

According to a report by ZOA Refugee Care, a Dutch nongovernmental organization focusing on education, between March and July 2000 in the seven registered temporary shelter areas in Thailand where the Karens were the majority, the Karen peoples made up roughly more than 80 percent of the registered shelter members (Maat and Taloung 2000, 1). For the 2000–2001 academic year, there were fifty-five primary schools (kindergarten [2 years] through fourth standard), eighteen middle schools (fifth through seventh standard), and thirteen high schools (eighth through tenth standard) in the seven registered temporary shelter areas, teaching 27,475 students with 917 teachers (ibid., 12). Primary school curriculum includes Karen language (especially *Sgaw* Karen, with the Eastern Pwo Karen dialect taught during the summer season), Burmese, English, mathematics, geography, science, and history.[34] In 1997, the Karen Education Project introduced five more courses—in the fields of music, sewing/knitting, drawing, carpentry, and first aid—into the high schools (Maat and Taloung 2000, 25–26). The head of the Karen education department pointed out that the education provided in temporary shelter areas is of a higher quality, from the Karen nationalist viewpoint, than that provided in schools in Burma's Karen State.[35] (Since the late 1980s, the languages of ethnic nationalities have rarely been taught beyond the fourth grade in Burma.)[36] In these security spaces, therefore, nationhood has been reenacted, forms of life have been reconstructed, and naked lives have been qualified. Nonetheless, as beings abandoned by sovereign power through the "inclusive exclusion," the Karens find their nakedness at times harrowingly revealed.

What was terrifying was that the atmosphere of fear I personally encountered in Rangoon between August and September 2000 also occurred in the shelter area. One time, while I was having dinner and a very friendly conversation with a group of students, whom I had taught and with whom I had spent time throughout the day, a student spoke in passing of violence committed upon shelter members by Thai authorities who were taking care of the shelter area. I asked that student to stop while I picked up a tape recorder, after which the student refused to say anything, no matter how much other students and I begged him to. He said he was afraid for his safety.

After that incident and others that I encountered, I found out that such violent incidents within temporary shelter areas had largely become practices of the past. Some infamous incidents had forced the Thai government to remedy the situation, especially after UNHCR was allowed to become involved in 1998. Nonetheless, many shelter members could not forget those violent events, especially those whom the atrocities were committed against—such as a girl who had been raped by a Thai local politician after she had set foot on Thai soil, but had decided to traverse through a "doorway" and became an illegal laborer outside the shelter area. At one point of our interview in 2001, the girl asked, "What will come out of this interview? . . . In the past couple of years I have been interviewed a couple of times, both in Bangkok and in this province, both by the Thai authorities and foreigners. But nothing better has happened to me, and nothing worse has happened to that man."

CONCLUDING REFLECTIONS

These violent incidents raise questions about the politics of authority: who speaks and who gets to listen to the memories of these "unqualified" voices abandoned before the law by the Thai sovereign power. Such politics of testimony implies hearing, which, as Ranajit Guha writes, "we know, 'is constitutive for discourse.' To listen is already to be open to and existentially disposed toward: one inclines a little on one side in order to listen."[37] One may still ask how one may *incline oneself* toward shelter members in restricted temporary shelter areas, being aware that hearing constitutes a person becoming: *being-open* as *being-with* for Others, in Heidegger's terms. Moreover, when Heidegger avers that one hears because one understands, and that one has "understood" only if one has heard "aright" (1962, 206), one can also inquire how one can hear the shelter members aright. These questions perhaps exemplify the imperceptibility attached to the status of shelter members.

Not only are lives in a "security space" not always secure, but the enunciation of their memories of violence is also usually not heard. Shelter members are caught in complex layers and frameworks of imperceptibility before the law. Those who would listen to such memories are often those who would not be authorized to enter the shelter areas. It is true that there are quite a few international relief agencies authorized to work in the string of temporary shelter areas along the Thai–Burmese boundary. Yet, these agencies' onus is not to listen to the forcibly displaced peoples' memories of past sufferings, but to remedy present ones. Such acts of remedy are so immediate that discerning the political dimensions attached to the work or to the act of listening has often been passed over. In this light, with a research project aimed at listening to shelter members' sufferings and the political ramifications of these sufferings, I needed an approval from the very sovereign power who had attempted all along to silence these voices; such is a moment of the paradox of perceptibility.

It is crucial that the moment of inscribing the shelter members as imperceptible before the law lays the foundation for the articulation of a political subjectivity that threatens the demarcation of what is to be perceptible and what not. As facets of imperceptible naked lives, the suffering of forcibly displaced peoples and their existence are conditions of the maintenance of sovereign power. Rendering these lives perceptible, as I have tried to do here, emphasizes not only that sovereign power is tenuous and dependent on the control (or the perception of control) of forcibly displaced peoples as naked lives. The attempt here to render perceptible that which is imperceptible is, in a Rancièrian sense, an attempt to expand the political. This is a Rancièrian politics that expands, disturbs, and deviates from the normal order of things by calling to mind, and making perceptible, that which has no part in the partition of the sensible of the Thai state. Equally important, I have here tried not only to shed light on the workings of sovereign power and thereby to expand the political; I have, relatedly, also taken note of the limits of sovereign power of the Thai state: the attempt to confine shelter members to the status of naked lives is not wholly successful; the shelter area provides a nationalist education unavailable in Burma.

As a way of ending my weaving of perceptibility of the forcibly displaced Karens, I would like to invoke again the stories of the silent boy (who refused to speak) and of the raped girl (who was not accounted for). The boy was inscribed by the statist discourse of a "temporary shelter area"; the girl, and her mother, originally passed through Thailand's "door." The boy was "before the law"; the girl was "beyond the law." The acts of *passing* and *being beyond* had rendered the girl extremely vulnerable, so that she became an *imperceptible naked life* with tremendous possibilities of having violence committed against her because of her lack of protection: having no one to whom to cry her agonies.

NOTES

I owe a debt of gratitude to the following people, especially my former advisor, Michael J. Shapiro, who left their marks in the earlier drafts of this article; others were Manfred Henningsen, Leslie E. Sponsel, Sankaran Krishna, Nevi Soguk, Brian Richardson, Jorge Fernandes, and Adam Sitze. I am grateful to Prem Kumar Rajaram for his friendship, patience, and helpful suggestions toward the article's drafts. I am also thankful to all the friends, especially Didier Bigo, who attended the "The Security Paradox of Open Borders: Control and Surveillance of Migrants" workshop in northern Thailand, for their exciting discussion on this article's draft. Needless to say, any errors are mine. Lastly, I am forever in the debt of the forcibly displaced Karens.

1. This article is part of a writing journey into the spaces of the forcibly displaced Karens, who have taken flight from within the Burmese nation-state. The fieldwork for my study was conducted between 2000 and 2001 and from 2005 onward.

The term *Karen* was originally used by outsiders, and its derivation is uncertain. The Karens were officially renamed by the ruling State Law and Order Restoration Council (SLORC) in 1989 as *Kayins*, a name that the Karen nationalist leaders have rejected "as strongly as they do the historic Burman term for their country, 'Myanmar'" (Smith 1999, 37). Under the name Karen, there are three major groups: the Pwo, the Bwe, and the Sgaw (see, e.g., Marshall 1922, 1). Size, persistence, and the fact that most of the peoples in the so-called temporary shelter areas on the Thai side are Karens are three major reasons why this study focuses on the Karens and not on the Shans, the Karennis, and the Mons, the other three major ethnic nationalities living in these Thai–Burmese border zones. Although Martin

Smith notes problems involved in conducting surveys within the territory of Burma/Myanmar, he states that the Karens are the second largest ethnic nationality after the Burmans (Smith 1999, 30).

Regarding the Karens' persistence, the Karen National Union (KNU) is one of the last remaining armed resistance organizations, and the longest-standing one, in Burma. Its members have been fighting against the Burmese government since the official announcement of their revolution (as they refer to it) on January 31, 1949. Since then, they have attempted to gain, origi-nally, a separate country for the Karens, and, later, an autonomous region for their Karen State under the Federal Union of Burma. Nonetheless, my study focuses mainly on civilians who have been forcibly displaced rather than on the political actors in the Thai–Burmese border zones.

2. Kafka 1984, 213–15. The term Burma is deployed throughout ex-cept when intending to convey the present officially recognized name of the country, Myanmar. The change from Burma to Myanmar has not been accepted by the opposition, and the use of the latter term has been politi-cally charged. For discussion, see, for instance, Taylor 2001, 1n1; Steinberg 2001, 41n1; Collignon 2001, 70n1; Silverstein 2001, 119n1. Moreover, there are three related terms that need clarification when employing the term Burma. First, Burma is a noun, the country's name. Burman is an adjective denoting an ethnic nationality living among other ethnic nation-alities. And Burmese is another adjective, signifying the discourse of state-hood of Burma; hence Burmese peoples, for instance, are peoples living within the territory of Burma.

3. Brian Richardson was greatly helpful in crafting this narrative.

4. Hence, the decision of the sovereign is not the expression of "the will of a subject hierarchically superior to all others" (Agamben 1998, 25–26). Neither is it the chaos that precedes legal order (Agamben, 1999, 162; 1998, 18). It is truly, according to its etymological root, "taken outside (ex-capere), and not simply excluded" (Agamben 1998, 18).

5. Agamben's project is unlike Michel Foucault's. For Foucault, bio-politics was a modern project, signifying the "threshold of modernity" (1978, 143, and see also 140–45); for Agamben, biopolitics was the origi-nary activity of the sovereign.

6. In this article, instead of using the term bare life, I follow the trans-lation of Vincenzo Binetti and Cesare Casarino (Agamben 2000b) and use naked life throughout, except in citing passages from Homo Sacer: Sov-ereign Power and Bare Life. As Binetti and Casarino emphasize, the term naked life translates the Italian nuda vita, which not only appears in the subtitle of Agamben's Homo sacer: Il potere sovrano e la nuda vita, but also throughout his work (see "Translators' Notes" in Agamben 2000b, 143n1).

7. The left column of the working agreements, which is "Steps to be taken by UNHCR," refers to "UNHCR as observer to Thailand's assess-

ment of the situation whether to grant temporary shelter on the ground of fighting and effects of civil war" (Department of International Organization 1998, 1; cf. Vitit 2005, 2, 7).

8. Nevertheless, the right column of the working agreements, which is "Steps to be taken by RTG," states: "Thai authorities have the right to grant or deny temporary shelter to arrivals. Those with valid claims must either enter designated temporary shelter areas, or otherwise decline the temporary shelter in Thailand and return to Myanmar" (Department of International Organization 1998, 1). In this light, when one considers the agreements against the horrendous effects of the Tatmadaw's Four Cuts strategy upon members of ethnic nationalities (effects not always reached through direct fighting), one recognizes that the Thai government's grant of temporary shelter to only peoples fleeing fighting is alarmingly narrow. The Four Cuts strategy was a so-called counterinsurgency program that the Burmese Army (the Tatmadaw) designed in the mid-1960s to cut the four main links—food, funds, intelligence, and recruits—between ethnic nationalities' soldiers, their families, and local villagers. The forcibly displaced peoples' sufferings that resulted from the Four Cuts strategy have thus become imperceptible. These peoples would be rejected at "the gate" and left outside the threshold of the sovereignty of the agreements, quintessentially situated *outside*, without existing in a corresponding *inside*. They would either have to attempt to traverse through other of Thailand's "doorways" or wander and hide in the war zones.

9. Kachadpai 1999, 3; see also Krit 1997, 2–3; Boonsak 1998, no. 23; Kachadpai 1999, 3–4; Kachadpai 2000, 4–5.

10. This is the third immigration act in the history of the kingdom. The first act was stipulated in 1950 and the second in 1954 (MOI 1999b). The fourth act, stipulated in 1980, was an addition to the 1979 act (see in, e.g., MOI 1999c, 108–9).

11. There are differing spellings (Moei or Moi) and names (Thaungyin or Moei/Moi) of this river; it is unclear which is more linguistically, historically, or politically appropriate; I choose to use Moei.

12. Kachadpai 1997, 68. On March 9, 1976, there was an announcement from Thailand's Ministry of Interior titled "Prohibiting Burmese National Aliens from Migrating into the Thai Kingdom" (MOI 1976). The announcement states that the Thai kingdom had, for humanitarian reasons, allowed Burmese nationals to *temporarily* take refuge in its provinces bordering Burma until conditions returned to normal (upon which, repatriation would occur). The announcement adds that since, at that moment in 1976, conditions inside Burma had returned to normal, there was no reason for further migration into Thailand from Burma; hence, from that day onward, forcibly displaced peoples would be prosecuted for illegal entry into the country (Kachadpai 1997).

13. Vitit Muntarbhorn, "Law and National Policy Concerning Displaced Persons and Illegal Immigrants in Thailand," unpublished paper (Bangkok: Institute of Asian Studies, Chulalongkorn University, n.d.), p. 7; quoted in Lang 2002, 92; see also Alexander 1999, 40.

14. The Thai state has a negative view of accession to the 1967 protocol and the 1951 convention because of perceived threats to national sovereignty and perceived legal conflicts. The state fears it would have to cede control to international instruments over how it deals with displaced peoples. The legal conflicts identified by the Ministry of Foreign Affairs' Department of International Organizations center on potential conflicts between the articles of the convention and Thai national law. Among the forty-six articles of the convention, twenty-nine correspond to the principle of humanitarian nondiscrimination in Thai laws, eleven are in conflict, and the rest are ambiguous (Department of International Organizations 2000; see more detail in Decha 2003, 198n43).

15. Kachadpai 1997; MOI 1999a, 2. These two books have, however, different classifications. Arguably, the difference has to do with the purposes of the classification: either as a response to "outside" threats, or for the purpose of governmentalizing practices, containment strategies, and maintaining "internal" order. The first ascribed purpose belongs to Kachadpai's book (the author worked thirty-seven years at the National Security Council), the second to the Ministry of Interior's book.

16. Kachadpai 1997, 68. The Ministry of Interior classified the post-1976 arrival as "illegal immigrants from Burma" (MOI 1999a, 2).

17. The narrative of this category, "peoples fleeing fighting," relies heavily on, in addition to Kachadpai (1997) and MOI (1999a), Alexander (1999) and Lang (2002).

18. Amnesty International 2002, 36. Read about the situations and plight of migrant workers from Burma in, e.g., ibid., 36–42; Amnesty International 2005; and a report of the Thai Action Committee for Democracy in Burma (TACDB 2002) titled "The Plight of Undocumented Burmese Workers in Thailand." This document covers interviews of forty-five migrant workers all over Thailand, in seven work categories and conditions: agriculture and plantation, construction sector, fishing, factory and small firms, sweatshop, housemates, and under police incarceration. For situations inside Burma regarding forced labor, consult the Web site of Federation of Trade Unions—Burma: http://www.tradenions-burma.org (accessed November 18, 2002).

19. These four "camps" are the manifestations of humanitarian assistance conducted by international relief agencies, but they could not be considered as preventive protection because of their lack of security.

20. The usage of the term *Karens* in this study is not meant to signify

the Karens as a "frozen" people. For, inasmuch as identities are contingent on the performative, I deploy the term as a signifier of those who enunciate, perform, and reenact *Karen-ness*. In this, I wish to retain Gayatri Spivak's notion that identities are strategically essentialized in encounters or political struggles (e.g., Spivak and Harasym 1990, 11). Following Rancière, such strategical essentializing of the Karens is an enactment of subjectification (the enunciative and performative acts attempting to render themselves perceptible and intelligible) with an intention to render them recognizable as qualified political subjects (e.g., Rancière 1999, 2001). Accordingly, to understand the sufferings and struggles of the forcibly displaced Karens in the war zones of Burma, or inside and outside the so-called temporary shelter areas on the Thai side, one must understand how crucial it is for the Karens to reenact themselves as Karens. For many illiterate, forcibly displaced Karens, Karen dialects are the only language, the only enunciative vehicle. For the civilians who have taken flight in the war zones, it is critical, after days, months, or years of running for their lives, to be able to trust that they belong to a community somewhere, a community that they believe can help them. For the members of the Karen National Union (KNU), after more than half a century of fighting in the name of Karen nationhood, it is crucial to be able to trust that the community is not nameless.

21. The interviewee even named Gen. Chavalit Yongchaiyut, the then prime minister and minister of defense, whose close relations with the ruling junta began in the 1980s.

22. Here, my student refers to another splinter group that broke from the Karen National Union in 1994. Since the focus of my study is on people in general and not political actors in the Thai–Burmese border zones, those interested in the latter issue should see, for example, Ball and Lang (2001).

23. According to Davide Panagia, the English term *political subject*(ivity) does not give an adequate sense of Rancière's *le sujet politique,* which refers both to the idea of a political subject and to the "proper" subject of politics (Rancière 2001, n2).

24. See also Rancière 2001, ¶12, 13, 14; ibid. 2004. When one views politics as such, Rancièrian political struggle is (as mentioned briefly here) about a special kind of "counting." And there are two contrasting ways of counting: one called *the police,* the other *politics.* The first only counts empirical parts, "actual groups defined by differences in birth, by different functions, locations, and interests that constitute the social body." The second counts "'in addition' a part of the no-part" (2001, ¶19). These two contrasting ways of counting depend on the "partitions of the sensible" (see more details in Decha 2003, 16–25).

25. Contrasting himself with Foucault, Rancière states that his own

concept of the partition of the sensible is his way of translating and appropriating Foucault's genealogical thought: ways of systematizing how things can be "visible, utterable, and capable of being thought" (Rancière et al. 2000, 12). Although Rancière's works could be expressed in terms close to those of Foucault's concept of *episteme*, they differ in some major aspects. Whereas Foucault thinks in terms of "limits, closure, and exclusion," Rancière thinks in terms of "internal division and transgression" (Rancière et al. 2000, 12). Foucault's concept claims to establish what is thinkable or not for a particular era, whereas Rancière's is more sensitive to "crossing-over, repetitions, or anachronisms in historical experience." For Rancière, Foucault's historicist's partition between the thinkable and unthinkable "seems to . . . cover up the more basic partition concerning the very right to think" (Rancière et al. 2000, 12).

26. See Jacques Rancière's critique of Giorgio Agamben in Rancière 2004. The expression "apparatus of recognition" belongs to Adam Sitze, to whom I am grateful.

27. Because the letter was not addressed to me, it is impossible to include it herein. But it can be viewed in Decha 2003, 212.

28. For my treatment of the incident, see Decha 2003, 235–48; the shorter version is in Decha 2004.

29. I am alluding to Walter Benjamin's sentence "There is a not-yet conscious knowledge of what has been: its advancement has the structure of awakening" (1999, 389[K1,2]). See the influence of Benjamin's treatment of the relationship between time and history, in Decha 2003, especially chapters 2 and 7.

30. Even the program report of the Thailand Burmese Border Consortium (TBBC) uses the term *camp* throughout (e.g., TBBC 2005). The name TBBC was changed from the Burmese Border Consortium (BBC) in 2004.

31. See the most recent report on food rations and supplementary feeding for shelter members by the TBBC (2005, 59–62). The following is an example of the monthly rations in 2000 (BBC 2000, 39–42; MOI 2001, 4):

Rice	35.2 lbs./adult; 17.6 lbs./child under 5 years
Fish Paste	2.2 lbs./person
Salt	2.2 lbs./3 persons
Yellow Beans	2.2 lbs./adult; 1.65 lbs./child < 5 years
Cooking Oil	1 lt./adult; 500 ml./child < 5 years
Dry Chili	0.55 lbs./person
Cooking Fuel	22 lbs./person for 2-person family;
	15.4 lbs./person for 5-person family
Blanket	1–2 persons/year*
Mosquito Net	1–3 persons/year*

Sleeping Mat   1–5 persons/year*
Cooking Utensils   1/family; 1 larger pot /family of over 5 persons
                         (cooking utensils first provided early in 2001)
Building Materials   for annual housing repair: bamboo and eucalyp-
                         tus poles (most camps), thatch or roofing leaves
                         (some camps) (starting early in 2000)
Clothing   warm clothing (starting with the cold season
                         of 2000)
(*or for a new arrival, if necessary)

32. According to the BBC, by the mid-1990s it had become necessary for the Refugee Relief Program to supply 100 percent of basic food needs (BBC 2000, 39).

33. From the dominant perspectives of many Thais, the forcibly displaced peoples as aberrants have not only inhabited the spaces of pollution but have also been sources of pollution, in many senses of the term. Below are two examples given of "environmental pollution":

First, in an interview I had with a village headman in a village adjacent to a temporary shelter area, he pointed to the stream that ran from the hill in the area, passed through the shelter, and flowed down to his village. In 1999, Thai villagers became sick and some of them died. They believed that the villagers' death had resulted from the sanitation problems in the shelter. The interview was conducted in May 2001.

Second, from a letter sent to a district chief in Kanchanaburi, a province where a temporary shelter area is located, one learns of a complaint made by a group of village leaders in that district. The villagers suspected that their reservoir, which had received water running downstream through a temporary shelter area at Ban Ton Yang, had been polluted. The Thai villagers had developed rashes and itching. According to the letter, these symptoms "may result from the water turning dirty due to the shelter members pollutes the upstream; now Thai villagers no longer use the water in the reservoir for either drinking or cleaning" (Villagers of Parainok 2001).

34. There were quite a few Burmese-speaking Muslims in these temporary shelter areas, and their children had to learn three to five languages starting from kindergarten. Besides the Burmese, Karen, and English languages, they had to learn Arabic and Urdu on a daily basis, in the evenings (Maat and Taloung 2000, 25).

35. Maat and Taloung 2000, 13. One day while I was talking with the family of the principal of the best high school in a temporary shelter area, a family of Thai villagers outside the shelter area came to meet the principal, asking for permission to send their teenage girl to study in the school. The parents of the girl told me, after being asked for their reasons, that English

teaching inside the shelter area was much better than that taught by the Thai local public high school near the shelter area. Since the family was also Thai Karen, the girl could also learn the Karen language should she be accepted. I was told by the principal that there had been quite a few cases like this at the school.

36. Smith 1995, 230. My Karen friends from Rangoon/Yangon mentioned to me, after learning about various dialects of the Karen language being taught in the temporary shelter areas, that his children had not been able to learn the Karen language because of the military junta prohibitive policy.

Alan Saw U, a Karen leader in Burma, writes a brief history of such a policy: "In the year 1922, during the British occupation of Myanmar, the Karen leaders worked hard for the Karen people to have access to information in their own language, the right to use their language in governmental educational institutions; and the right to have adequate provision created for the use of the Karen language where needed. . . . Consequently, a bill was declared for the Sgaw Karen dialect to be used in the High School up to 10th Grade; and the Delta Pwo [i.e., Western Pwo] dialect to be used up to the Seventh Grade. Primers and textbooks were published and the Karen people were able to fully enjoy the right to their own language for quite a number of years. . . . Unfortunately that right to language was short lived. The Myanmar Government started the process of *ethnic linguicide* soon after it gained independence from the British in the year 1948" (2000, 1). According to Alan Saw U, during the session in the parliament in 1958, the bill for the right of the Karen people to their own language was rejected without any reason being given.

37. Guha 1996, 9; the quotation is from Heidegger 1962, 206.

WORKS CITED

Agamben, Giorgio. 1998. *Homo Sacer: Sovereign Power and Bare Life.* Trans. Daniel Heller-Roazen. Stanford, Calif.: Stanford University Press.

———. 1999. "The Messiah and the Sovereign: The Problem of Law in Walter Benjamin." In *Potentialities: Collected Essays in Philosophy,* ed. and trans. Daniel Heller-Roazen. Stanford, Calif.: Stanford University Press, 160–74.

———. 2000a. "Form-of-Life." In *Means without Ends: Notes on Politics,* trans. Vincenzo Binetti and Cesare Casarino. Minneapolis: University of Minnesota Press, 3–12.

———. 2000b. *Means without Ends: Notes on Politics,* trans. Vincenzo Binetti and Cesare Casarino. Minneapolis: University of Minnesota Press.

Alan Saw U. 2000. "Language Rights of Refugees: A Personal Reflection." Unpublished paper.

Alexander, Michael. 1999. "Refugees in Thailand." *Asian Migrant* 12, no. 1 (January–March): 37–44.

Allen, T., and D. Turton. 1996. "Introduction: In Search of Cool Ground." In *In Search of Cool Ground: War, Flight, and Homecoming in North-east Africa,* ed. T. Allen. London: James Currey.

Amnesty International. 2002. "Myanmar: Lack of Security in Counter-Insurgency Areas." Report—ASA 16/007/2002, July 17. http://web.amnesty.org/ai.nsf/recent/ASA160072002?OpenDocument.

———. 2005. "Thailand: The Plight of Burmese Migrant Workers." Report—ASA39/001/2005, June 8. http://web.amnesty.org/library/pdf/ASA390012005ENGLISH/$File/ASA3900105.pdf.

Appadurai, Arjun. 1996. "Sovereignty without Territory: Notes for a Post-national Geography." In *The Geography of Identity,* ed. Patricia Yaeger. Ann Arbor: University of Michigan Press, 40–58.

Ball, Desmond, and Hazel Lang. 2001. "Factionalism and the Ethnic Insurgent Organisations." Working Paper No. 356, Canberra: Strategic and Defence Studies Centre, Australian National University.

Benjamin, Walter. 1999. *The Arcades Project,* trans. Howard Eiland and Kevin Mclaughlin. Cambridge, Mass.: Belknap Press, Harvard University Press.

Boonsak, Kamheangridhirong. 1998. "Statement by General Boonsak, Secretary-General of the National Security Council, Leader of the Thai Delegation at the 49th Session of the Executive Committee of the Programme of UNHCR." Geneva, October 5.

Burmese Border Consortium Relief Programme (BBC). 2000. "Programme Report for the Period July to December 2000 including Revised Funding Appeal for 2001." [Bangkok] Thailand: Burmese Border Consortium Relief Programme.

Collignon, Stefan. 2001. "Human Rights and the Economy in Burma." In *Burma: Political Economy under Military Rule,* ed. Robert H. Taylor. New York: Palgrave, 70–108.

Decha Tangseefa. 2003a. "Imperceptible Naked-Lives and Atrocities: Forcibly Displaced Peoples and the Thai-Burmese In-Between Spaces." Ph.D. diss., University of Hawai'i at Manoa.

———. 2003b. "Imperceptible Naked-Lives: Constructing a Theoretical Space to Account for 'Non-statist' Subjectivities." Paper presented at the CEEISA/ISA International Convention, Budapest, Hungary, June 26–28.

Department of International Organizations, Ministry of Foreign Affairs of Thailand (DIO). 1998. Second RTG-UNHCR Brainstorming Session, Bangkok, May 15, 1998.

————. 2000. "An Accession to the 1951 Refugee Convention" (document outlining the reasons why Thailand has not acceded to the 1951 convention). June 9.

Foucault, Michel. 1978. *The History of Sexuality*. Vol. 1, *An Introduction*, trans. Robert Hurley. New York: Vintage Books.

————. 1986. "Of Other Spaces." *diacritics* 16, no. 1 (Spring): 22–27.

Guha, Ranajit. 1996. "The Small Voice of History." *Subaltern Studies* 9: 1–12.

Heidegger, Martin. 1962. *Being and Time*, trans. John Macquarrie and Edward Robinson. New York: Harper and Row.

Kachadpai Burusapatana. 1997. *Burmese-National Ethnic Minorities [Chonklumnoi sanchat phama]*. Bangkok: Phaephitthaya Publisher.

————. 1999. "Statement by Mr. Kachadpai Burusapatana, Secretary-General of the National Security Council, Head of the Thai Delegation at the 50th Session of the Executive Committee of the Programme of UNHCR." Geneva, October 4.

————. 2000. "Statement by Mr. Kachadpai Burusapatana, Secretary-General of the National Security Council, Head of the Thai Delegation at the 51st Session of the Executive Committee of the Programme of UNHCR." Geneva, October 3.

Kafka, Franz. 1984. *The Trial*, trans. Willa and Edwin Muir. Revised, and with additional material translated by E. M. Butler. New York: Schocken Books.

Krit Garnjana-Goonchorn. 1997. "Statement by H. E. Mr. Krit Garnjana-Goonchorn, Permanent Representative of Thailand to the UN, Leader of the Thai Delegation at the 48th Session of the Executive Committee of the Programme of UNHCR." Geneva, October 13.

Kritaya Archavanitkul and Pramote Prasatkul. 2005. *Raingan prachakorn lae sangkom* (Population and Social Report). Nakhonprahtom, Thailand: Institute for Population and Social Research, Mahidol University.

Lang, Hazel J. 2002. *Fear and Sanctuary: Burmese Refugees in Thailand*. Ithaca, N.Y.: Southeast Asia Program Publications.

Maat, Hendrien, and Loytee Taloung. 2000. "Education Survey." Survey carried out in Karen Refugee Camps on the Thai–Burmese Border. Mae Sot, Thailand: ZOA Refugee Care.

Malkki, Liisa H. 1995. *Purity and Exile: Violence, Memory, and National Cosmology among Hutu Refugees in Tanzania*. Chicago: University of Chicago Press.

Marshall, Harry Ignatius. 1922. *The Karen People of Burma: A Study in Anthropology and Ethnology*. Columbus: Ohio State University.

Ministry of Interior (MOI) of Thailand. 1976. *"Ham khontangdao sanchat phama opphayop khaoma nai ratcha'anachak thai"* (Prohibiting

Burmese National Aliens from Migrating into the Thai Kingdom). An announcement from the Ministry of Interior, March 9, 1976.

———. 1999a. *Pramuan kotmai rabiap kieokap chonklumnoi nai prathetthai* (A Compilation of Laws and Regulations Regarding Minority Groups in Thailand). Bangkok: Department of Local Administration.

———. 1999b. *"Phraratchabanyat khonkhaomuang pho so 2522"* (Immigration Act 1979). In *Pramuan kotmai rabiap kieokap chonklumnoi nai prathetthai (A Compilation of Laws and Regulations Regarding Minority Groups in Thailand)*. Bangkok: Department of Local Administration, 89–107.

———. 1999c. *"Phraratchabanyat khonkhaomuang (chabap thi 2) pho so 2523"* (Immigration Act 1980). In *Pramuan kotmai rabiap kieokap chonklumnoi nai prathetthai (A Compilation of Laws and Regulations Regarding Minority Groups in Thailand)*. Bangkok: Department of Local Administration, 108–9.

———. 2001. *Banyai sarup: phunthi phakphing chuakhraw ban mae la (Briefing: The Temporary Shelter Area at Mae La]*.

Rajaram, Prem Kumar. 2002. "Humanitarianism and Representations of the Refugee." *Journal of Refugee Studies* 15, no. 3.

Rancière, Jacques. 1999. *Disagreement: Politics and Philosophy*, trans. Julie Rose. Minneapolis: University of Minnesota Press.

———. 2001. "Ten Theses on Politics." *Theory and Events* 5, no. 3.

———. 2004. "Who Is the Subject of the Rights of Man?" *South Atlantic Quarterly* 103, nos. 2–3 (Spring/Summer): 297–310.

———, Solange Guenoun, and James H. Kavanagh. 2000. "Jaques Rancière: Literature, Politics, Aesthetics: Approaches to Democratic Disagreement." Trans. Roxanne Lapidus. *SubStance* 29, no. 2: 3–24.

Shapiro, Michael J. 1994. "Moral Geographies and the Ethics of Post-Sovereignty." *Public Culture* 6, no. 3: 479–502.

Silverstein, Josef. 2001. "Burma and the World: A Decade of Foreign Policy under the State Law and Order Restoration Council." In *Burma: Political Economy under Military Rule*, ed. Robert H. Taylor. New York: Palgrave, 119–36.

Smith, Martin. 1995. "A State of Strife: The Indigenous Peoples of Burma." In *Indigenous Peoples of Asia*, ed. R. H. Barnes, Andrew Gray, and Benedict Kingsbury. Ann Arbor, Mich.: Association for Asian Studies, Inc. Monograph and Occasional Paper Series no. 48: 221–45.

———. 1999. *Burma: Insurgency and the Politics of Ethnicity*. 2nd ed. London: Zed Books.

Steinberg, David I. 2001. "The Burmese Conundrum: Approaching Reformation of the Political Economy." In *Burma: Political Economy under Military Rule*, ed. Robert H. Taylor. New York: Palgrave, 41–69.

Taylor, Robert H. 2001. "Introduction: Stagnation and Stalemate." In *Burma: Political Economy under Military Rule,* ed. Robert H. Taylor. New York: Palgrave, 1–4.

Thai Action Committee for Democracy in Burma (TACDB). 2002. *The Plight of Undocumented Burmese Workers in Thailand.* Bangkok: TACDB.

Thailand Burma Border Consortium (TBBC). 2005. "Relief Programme: January to June 2005." [Bangkok] Thailand: Thailand Burma Border Consortium.

United Nations High Commissioner for Refugees (UNHCR)–Thailand. 1998. "UNHCR's Role on the Thai–Myanmar Border." Bangkok: UNHCR.

Villagers of Parainok. 2001. "Chotmai thung naiamphoe sangklaburi: ruang nam nai 'angkepnam bo. Parainok sokkaprok" (letter to the district director of Sangklaburi, Kanchanaburi: The water in the village's reservoir has become dirty). Dated January 2001.

Vitit Muntarbhorn. 2005. "Refugee Law and Practice in the Asia and Pacific Region: Thailand as a Case Study." Unpublished paper.

———. No date. "Law and National Policy Concerning Displaced Persons and Illegal Immigrants in Thailand." Unpublished paper. Bangkok: Institute of Asian Studies, Chulalongkorn University.

11

Locating Political Space through Time: Asylum and Excision in Australia

PREM KUMAR RAJARAM

On November 4, 2003, the vessel *Minasa Bone* landed without authorization on Australia's Melville Island, some twenty kilometers off Darwin. On board were ten male Kurdish individuals plus four Indonesian crew. The boat had arrived from Indonesia; the passengers on board were Turkish nationals. The *Minasa Bone* appeared to be the latest of a series of unauthorized vessels carrying asylum seekers to Australia. In chapter 9 of this volume, Suvendrini Perera has described an Australian Pacific borderscape of moving bodies, bodies written out of the Australian sovereign landscape.

All unauthorized arrivals in Australia have been subject to mandatory detention while asylum claims are being processed on the presumption that, if they are found to have a case for asylum under the relevant conventions, they will be eligible only for "temporary protection visas," which may be revoked after a period; or else they will be subject to removal from Australian territory, without consideration of asylum claims, either to offshore processing countries or, under regional agreements, to neighboring countries, without any guarantee that if found to be refugees they would be allowed to return to Australia.

These are the principal possible outcomes of unauthorized boat arrivals in Australia, following the passage of legislation in September

2001 that amended the 1958 Migration Act. All interpret the immigrant as a security threat to the Australian polity, and thus insist on punitive or exclusionary measures. I suggest that the varied responses by the Australian government to unauthorized boat arrivals are performative acts of sovereignty, where the outline and limits of territorial sovereignty and the state itself are drawn. What is particularly interesting about the *Minasa Bone* example is that it shows that the performance of territorial sovereignty is not only about assertion of control over a space of land, but also, as I will show, about the control of time. On November 9, 2003, Australia's immigration minister, Amanda Vanstone, in a press release headed *"Minasa Bone* Returns to Indonesia," blandly stated, "The *Minasa Bone,* the boat that arrived illegally near Melville Island on 4 November, has returned to Indonesia following its escort back to international waters near Indonesia.... The passengers on the *Minasa Bone* did not claim asylum in Australia. If they were to claim asylum in Indonesia, the regional cooperation arrangements in place in Indonesia would provide the basis for their care and consideration of any claims" (Vanstone, 2003). Among the amendments made to the 1958 Migration Act was the Migration Amendment (Excision from Migration Zone) Bill 2001. This allows for the excision of territory from Australia for the purposes of the Migration Act 1958, and was the piece of legislation enacted as a response to the entry of the *Minasa Bone* and its passengers into Australian territory. The excision amendment allows for chunks of Australia to be removed from a "migration zone"; obligations toward asylum seekers in these zones are different from those in nonexcised parts of Australia. The excision amendment makes arrival, at designated offshore territories, insufficient grounds for claiming asylum in Australia.

Vanstone's press release is made up of a series of definitive statements: the boat "arrived illegally"; "the passengers on the *Minasa Bone* did not claim asylum in Australia"; "if they were to claim asylum in Indonesia, regional cooperation arrangements . . . would provide the basis for . . . consideration of any claims." There are different ways of seeing the events preceding and following the arrival of the *Minasa Bone.* The terse language of the press release favors a conception of reality (and of geography) set in motion by the Migration Amendment Act. The legislation provides a means of responding to political uncertainty. It is a means of response enabled by a prior conceptualization of the meaning of the event. The response to unauthorized boat arrivals is predetermined; it is set on a particular trajectory by the relevant legislation. Such legislation renders imperceptible asylum claims that, the passengers were to claim, had actually been made. Political legislation provides a template for responding to the future by structuring external certainty. This is a structuring centered on clarifying the present, the political order afforded by the territorial state, as the basis for facing the future. This is, in sum, the "integration of the future into the horizon of the present" (Chowers 2002, 668). It is the establishment of the structure of recognition of the territorial state as a means of making external uncertainty knowable and of responding to the uncertainty of the future. It is a means of ensuring the stability of territorial politics by attempting to incorporate the unknowable future into the horizon of existence allowed for by the state. Vanstone's terse and certain language, particularly in identifying the *Minasa Bone* as illegal and stating that no asylum claims were made in Australia, is a reflection (indeed is made possible) by the legislation that establishes a dominant apparatus of recognition and is centered on the excision of a part of Australia from itself.

On November 14, the Australian government admitted that individuals on the boat made repeated claims for asylum. By then, the boat, and its passengers, had been forcibly sent to Indonesia. Amanda Vanstone noted, "some people did say things referring to human rights and refugees." Such things, however, were said in a vacuum, Vanstone continues: "It's not news or relevant whether they did or didn't make certain remarks because they were never in the migration zone." Melville Island was retrospectively excised from the migration zone. Excision of the island was legally confirmed some hours after the *Minasa Bone* landed.

A POLITICS OF PLACE

The amended Migration Act has a broader history, which links it first to Australian immigration policy and then to the exclusionary practices of identification of settler colonialism. Vin D'Cruz and William Steele suggest that the Australian political imagination is run through with a fear of the "world's most dispossessed people (invariably people of colour) whether refugees or Aborigines" (D'Cruz and Steele 2003, 279). Kathryn Manzo notes that, though the overt racial aspects of immigration policy ended in the 1970s, Australia continues to define itself against externalized others, notably Asians and "boatpeople" (Manzo 1996).

The garrison mentality of Australian immigration policy is notable for its externalization of danger and for its capacity to represent a threat to the state as a threat to an Australian way of life. The manner of territorialization leads to particular societal technologies of the self that, in return, bind the individual to the territorialized space. Thus a history of nation building, of building national identity, has been "marked by class conflict, an uneasy stalemate between Capital and Labour, the dispossession of Aborigines and a visceral fear of Asia" (Burke 2002, 24). Australian national identity, Burke continues, has been "based on exclusion and security, consistently purchased at the expense of the Other." The way of life guaranteed by the state is under threat by deviant elements, whether refugees, other migrants of certain hues, or Aborigines, that would question the limits of community, justice, and belonging delineated by the Australian Commonwealth. Illegal immigration is not just a problem perceived by a state; it becomes a general moral problem to be addressed by all Australians.

The stranger, or the refugee, or the unauthorized migrant, is projected as "a symptom of (controllable) uncertainty" (Burke 2002, 24). A community premised on maintaining commonality as a goal is readily fearful of the uncertainty that the migrant poses. Anthony Burke argues that this fear of uncertainty is instrumentally used in Australian nation building, "projections of the stranger/refugee as . . . uncertainty are calculated and wilful displacements of larger and more complex social phenomena" (ibid., 24). In other words, the deployment of this fear or anxiety is an instrument by which the state vindicates or reterritorializes its space. This reterritorialization of space occurs through the identification of a threat only discernible from within the logic of a national order of things. The identification and promotion of this threat thus further binds a population to the structure of order and justice given by the state. The "circulation of anxiety" (Bigo 2002) is a form of governmentality, a means by which the state emphasizes its embrace of society. This power is also visible in those rendered "outsiders." The making of particular forms of subjects with particular forms of relations perceptible connects these subjects to those who no longer have reason to be seen, who are imperceptible. This arrangement of home is not a static one: it is involved fundamentally in an ongoing encounter with otherness, with those rendered imperceptible.

Burke describes further the deployment of the concept of *home* in Australian political discourse. He cites Prime Minister John Howard responding to the uncertainties of global structural change by advocating a concept of home that provides a "sense of security": "The loss of security challenges traditional notions of home and people feel the need to react to alienation. Part of the job of a Prime Minister in these contemporary times is, whilst enthusiastically embracing change and globalisation, he or she must embrace what is secure, what people see as 'home.' I want to provide Australians with this security as we embrace, as we must and will, a new and vastly different future" (Burke 2001, 186–87; Burke 2002, 24). Burke notes how the desire for a home may be also seen as a desire to displace attention from the failure of public policy to deal with structural change. This desire for some sort of home is a response to a failure of public policy to provide people with "a rightful and secure position in society, . . . a space unquestionably one's own . . . where the rules do not change overnight without notice" (Burke 2002, 24, citing Bauman 1997, 26). In Rancière's terms, the failure of public policy may be read as a failed mapping of the social. A mapping that defines individuals in terms of particular purposes or functions is proven inadequate to address how society is affected by long-term global structural change.

Ian Duncanson argues that this conception of home serves as the basis for thinking in exclusionary and demonizing ways about unauthorized arrivals (Duncanson 2003). He outlines a "trope of the ordinary Australian" that serves as the basis for interpreting and judging refugee stories without actually needing to listen to these stories. The "ordinary Australian" occupies a space of "sameness and greatness" (Duncanson 2003, 31). Duncanson cites John Howard, "One of the great things about living in Australia is that we're essentially the same. We have a great egalitarian innocence" (ibid., 31). Repeating the trope of invasion anxiety pursued by Burke, Manzo, and D'Cruz and Steele, Duncanson suggests that this space of sameness and greatness is animated and outlined by another trope: that of a great horror. Internal bliss is contrasted with external fear and horror; there is a desire thus to structure that external uncertainty, whether it is the hostility of the land, Aboriginal threats at the time of conquest, or racialized migrants and asylum seekers: "The refugees, fragmented, dislocated, 'out there,' confirm my being as unified, placed, 'in here,' at the same time as they threaten my identity with their implied numbers and sheer otherness" (ibid., 32–33). The emplacement of identity in a territorialized home is both confirmed and placed under threat by the "sheer otherness" of refugees.

Duncanson notes a class divide here, one apparent in recent governmental attempts to outline a real or proper Australian people *(that great egalitarian innocence)*. In the Rancièrian terms pursued here, the "ordinary Australians" are those subjectivities that are instrumentally and functionally perceptible in the dominant political formation. Rather than focus on cosmopolitan élites who engage in economic or cultural arguments both in favor of and against migration, Duncanson (and Burke) note that the current Australian government has outlined a figure of "ordinary Australians" sitting in judgment against refugees. This figure of sameness and greatness provides the basis for thinking and judging refugee movements into Australia, while precluding the need to investigate individual stories—favoring a visceral reaction against a threat by sheer otherness to home and a way of life.

A politics of place thus has two aspects. As Ahmed and Fortier note, these are the creation of a sense of home and the maintenance of borders against outsiders. Regarding the former, the disciplining role of landscapes in imparting norms of behavior is important. Allaine Cerwonka traces the disciplining of minority ethnic groups in two sites in Australia:

> Bodies are managed and nations [are] physically constructed in everyday constructions and cultivations. The spatial practices in this study indicate that in many ways the bodies and social position of so-called ethnic groups in Australia are being managed, despite popular support for multiculturalism. Through police regulations and surveillance, for instance, and through community regulations aimed at preserving heritage, "ethnics" are assigned and reminded of their place in national community. (Cerwonka 2004, 232)

Cerwonka studies spatial practices in the East Melbourne Garden Club and in Fitzroy Police Station that recenter white culture through processes of police surveillance and discourses on "heritage" preservation. *Home* is being recreated through the dual disciplinary practices of control and social memory in the face of perceived threats posed by "ethnics" to a population perceiving itself "on the margin of what it considers its own nation" (Cerwonka 2004, 232). This is thus a practice of clarifying a place-based politics; it is based on attempts at sharpening the border between belonging and non-belonging. The nature of landscapes is such that dominant meanings can only be perceived in terms of other meanings, meanings that it considers deviant, in the face of its own normalcy, and upon which it relies for its clarity and vindication (Bunnell 2002). The attempt at clarifying the border between insider and outsider itself constitutes a fundamental relation between *inside* and *outside*.

Different imaginings of Melville Island also point to attempts to draw borders between insiders and outsiders and to the different imaginings of an Australian people placed within the community. Senator Andrew Bartlett, leader of the Australian Democrats, highlighted a concern about a "logical extension" of excising parts of Australia: "A logical extension of excising islands is excising everywhere. There is nothing special about an island compared to the mainland in a legal sense and they already have rules that prevent some people who are on the mainland from being able to apply to seek protection" (SMH 2003). In an editorial titled "Diminishing the Land of Australia," the *Age* newspaper wrote: "Excising almost 4000 islands is an extreme reaction to the threat posed by the 14 passengers on the Minasa Bone. The Government's response plays to unfounded domestic fears, encourages xenophobia and should be condemned" (*Age* 2003b). The reactions of the *Age* and of Senator Bartlett point to more expansive responses in the undercurrent of Australian public life toward asylum seekers, responses that Howard might dismiss as those of elites out of touch with "ordinary Australians."

Senator Bartlett's "logical extension" argument, and the title of the editorial in the *Age,* invoke a tradition of nationalism centered on the wholeness and integrity of a unique land. Sneja Gunew writes of how postcolonial anglophone nationalism focused on the strange land and landscape of Australia to promote ideas of national uniqueness: "What, after all, differentiates a postcolonial Anglophone national culture if not 'the' land, the uniquess of the landscape" (Gunew 1990, 99).

The geography of postcolonial anglophone Australia is, however, a contested one. The land and landscape was not won without some brutality. Jane Jacobs looks at the contest over the meaning of places in contemporary Australia, taking note of the resistance of the state of Victoria to allowing an aboriginal mosaic to stand near the law courts: "The mosaic at the Law Courts did not address an injustice safely contained in a historical event. It spoke to a far more persistent and ubiquitous notion of injustice—one that extends to the present and right into the ordered confines of the Law Courts themselves. Being reminded of the 'injustice' of the law was possibly not what the modern Austrailan legal fraternity wanted to hear" (Jacobs 1997, 214). Australian histories of forceful dispossession are transformed through narratives and paeans that focus on the land. Heroes of such songs to the nation are invariably white and male, with a preternatural affinity for "the bush" (Manzo 1996, 202). Sneja Gunew writes that "the land . . . 'speaks' most authentically through the oral literature of the indigenous nomads: in translation" (Gunew 1990, 99). The appropriation of aboriginal cultural symbolism and of Aboriginal ways of living in the bush by anglophone Australian literature in the early years of the Commonwealth entrenched the priority of white Australia. This is reflected, politically, in the Australian Natives Association, founded in 1871 and important in the popularization of the Federation (or independence) project (Irving, 1999). The Natives Association comprised solely white, Australian-born men, and rested on the doctrine of *terra nullius* to entrench the dissociation of aboriginal peoples from the land.

The contrasting response of Shadow Immigration Minister Stephen Smith and of indigenous Tiwi Islanders to the excision of Melville is telling. Tiwi Islanders are inhabitants of Melville Island (part of the Tiwi group) and, reportedly, were the first Australians to make contact with the *Minasa Bone*. The Shadow Minister railed at the government's surrendering of parts of Australia:

> Senator Vanstone has tried to make a point, of . . . if you excise Christmas Island and Ashmore Reef, then why not Melville Island. Well, the truth is, Melville Island is a stone's throw from Darwin and there is a limit . . . to how much you can rely upon legal technicality. The Government says, 'Let's excise essentially, every island from half-way up the north coast of Western Australia to the Coral Sea.' That is essentially a surrendering of our borders to people smugglers.

It's the modern day equivalent to the Brisbane Line. We might just as well . . . excise Rottnest and Tasmania or really go the whole hog and say, "Let's excise the mainland island and leave Tasmania only as part of our migration zone." (ABC 2004)

Smith's dismay at the excision by the Howard government is one that repeats and reinforces the sovereign right of the Australian Commonwealth to jurisdiction over the land. The unnamed (at his or her request) Tiwi Islander, in an interview with the *Green Left Weekly*, points to the nub of the matter, which Smith cannot recognise. "We watch the news and read the paper. We're not stupid people, we're educated. We know what it means to be non-Australians. If that boat comes back, we'll welcome them and give them food and water. You know why? Because we're all one group—non-Australians" (Hodson 2003; Perera, chapter 9 in this volume).

The above examples perhaps point to the dispute inherent in establishing a place-based politics. The politicians cited speak, in different ways, to a community of care and responsibility, one overwhelmingly national in character; there is a sense of common care and responsibility for one another and for the integrity of the landscape and territory of Australia. However, as noted by Ahmed and Fortier, such concepts of care and responsibility are conjoined with a political performance by the citizen that demonstrates his or her likeness with his or her fellows as well as his or her fundamental difference before outsiders. The demand is not only that care and responsibility be demonstrated, but also that belonging be performed to cohere the community against the threat posed by outsiders. The descriptions of the landscape of Australia by both Bartlett and Smith act as disciplinary technologies of subjectification (that enact particular performances of identity). There is a demand inherent in their descriptions of Australian territorial integrity that belonging be performed and that it be performed through the mantle of citizenship. Taking up such a mantle connects individual subjectivity, national or civilizational identity, and the state in a symbiotic relation. Such a multiscalar linkage, which provides a complex idea of home and a politics of place, must also, however, be premised on the implicit or explicit disavowal of other conceptions of the landscape of Australia, ones that precisely break the links among individual, community, and geopolitical power, such as those of the Tiwi Islander interviewed.

In this section, I have tried to show how a politics of place in Australia is developed through a complex interaction among individual, community, and state. I have noted that this sense of place is premised on creating a foundation of certainty amid the threats perceived to be posed by internal or external outsiders. I have noted that this politics of place emanates a structure of recognition that leads to certain subjects being recognized as political, and certain issues becoming political questions. I have noted, but not in detail, that this place-based politics is not viable, because the border between inside and outside is called into question by the very act of constituting place and home. In the next section, I will explore this idea further by looking at Rancière's conception of place and politics, noting in particular that society cannot be understood as static. Place politics is always premised on an engagement with otherness, an engagement that leads to any particular vision of society being episodic, malleable, in flux. The consequence is that restrictions on belonging and the temporal controls that these require (the cultivation of a particular all-encompassive social memory) are untenable. Practically, what is untenable is the exclusion of the migrant or refugee (from the networks and practices of care and responsibility) because of a prior conception of a foundational place for conducting a limited politics. This place, as I shall argue, is always premised on a fundamental dispute about limits and borders on identity and belonging.

RANCIÈRE'S PLACE

In his analysis of the spatial location of politics, Rancière talks about "police" and about "politics": both have particular but different relations to place; they are similar, though, in that both focus on the distribution of elements within space. Both police and politics see space as constituted by a particular distribution of elements and functions that make sense and are logically perceptible within the space (and are not necessarily sensible outside of that space). The essential difference between the two is that *police* involves a form of counting that sees political community as a sum of its parts. Each part has a particular definitive function that makes sense in terms of the aggregate community. For Rancière, politics, or politics as process, is one that counts the "no-part," that which has no part in political community; politics for Rancière is about the process of dissension that distances the community from itself by making visible that which has no reason to be seen. If *police* is about the clarification of the spatial location of politics, then *politics as process* asks questions about the constitution of a unitary world of politics. The essence of politics, Rancière says, is to manifest *dissensus,* "the presence of two worlds in one" (Rancière 2001).

Although politics as police is centered on the clarification and securing of place, politics as process is not place-centered. That is, it is not a particular locale that gives us politics. Dissensus for Rancière is about the clash of different spatial formations or different ideas about the distribution of elements within space. This is the partition of the sensible that Rancière speaks about: the definition of the forms of partaking in politics by, first of all, defining the modes or manner of perception within which forms of partaking must function.

The partition of the sensible is the operator of a particular mode of subjectification and political partaking. Politics as police is a form of counting that denies the contingency of this arrangement. The partition of the sensible undertaken by politics as police is characterized by the constitution of groups "dedicated to specific modes of action, in places where these occupations are exercised, in modes of being corresponding to these occupations and these places" (Rancière 1999). What is missing is what Rancière calls the void, or supplement: an accounting of the no-part, that which has not been given a function within the partition of the sensible.

The vindication of the place-based politics of the police rests on its ongoing capacity to deny the contingency of its partition of the sensible (in other words, the contingency of its concept and politics of place). This denial itself rests on a form of counting that ignores or passes over that which has no reason to be seen. The partition of the sensible defines the modes of perception in which politics must operate. The vindication of this participation is the performance of political subjectivities in accordance with the rules and functions attributed to them by the partition. Thus, it is the ongoing performance of subjectivities of the functions attributed to them that, teleologically, maintains political space.

More fundamentally, the functionality of subjects rests on their continued representation as unitary subjects who can be understood, above all, in terms of the clarity of their allotted functions. This representation rests on a form of counting that refuses to perceive the imperceptible (that which has no reason to be seen). The sense that subjectivity may be other than it is, and that politics may thus also be other than it is, is policed against by a form of control that refuses to see certain acts or groups as politically consequential.

> Police intervention in public spaces does not consist primarily in the interpellation of demonstrators, but in the breaking up of demonstrations. The police is not that law interpellating individuals (as in Althusser's "Hey, you there!") unless one confuses it with religious subjectification. . . . It is, first of all, a reminder of the obviousness of what there is, or rather, of what there isn't: "Move along! There is nothing to see here!" The police says that there is nothing to see on a road, that there is nothing to do but move along. (Rancière 1999)

For Rancière, a place-bound politics is one premised on a structuring of society where a form of "identificatory distribution (naming, fixing in space, defining a proper place) is an essential component of government" (Dikec 2005, 186). This creates a sense of the visible and the sayable. Certain things and certain acts may be recognized as political, and those that have no basis from which to speak or to be seen are imperceptible. The maintenance of a place-based politics, (a politics of home, in Connolly's terms), rests on the continuation of the system of recognition given by a particular purposive structuring of space. "If there is someone you do not wish to recognize as a political being, you begin by not seeing them as the bearers of politicalness, by not understanding what they say, by not hearing that it is an utterance coming out of their mouths" (Rancière, 1999). The space of politics becomes a static space. It is one that is performed and thus maintained within a closed order. That is, it relies on the ongoing performance by subjects of the functions allotted to them, and this itself depends on a policing of the perceptible. This is a policing that controls what may be seen within the different segments of political space; this is a way of seeing that recognizes something in terms of its functionality and proper place within the wider partition of the sensible. That which is afunctional, or has no particular reason for being seen, is imperceptible within this closed system that only recognizes that which is sensible to it: only that, in other words, that contributes to the sum of the greater community is counted as political.

The stasis of society, the control over what can be seen, involves then a temporal forgetting. The structuring of a politics of place involves an appropriation of history, as well as the spatialization of time, where duration is experienced within the functions created by the partition of the sensible. Social subjection, argues Rancière, is sustained by a particular partitioning, or ordering, of how time is experienced (Rancière 2003, 5–6). The obvious partitioning of time, says Rancière, is that workers work during the day and rest during the night. Time is experienced in terms of the functions attributed to particular subjects. The ordering of time is vital for the maintenance of the static social space of the police. There is no place for a void—that which has not been given an order or a structure; there is no place, in politics as police, for an unconquered time within which all sorts of subjectivity may compete for perceptibility.

The retrospective excision of Melville Island was a reassertion of statist control over time, a reassertion designed to prevent or suppress the perceptibility of the problematic political subjectivity expressed by the *Minasa Bone*'s passengers. The retrospective excision, in John Howard's terms, rendered irrelevant the speech of the passengers, including any asylum claims that may have been made. It is a forceful location of the place of politics; it is a reassertion of the temporal order that gives the possibility of not recognizing someone as political.

EXPELLING THE *MINASA BONE*

The Howard government had initially suggested that the passengers on the *Minasa Bone* did not claim asylum. Ten days later, on November 14, the government backtracked and admitted that a number of the passengers communicated their intention to claim asylum—by pointing to the word *refugee* in a dictionary and by saying "human rights" (Forbes and Shaw 2003). The government's response to this is telling. The prime minister, John Howard, in London at the time, became embroiled in a discussion over whether or not it was relevant that a claim for asylum had been made. Speaking on national radio, Howard said, "It doesn't really matter. The key thing here is that at the time any so-called application for asylum might have been made, the islands had been excised" (Grubel 2003). At the time asylum claims were made, the islands were no longer a part of "Australia" (for purposes of the Migration Act).

The curtailing of political space that leads to the possibility of seeing some individuals as irrelevant to the politics of place of Australia
rests on a concrete and visceral *cutting* of Australia. That this cutting takes place retrospectively demonstrates the control of time to
which the state aspires. It is a control centered on the dismissal of a
void where subjectivity unstructured by territorial time may be apparent. It demonstrates that the control of the way time progresses is
perceived as integral to the maintenance of statist domination over
society. In another sense, the capacity of the state to look back and
change the course of events by rendering imperceptible what was, momentarily, perceptible points to Rancière's politics as police, a form
of counting that rests on a particular ways of naming. This naming
is based on legislation that institutes, in Rancière's terms, a *wrong*,
a restriction of rights. The restricted political space of Australia is
grounded in legislation that imposes an apparatus of recognition.
This is an apparatus centered on a particular structuring of society
and its containment within a particular landscape. This society is
taken as a definitive adjudicator of political meaning.

The people on the *Minasa Bone* were escorted back to Indonesia
only four days after landing. Forty-eight hours after the foreign
minister, Alexander Downer, justified this expulsion of would-be
asylum seekers by saying, "Now, we don't know a great deal about
these people; they didn't claim asylum in Australia while they were
in Australian waters" (Gordon 2003), Australian television interviewed one of the boat people, Abuzer Goles, in a detention center
in Jakarta. In response to a question whether he had asked for asylum, he replied, on television: "Thousands of times, thousands . . .
I begged them, I pleaded down on my knees. They sent a Turkish
interpreter and I pleaded with him saying I'll do anything not to be
sent back. We spent four days on the water, ten days without sleep,
it nearly killed us. I'm human, I'm a human being. I'm a refugee"
(Gordon 2003).

Claims for asylum, claims that were in fact made, are dismissed
retrospectively. Actual claims are now "so-called," suddenly lack
legitimacy; indeed, it is as if they had never been made. Howard's
statement draws us, again, to the nub of the matter; it draws us into
considerations of place, of speech, and of legitimacy. It takes us further than a construction of not-Australians; indeed, it draws us into
a consideration of what it is to be human.

The right to retrospective excision preempts the chaos of time. Such a right invests the state with the privilege to recast how time has passed. Walter Benjamin argues that history is imbued with the "presence of the now" (Benjamin 1999, 252–53). Images of the past are conceptualized in ways that make sense to, and reemphasize the concerns and limits of, the hegemonic view of the present condition. Those images that appear counter to a state-centric vision of history and humanity need to be reconceptualized or forgotten. History is an exercise in representation and interpretation; but it is rare that such an instance of representation and reinterpretation, replete with hegemonic conceptualizations of belonging and nonbelonging, becomes as clear as it does in the case of the *Minasa Bone* asylum seekers. History, as the interpretation of events, can potentially hold time in stasis, as the retrospective exclusion of Melville Island demonstrates. History is to be codified into meaningful images that fit the ends of state-centric histories, geographies, and politics. The image of legitimate asylum claims coming from an illegitimate entry into Australian sovereign space does not square with other images of boundaried and controlled sovereign space. There is a vested interest in erasing from images the palimpsest of different times and geographies and the plethora of different interpretations that ensue therefrom (Esch 1999, 2). As Benjamin says, "Every image of the past that is not recognized by the present as one of its own concerns threatens to disappear irretrievably" (Benjamin 1999, 247).

The legal right to retrospectively excise, to retrospectively make judgments on what has gone before, and to reserve the right to pretend that events have never occurred is a manifestation of the desire to contain history and "a people" within territorial bounds. Indeed, it is a manifestation of the sense that the right to speak, the right to *be,* is given or not given by the state and the structures of legitimacy it possesses. To speak is to accept and be accepted by these structures and their temporal and spatial horizons. The control of time investigated here involves a control over both the integrity of the social space and the definitive political subject upon which the Australian state's claim to legitimacy rests. In Rancière's terms, the static space of politics as police rests on the maintenance of structures of recognition that allow *some* utterances and some forms of life to be seen as political. These are utterances and subjectivities that have a functional space within the partition of the sensible.

The territorial partition of the sensible would be under threat from a time consciousness that refused containerization within the state and manifested this refusal in extraterritorial communities and solidarities: that expressed itself in Goser's lament "I'm a human, I'm a human being." A human being without an address who calls to mind the restrictions on being that underline the territorial enterprise.

But the possibility of this extraterritorial threat has been, at least temporarily, precluded by the Australian government, who can act in retrospect. Such is the nature of the preemptive strike, weapon of choice in our suddenly insecure world. The Australian government makes a preemptive strike on the border communities and solidarities unfolded by the advent of the *Minasa Bone* and the bare claim to human rights. It is a slightly delayed preemptive strike, but is probably more honest than other "preemptive" strikes when we never know for sure that there is or will be something to preempt. Acting retrospectively, the Australian state fantastically undertakes a preemptive strike on that which has already taken place.

CONCLUSION: A POLITICS OF BECOMING?

The creation of an affective connection between a people and a particular landscape of home rests on a particular form of political partaking. This is a partaking that implies both division and participation. The landscape of Australia is made out to make sense to a generalized community of "ordinary Australians." This community is run through with relations of care and responsibility designed to maintain the community, which becomes an end in itself. The landscape of Australia is thus the subject of a division, one that creates a dominant spatial formation sitting instead of other potential spatial formations, and that serves as the basis to identify and judge outsiders, like unauthorized boat arrivals. The dominant landscape of Australia reflects a particular politics of place; as it divides to create a spatial formation that provides an affective link between "a people" ("ordinary Australians") and the landscape, so too does it generate exclusion.

This exclusion, as I have tried to demonstrate, relies on a control of time. It relies on a sense that subjectivities of "ordinary Australians" are preformed and discoverable repositories of rights and identity. It is a containment of what it is to be within a temporal and spatial horizon. In William Connolly's terms, it is the institution of a politics of Being in the stead of the possibility of a politics of becoming. This institution is centered on a "politics of realisation of an essence or universal condition already known by reasonable people" (Connolly 1999, 130), for example Ian Duncanson's trope of ordinary Australians.

A politics of becoming, by contrast, is a politics that disavows a search for essences and homes. It is a politics of movements and flows, of identity games and fragments. It is a politics centered on the possibilities of and at the border, from whence disruptive claims, communities, and identities may extrude onto the happily settled home: "the politics of becoming . . . sows disturbance and distress in the souls of those disrupted by its movement" (Connolly 1999, 136).

Rancière suggests that the goal of politics as process is to separate the community from itself. That is, its goal is to ask questions about the functional distribution of identity over space, a distribution that vindicates the enforced location of a so-called proper politics and the identification of those who are always already outsiders. This distribution involves, most importantly, the refusal to think *the void*. It is a refusal to think about how a partition of the sensible involves the noncounting of those deemed to have no part in the partition.

Rancière's politics of process and Connolly's politics of becoming are ways of thinking the no-part that strives to expand political space. The asylum seeker Abuzer Goles makes a claim to political solidarity and empathy based on a common humanness. Such claims cannot be heard within the Australian partition of the sensible that hears only the speech of those who have a part in it. This result points to affixing of a temporal horizon on the meaning of humanity.

Such representations may tend toward the despairing: if the meaning of humanity has been circumscribed, what hope is there for political expansion? Rancière, however, argues that the restrictive form of a partition of the sensible, the "police" form, depends for its ongoing validity on a static sense of social space. Such stasis only comes about if we see the repositories of political power and political rights as preformed individuals, existing almost metaphysically above society and its changes (such as the "ordinary Australians" trope). Rancière suggests that political rights exist in two senses, written and exercised. In their written form, political rights are not endowed or vested in a particular individual; they are, rather, "open predicates." Rights, such as the right of participation, are not "predicates belonging to definite subjects. Political predicates are open predicates: they open up a dispute about what they exactly entail and whom they concern" (Rancière 2004, 303). Rights are open predicates because the subject in whom they are supposedly vested is similarly open: "Man and citizen do not designate collections of individuals. Man and citizen are political subjects. Political subjects are not definite collectivities. They are surplus names, names that set out a question or a dispute . . . about who is included in their count" (303).

The stasis of social and political space depends then on a misreading of the political subject. That subject is not definitive but is constituted and reconstituted over time by different means of exercising or using rights. The borders of the space of politics, community, justice, and belonging are not closed; they are open to dispute. Parallels may be found in the cultural geography literature, where space is understood as a landscape whose meaning is always under contest (Bunnell 2002; Kong and Law 2002).

Politics as process is thus the deployment of an alternative partition of the sensible against the dominant police form. It is a deployment that, most important, separates the community from itself, from the foundational points of definitive subjects that locate political space. Politics as process seeks to expand political space and to question its borders not by positing an alternative definite subject of politics against the police form, but by investigating the border between inclusion and exclusion. Rather than positing an alternative definite subject, politics as process seeks to count the no-part: those who have no proper place or function in the partition of the sensible. In this counting, the onus is on using the language of rights in imaginative ways to question the foreclosure of the ambit of justice, community, and politics.

The unnamed Tiwi Islander, in the previously given quote, points to a solidarity of those who have no part in the Australian partition of the sensible. She or he points to a solidarity of those who have been read out of the temporal and spatial foreclosures of the particular spatial formation and understanding of the home of the Australian landscape. From this solidarity, in a place of negativity, comes the basis to think to and against the premature foreclosing of what it is to belong and partake in Australia. This basis asks questions about the validity of a territorial foreclosure of rights and of the control of time that runs through it.

WORKS CITED

Age. 2003a. "Asylum Claim Irrelevant: Howard." November 14.

———. 2003b. "Diminishing the Land of Australia." November 7.

Ahmed, Sara, and Anne-Marie Fortier. 2003. "Re-imagining Communities." *International Journal of Cultural Studies* 6, no. 3: 251–59.

Australian Broadcasting Corporation (ABC). 2004. *Insiders: Transcript of Interview with Stephen Pile,* ABC TV 7, March 2004. http://www.abc.net.au/insiders/content/2004/s1060560.htm (accessed November 21, 2005).

Bauman, Zygmunt. 1997. *Postmodernity and Its Discontents.* London: Polity.

Benjamin, Walter. 1999. *Illuminations.* London: Pimlico.

Bigo, Didier. 2002. "Security and Immigration: Toward a Critique of the Governmentality of Unease." *Alternatives: Local, Global, Political* 27 (Special Issue): 63–92.

Bunnell, Tim. 2002. "*Kampung* Rules: Landscape and the Contested Government of Urban(e) Malayness." *Urban Studies* 39, no. 9: 1685–1701.

Burke, Anthony. 2001. *In Fear of Security: Australia's Invasion Anxiety.* Sydney: PlutoAustralia.

———. 2002. "Prisoners of Paradox: Thinking for the Refugee." *Social Alternatives* 21, no. 4: 21–27.

Cerwonka, Allaine. 2004. *Native to the Nation: Disciplining Landscapes and Bodies in Australia.* Minneapolis: University of Minnesota Press.

Chowers, Eyal. 2002. "The Physiology of the Citizen: The Present-Centered Body and Its Political Exile." *Political Theory* 30, no. 5: 649–77.

Connolly, William. 1999. "Suffering, Justice, and the Politics of Becoming." In *Moral Spaces: Rethinking Ethics and World Politics,* ed. D. Campbell and M. J. Shapiro. Minneapolis: University of Minnesota Press.

D'Cruz, Vin, and William Steele. 2003. *Australia's Ambivalence towards Asia.* Clayton: Monash Asia Institute.

Deranty, Jean-Philippe. 2003. "Jacques Rancière's Contribution to the Ethics of Recognition." *Political Theory* 31, no. 1: 131–56.

Dikeç, Mustafa. 2005. "Space, Politics, and the Political." *Environment and Planning D: Society and Space* 23, no. 2: 171–88.

Duncanson, Ian. "Telling the Refugee Story: The 'Ordinary Australian,' the State of Australia." *Law and Critique* 14, no. 1: 29–43.

Esch, Deborah. 1999. *In the Event: Reading Journalism, Reading Theory.* Stanford, Calif.: Stanford University Press.

Forbes, Mark, and Meagan Shaw. 2003. "Kurds Sought Asylum: Vanstone." *Age,* November 14.

Gordon, Michael. 2003. "Another Case of Truth Overboard?" *Age,* November 15.

Gunew, Sneja. 1990. "Denaturalizing Cultural Nationalisms: Multicultural Readings of 'Australia.'" In *Nation and Narration*, ed. Homi Bhabha. London and New York: Routledge.

Hodson, Michael. 2003. "Tiwi Islanders: 'We're All Non-Australians.'" *Green Left Weekly*, November 19, 2003.

Irving, Helen. 1999. *To Constitute a Nation: A Cultural History of Australia's Constitution*. Cambridge: Cambridge University Press.

Jacobs, Jane. 1997. "Resisting Reconciliation: The Secret Geographies of (Post)-Colonial Australia." In *Geographies of Resistance*, ed. Steve Pile and Michael Keith. London and New York: Routledge.

Manzo, Kathryn A. 1996. *Creating Boundaries: The Politics of Race and Nation*. Boulder, Colo.: Lynne Rienner.

Rancière, Jacques. 2001. "Ten Theses on Politics." *Theory and Event* 5, no. 3.

———. 2003. "The Thinking of Dissensus: Politics and Aesthetics." Paper presented at Fidelity to the Disagreement: Rancière and the Political, conference at Goldsmiths College, September 16–17, 2003. http://homepages.gold.ac.uk/psrpsg/ranciere.doc (accessed May 8, 2006).

———. 2004. "Who Is the Subject of the Rights of Man?" *South Atlantic Quarterly* 103, no. 2–3: 297–310.

Shapiro, Michael. 2003. "Radicalizing Democratic Theory: Social Space in Connolly, Deleuze, and Rancière." http://homepages.gold.ac.uk/psrpsg/shapiro.doc (accessed May 8, 2006).

Sydney Morning Herald (SMH). 2003. "Island Excision Thrown Out: Hunt for New Plan." November 25.

United Nations High Commissioner for Refugees (UNHCR). 2004. "UNHCR's Views on the Concept of Effective Protection as It Relates to Indonesia." *UNHCR Position Papers* (update January 2004a). http://www.unhcr.org.au/UNHCR-protlegal-EPIndonesia.shtml (accessed November 22, 2005).

Vanstone, Amanda. 2003. "Minasa Bone Returns to Indonesia": Joint Media Release with the Minister for Foreign Affairs, Alexander Downer. VPS 006/2003. Minister for Immigration and Multicultural and Indigenous Affairs. http://www.minister.immi.gov.au/media_releases/media03/v03006.htm (accessed May 8, 2006).

Warner, Daniel. 1999. "Searching for Responsibility/Community in International Relations." In *Moral Spaces: Rethinking Ethics and World Politics*, ed. David Campbell and Michael J. Shapiro. Minneapolis: University of Minnesota Press.

12

Border's Capture: Insurrectional Politics, Border-Crossing Humans, and the New Political

NEVZAT SOGUK

> I carry two worlds within me
> but neither one whole . . .
> the border runs
> right through my tongue.
> —ZAFER SENOCAK, "DOPPLEMANN"

Borders have lives of their own. They move, shift, metamorphose, edge, retract, emerge tall and powerful or retreat into the shadows exhausted, or even grow irrelevant. They are not simply fences, walls, and chains that divide the earth's surface into sovereign territories, simple in purpose and function as they appear on a world map.

True, they exist seemingly lifeless in the vast openness of geographical landscape, not shifting, always immobile, always steady in time and place. However, their appearance belies the dynamism that underlies the calm of their surface appearances. Borders can come alive and either make way or make trouble for the sojourner or traveler. They come alive through the intentionalities applied to them. They are imbued through and through with intentionalities that not only construct and empower borders but also open holes in them and violate them.

Yet, strangely, even as they exist out there, borders' existence

283

appears all but devoid of consequence and meaning without the
support of the intentionalities that erect them in place and attend
to them through time. When enforced, borders can grab you. Left
unenforced, they are swallowed up and smoothed into the transver-
sality of space. They are recouped and recast in ways useful to the
dreams and desires of those who navigate them. David Chorlton
(1984) captures this political economy in border poetics:

> A refugee has crossed
> so many borders, he becomes
> Invisible where countries change
> their names. When he stops
> in the shadows to catch
> his breath, pieces
> of a border lace his shoes

The border laces the refugee's shoes and makes him invisible. It be-
trays its primary raison d'être in the momentary relationship with
the refugee. Recast and appropriated in the refugee's survival intentionality, it serves as an instrument of transition, not of obstruction.
It supports fugitive feet; it becomes a resource.

In short, borders acquire their meanings always contingently,
through the activities and practices undertaken around and through
them. They are consequential only where and when border practices
are at work, making a border out of a fence or digging a border out
of a ditch. Thus understood, borders are always ephemeral, never
eternal.

If there is any constant to borders in time and place, particularly
in the order of the national territorial state, it is the logic of the statist and territorial governmentality, which political borders are compelled to reflect and embody. Often overlooked in those studies of
borders that take actual fences and walls as sites of a border's enunciation and actualization, this statist logic is nevertheless central to
border practices as limit markers. It is this logic that lurks behind
fences and walls. More important, it is the same logic that shows us
how fences and walls are not the only borders that can be deployed
as borders. This logic shows us how fences, walls, and ditches can
be transfigured, moved, mobilized, and deployed as limit markers in
political space. They are camouflaged and concealed in other forms.
Culture, race, class, gender are all appropriated as camouflage in the
reproduction of borders creatively and resourcefully in unexpected

places and surprising forms. Cultural fences, class walls, and gender ditches emerge as new and powerful borders even as they never announce themselves as borders.

Many of the essays in this book highlight these dynamics now prevalent across the world. Among others, the chapters by Didier Bigo, Elspeth Guild, Alice Nah, Suvendrini Perera, and Decha Tangseefa explore border lives and border geographies that are increasingly conditioned in paradoxes and contradictions reflective of the old and the emergent modalities of territoriality and political community. Significantly, despite all of their differences in theory, method, and substance, these works point to remarkable experiential parallels in contemporary borderscapes that are ever more transversalized, yet still remain within the ambit of statist governmentality working to impose new regimes of control and hierarchy. Whether in Bigo's work on *Banoptican* and the detention of foreigners or in Decha's study of the paradoxes of im/perceptibility attendant to Karens' life worlds, the statist territorial logic proves both resilient, inventing and erecting new camouflaged borders, and forever insufficient to the challenges and pressures relentlessly issuing the political anew.

So borders proliferate burdened to reflect the statist territorial logic of not only territorial confinement but also bipolitical regimentations. They emerge in a shifting mélange of political, administrative, sociocultural, aesthetic-poetic, and political-economic interventions. They constantly "unfold" in time, and across place and space, in form and content. States' borders are not simply found in a fence or a ditch but also in the resourceful and ever-shifting border practices permeating space both within the confines of fences and across the barbed wires in everyday sites. A fence, as border, can shift and move in multiple directions and metamorphose into practices that capture people in labyrinths of political regimentations as if encircled by a fence. A border can move inward and become a policy of denial of rights to migrants and refugees. Or it can fold outward and translate into a policy of intercepting refugee ships and forcing them to return to worlds of insecurities. It is in this sense that I say borders are alive, mobile, resourceful, and operating to multiple rhythms under different temporal and spatial conditions. They are practices that work to capture and regulate contingencies. This is their strength.

Still, as we are reminded in the Sucker's "Journey to the Sky and Back Down" in the aboriginal Salish story—"Watch what you grab, it might grab you"—border practices, which work to capture contingencies, always run the risk of themselves being captured in these very contingencies. They relate—that is, border to and border on—as much as they separate and disconnect. Therefore, they not only create or highlight existing differences but also reveal the extant commonalities, as well as point to future convergences of relations yet to be formed. Ultimately, borders inexorably reveal as they are revealed, contributing to the opening and maintaining of a field of borderizations. In this dynamism, borders become *borderizations*—practices of relationality—made possible in tensions, conflicts, and contradictions as well as unexpected convergences of intentionalities. In borderizations, they grow ambiguous. Their ambiguities cultivate the new political.

It is borderization as a location, position, and condition that emerges as the site of the "political" not only in terms of constraints but also in terms of promises. In this paper, I turn to borderizations in refugee, asylum seeker, and immigrant worlds. Specifically, I turn to refugee or asylum seeker sites as they engender movements that target the prevailing hierarchies in national and international politics. I call such contra movements "insurrectional" movements born in borderizations. I argue that a border's capture of humans in political, economic, and cultural regimentations are also simultaneously a capture of the border in insurrectional politics. As border practices unfold so does the insurrectional politics, in what Michel de Certeau calls the "illusory inertia" (de Certeau 1988) of the border.

As you see in reading this volume, I am not alone in daring to think that this is the case and that migrant movements are growing ever more insurrectional in unexpected ways. Suvendrini Perera, James Sidaway, Prem Kumar Rajaram, and Decha Tangseefa richly demonstrate contemporary borderscapes as the new political horizon that pressures the familiar political orders of states, territories, and borders. The old may not yet be dying, but the new is obstinately being born in insurrectional movements, especially those cultivated in migrant worlds.

BORDER AS "ILLUSORY INERTIA":
MICHEL DE CERTEAU ON BORDERS

De Certeau situates his critical interrogation of the border by appealing first to the map. The map, de Certeau argues, is an "operation" marking out boundaries and borders in modern politics. It is a sine qua non in imagining the world through the nation-statist territorialities. The map issues forth and reifies borders. In turn, borders underlie the map's authority, helping to express it in its sterilized form. "The map" de Certeau argues, "colonizes space" (de Certeau 1988, 121). It imposes a narrow, uniform code of comprehension by eliminating the crisscrossing, itinerant intelligibilities, which in reality mark the grounds of operations. The map produces abstractions where concreteness calls for recognition and registry. The map projects an insular order where transitions, translations, and openings characterize social, economic, and political landscapes. Finally, it is deeply ahistorical, where history in temporal and spatial happenings appears most manifest and dynamic. As de Certeau puts it, the map "wants to remain alone on the stage" (ibid.). The map wants to remain central to modern imagination.

Paradoxically, for all the work and activity, the map has to be silent about the work that produces it. "The map," writes de Certeau, "functions as a theater, a totalizing stage, on which elements of diverse origin are brought together to form the tableau of a 'state' of geographical knowledge [but only by] push[ing] away into its prehistory or into its posterity, as if into the wings, the operations of which it is the result or the necessary condition" (ibid.).

Working to obscure the traces of its own genealogy as a political intervention in physical and human landscapes, the map confidently conveys a status of finality and stability about what it marks as the borders in the otherwise transversal geography, as if nothing is moving, changing, or shifting. Borders on a map appear not only stationary but also as if "lifeless": they simply stand there, in a certain "inertia," mechanically expressing the fundamental separation of presumably already existing distinct cultural, political, and economic life spaces. The inertia, argues de Certeau, is "illusory."

In reality, the grounds of the map and of the borders they express through cartographic stylizations are conditioned and energized through a thousand itineraries and worldly dreams and desires. It is dynamism, not inertia, that characterizes border landscapes. It is a dynamism both permissive and constraining; it affords opportunities, but never guarantees; it poses challenges yet is also productive of openings. That is why, in spite of its central place in modern imagination, de Certeau argues, maps and borders remain unstable operations. They can never fully hide their limits or control the openings their operations facilitate.

Ultimately, de Certeau maintains, what is inexorable in border practices is the relationality, which is revealed even in attempts to control and obscure it. Its logic is one of ambiguity:

> "Stop!" says the forest the wolf comes out of. "Stop!" says the river, revealing its crocodile. But this actor, by virtue of the very fact that he is the mouth piece of the limit, creates communication as well as separation; more than that, he establishes a border only by saying what crosses it, having come from the other side. [He] articulates it. . . . The frontier functions as a third element. It is an "in-between"—a space between. . . . *It privileges a logic of ambiguity through its accounts of interaction. It turns the frontier into a crossing, and the river into a bridge. It recounts inversions and displacements:* the door that closes is precisely what may be opened; the river is what makes passage possible; the tree is what marks the stages of advance; the picket fence is an ensemble of interstices through which one's glances pass. (emphasis mine; de Certeau 1988, 127)

In reality, de Certeau suggests, the border is an intimate coproduction or composition of "smooth" places and places of rupture, even though on the surface it may look highly striated. "In reality, in its depth [border] is ubiquitous." It is everywhere. It is "a piling up of heterogeneous places. Each one, like a deteriorating page of a book, refers to a different mode of territorial unity, of socio-economic distribution, of political conflicts and of identifying symbolism" (ibid., 201).

The logic of ambiguity, which in interaction resists and even inverts the logic of the statist territoriality, produces an ambiguous "spatial syntax" of elements and forces. Ambiguity and indeterminence, born in the "intersections of mobile elements" (ibid., 115), condition and mediate the political space in which the real struggles transform their grounds into specific places of living. For de Certeau, the status quo is never stable or totally dominant. Even the most dominant order can be expropriated in the service of resistances (ibid., 115; Thacker 2003, 31). Ambiguity is productive of transformative politics.

In much the same way, though the border as a marker of sovereign territorial and statist politics in the world may be the predominant claim, the border can neither privilege nor empower separations and exclusions alone. Simultaneously and inescapably, it also reveals the multiple ontologies and knowledges of translations, flows, and transformations. Hear de Certeau again:

The whole made up of pieces that are not contemporary and still linked to totalities that have fallen into ruins, is managed by subtle and compensatory equilibria that silently guarantee complementarities. These infinitesimal movements, multiform activities, are homologous to that "boiling mass of electrons, protons, photons, . . . all entries whose properties are ill-defined and in perpetual interaction" . . . these movements give the illusion . . . of immobility. *An illusory inertia.* (emphasis mine; de Certeau 1988, 115)

Inertia is therefore political. Behind its façade, it hides the border's movements. Behind the surface certainties of the façade, the border is revealed as a dynamic field like that of a "boiling mass of electrons, protons, photons"; its "properties are ill-defined and in perpetual interaction." It is in this field that a host of border activities are orchestrated to capture and organize movement in the service of the dominant governmentalities. Yet it is also in this field that insurrectional movements emerge through people's stories in movement, engendering novel relations that capture and open the borders of confinement. I turn to refugees, asylum seekers, and illegal immigrants to explore insurrectional politics.

CAMOUFLAGED BORDERS AND INSURRECTION STORIES

Writing in an article in the annual report of the United State Committee for Refugees, the United Nations High Commissioner for Refugees described the prevailing conditions of inhospitality confronting refugees and asylum seekers the world over: "Since September 11, refugees and asylum seekers have had even more difficulty than before in finding safety. No corner of the globe has been immune. . . . Increasingly governments exclude (refugees and asylum-seekers) from protection and detain them" (*Migration News,* October 2003).

The commissioner was not exaggerating. Nowadays, "becoming" a refugee or an asylum seeker through legal openings is almost an impossibility, while illegal immigration translates into experiences of overwhelming estrangement from basic rights. When human beings are violently ushered into the twilight zone between their citizen selves and their displaced selves, they are also ushered from their "proper" subjectivity within the fold of a state into the uncertain lots in displacement. In the words of philosopher Giorgio Agamben (1995, 114), movements of the displaced open into the zone of "bare life" in the order of the nation-state system, in which the displaced appear "sacred" but "worldless" bodies, bodies ideal-ized and sacralized in discourses of protection, yet marginalized and excluded from the full possibilities of life in acts of regimentation. Characterized by their bare lives in displacement, refugee, asylum, and illegal migrant movements are inevitably mobilities through insecurity and loss before they can become openings to new lives. Borders in camouflage proliferate in their lives.

Consider the story of Yaguine Koita and Fode Tounkara, two teenagers from Guinea, who were found dead in the landing gear of a plane when it landed in Brussels in 1999. A note was found on them, in which they had anticipated the risky passage they were embarking on: "Excellencies. We . . . write this letter to you to talk about the objective of our journey and the suffering of us, the chil-dren and young people of Africa. If you see that we have sacrificed our lives, it is because we suffer too much in Africa and need you to struggle against poverty and war. Finally, we appeal to you to *excuse us very, very much for daring to write this letter"* (emphasis mine; Sullivan and Casert 2000). Never before told this dramati-cally, their message of life after death arrested imaginations across the world for its daring. Yaguine and Fode were heard only in death, for while alive they were expected to undertake "living," with its grinding poverty and devastating wars, quietly. Instead, their au-dacity spanned the borders that slice through humanity, and ex-posed the lie of human solidarity. Yaguine and Fode knew that they were "bare" and expendable bodies. So they asserted their voice by recording their dying.

The "list of death" is a living document attesting to border prac-tices. New names are daily added to its list of documented dead refugee migrants in the borders of Fortress Europe since 1993. It is a strangely captivating list, paying homage to the dead human beings through the fragments of their journeys of hope cut short:

8/10/02 14 N.N. (7 women, 7 men) Sub-Saharan Africa, presumed drowned, boat capsized near Barbate avoiding detection by SIVE.
22/9/02 14 N.N. (men) Tunisia drowned, forced by smugglers to swim ashore near Scoglitti (south Sicily, I).
2/8/99 1 Koita Yaguine (boy 14) Guinea stowaway, frozen to death in undercarriage of Conakry (Guinea) to Brussel.
2/8/99 1 Tounkara Fodé (boy 15) Guinea stowaway, frozen to death in undercarriage of Conakry (Guinea) to Brussel.
(UNITED 2004)

This remarkable document is a border marker in all directions. It records not deaths alone, but also the borders' shifts and metamorphoses forward and backward, upward and downward. Now a trailer, now an airplane's wheel bay, now a train or a boat—borders take on many ambiguous forms far away from where they lie on maps. "Watch out!" The "French border can reach and grab you in the air over Guinea, or the Spanish border can capture you in a trailer even when inside Spain."

Surely, where refugee, illegal migrant, or asylum seeker bodies fall, they mark borders in their resourceful and rich unfolding temporally and spatially. Bodies fallen, drowned, frozen, mangled, and suffocated highlight borders' capture of people daring to move unauthorized. On the other hand, they also point to the trails through which border-crossing people turn insurrectional, capturing borders and harnessing them to their movements. In this way, while reflecting the dead certainties, stoppages, and terminations effected by the border, they also point to the ambiguities energized through border practices, which manifest passages, continuities, and interactions. Even the dead bodies, to recollect de Certeau, "mark the stages of advance" (de Certeau 1988, 127).

The contours and contents of these migrant movements are shaped largely by the politico-administrative regimentations of human migrancy, which is increasingly state-fundamentalist in spite of, and perhaps due to, the transversal challenges to states' privileges in politics. Aware of the new global imperatives, most states are attempting to smother human migration by erecting new borders in camouflage.

As early as 1995, in its annual report on the state of the world's refugees, the UNHCR observed this phenomenon: "States are increasingly taking steps to obstruct the arrival of asylum seekers, to contain displaced people within their homeland, and to return refugees to their country of origin" (UNHCR 1995, 16). Not surprisingly, such draconian closures of legal openings are compelling many millions into underground networks, into illegal immigratory circuits. A sort of war of positioning ensues, activating migrants' insurrectional desires and capacities.

Reflecting this conjuncture, many European governments have been devising common, integrated asylum policies within the context of the European Union (EU) since the mid-1980s to curb the right of asylum in Europe. From the Dublin Convention to the Schengen

Accords, from the Maastricht Treaty to the Amsterdam Treaty, to numerous ministerial agreements since the 1990s, the European Union policies have made it virtually impossible to receive asylum in Europe, while also severely restricting legal migratory routes. With EU membership now at twenty-five countries, Europe's common policy is set to extend its reach, its camouflaged borders, into greater distances.

Similar sentiments have also swept the United States, particularly after September 2001, leading to the enactment of unprecedented border controls and immigration policies. Initiated as part of a larger effort to fight a "war on terror," some of the policies have come to engender broader implications the world over for the ways in which immigrants, legal or illegal, are cast and treated. However understandable, these policies have intensified the feelings around the figure of immigrant, in effect, narrowing the field of legal migratory possibilities for many.

Observers also argue that in some cases the prevailing attitudes have emboldened governments already predisposed to casting all migrants, not just specific persons, through a language of fear and danger, even of terror. A strange case of such an enabling, a camouflaged border practice going astray, took place in Macedonia in 2002. Seven Pakistani and Indian "illegal migrants" were reportedly on their way to Greece, where they had hoped to find work in Greece's Olympic construction industry. However, they were apprehended by the Macedonian police, "taken to a spot en route to the U.S. Embassy and executed." Subsequently, they were presented to the world as terrorists on the way to strike the U.S. embassy (Bearup 2004). One government later, and after much international pressure, Macedonian officials admitted that the murders were staged as "part of a clumsy plot to try to impress the U.S." What is instructive is that "Macedonia's quest to join the war on terror" could be initiated this way, on migrant bodies seen and treated as bare lives whose killing or abuse did not necessarily constitute a crime in the minds of many.

Across the ocean in Australia, the disdain for asylum seekers and refugees has reached even greater heights. The government not only openly flouts the very foundations of protection regimes by denying asylum seekers the right to be heard, but also relentlessly dehumanizes them through fabricated charges of barbarism. Among the charges was the accusation, now discredited, that asylum seekers and refugees tossed their children into the ocean to force the Australian government to receive them as asylum seekers (Perera, chapter 9 in this volume). Sadly, such charges cast a thin veil of legitimacy over the duplicity of the Australian government in rendering refugees as bare bodies who can be denied the right of refuge and cast into oblivion before the eyes of the world. As Jacques Rancière argues, visibility does not necessarily effect "perceptibility" (Rancière 1998). It effects a shift in the strategies of border making. It makes borders move. For example, "excision" of refugees from the territorial body politic of Australia emerges as the new border to be negotiated.

In an effort to stop asylum seekers and refugees from coming to Australia, the Australian government "excised" thousands of small Australian islands from Australia's sovereign territory, for migration and asylum purposes, thus delegitimizing any appeals on those islands to the U.N. Geneva Convention. Shortly after the excision decision, an Indonesian vessel carrying fourteen Turkish Kurdish asylum seekers arrived at Melville Island, paradoxically at once a part of Australia and removed from it after excision. They asked for asylum, but were told that Australia had shifted its border, contracting inward in terms of asylum space. Kurdish asylum seekers were forced back out to international waters. Excision as the new border holds back the refugee hordes, the new barbarians—all fourteen of them.

Ironically, the Australian government hides its transgressions by discarding asylum seekers onto islands like Nauru to be heard from no more, or by capturing them in camps in the Australian wilderness that border them away from Australia while inside Australia. Only poetic whispers escape incarceration:

> My name is asylum
> I was born here
> Here is the detention centre . . .
> The border between me and Australia
>   (Angel Boujbiha, in Evans 2003, 163)

That behind this border fence the incarcerated may immolate themselves or sew their lips to protest the silence of the outside world barely makes a ripple in the desert-sea in which they are confined and abandoned. They are at once visible as bodies yet imperceptible as human beings.

That is, as bare and strange bodies, they are visible even in their relative obscurity through representations that characterize them as disruptive externalities to host communities, posing political, economic, and security challenges. But as human beings, they are imperceptible and inaudible even when they speak, because of the condition of voicelessness imposed on them in formal and popular representations. No matter, their presence is reinscribed in terms that escape their control and is made to stand for specific culturally and politically encoded images, identities, and subjectivities.

In most European countries, refugees, asylum seekers, and illegal immigrants are made visible beyond their actual numbers to stand as a *deluge* (of strangers) that the countries' cultural, social, and aesthetic terrains—not to speak of political-economic welfare geographies—cannot absorb. Thus indicted and rejected, the unexpected and clandestine migrants' very existence reveals the state-fundamentalist rhetoric employed against refugees, asylum seekers, and illegal immigrants, and exposes the hollow myth of universal humanism. In the final analysis, even if no "host" cares to recognize the fact, such migrants show the inexorable borderizations that sharpen migrants' will to defy expectations. As the following poem demonstrates, an expression of a migrant's regrets is not followed with a promise to leave. The refusal to do so emerges as a moment of insurrectional politics.

> Sorry
> Sorry that we are here
> That we take your time
> Sorry
> Sorry that we breathe in your air
> That we walk on your ground
> That we stand in your view. . . .
> And sorry that we brought nothing
> The only thing we have is a story
> Not even a happy story.
> (cited by Alibhai-Bown, October 2002)

In a relational sense, such an insurrectional migrancy is born in the ontopolitics of territorial nation-states where the dominant institutions, relations, and subjectivities of membership in community are anchored in, and defined by, exclusivist border practices. In other words, territorial national states owe their practical authority to the successful production and policing of borders of difference across political, cultural, and aesthetic lines where none exist naturally or as a given. Much of the contemporary politics around migrants is energized by this imperative of helping to produce and empower the national communities that can then be referred to, and represented by, states as if they exist as given authentic communities. Migrants are caught up in this ontopolitics. The predominant norms and values of the territorial national states as political communities work against their moving subjectivities not only as migrant human beings who for a variety of reasons are compelled to move, but also as migrant political subjects who are inexorably constitutive of the host communities of their residence.

Not surprisingly, efforts are undertaken to orchestrate and condition migrant lives so that migrants figure as externalities to the "receiving" communities even as they are internal to the practices by which the communities become possible and members recognize themselves as proper members. Rendered liminal figures, migrants seem bound permanently to a search for community even as they reside in a community.

Yet, paradoxically, as the borders that attempt to capture migrants proliferate, so do the inevitable ambiguities and openings that border practices lead to. Migrants' defiance against accepting their imposed locations intensifies the ambiguities and openings. Certainties, closures, and terminations are thrown into the migrant winds. In uncertain yet open migrant horizons, and because of the sheer imperatives of living, migrants begin to imagine, cultivate, and further enact insurrectional politics—a transformative praxis that works to shape through migrant imagination the worlds that migrants crisscross yet never permanently inhabit. In migrant lives, insurrectional politics find fertile grounds.

HIERARCHIES, DESIRES, AND THE POLITICS OF INSURRECTION IN MIGRANT MOVEMENTS

Recollect Michel de Certeau in "Spatial Stories," where he highlights the inevitable paradoxical transnational and translational dynamics of border practices: "the mouth piece of the limit always creates communication as well as separation. . . . It turns the frontiers into a crossing, the river into a bridge. It recounts inversions and displacements: the door that closes is precisely what may be opened" (de Certeau 1988, 127).

Ironically, the impossibility of return to previous alignments sharpens and focuses the determination of insurrectional migrants to capture back that which regiments their lives. In surprising ways, what seem to facilitate migrants' resistant mode of life are the productive ambiguities border practices necessarily engender. The dynamics of border practices are such that they create borders but also create inevitable borderizations. Borders, after all, become meaningful by expressing what they cross and separate. As de Certeau suggests, borders separate the other side only by articulating it. In articulating it, they relate to it, they rely on it, and they thus internalize it even as they describe it as external. This ontological relation operates in border practices that attend to insurrectional migrant lives and work to externalize them. Just as migrants are made subjects of border practices, migrants relate to and recast such practices. As much as being regimentations of migrant lives, border practices turn ambiguous in their dynamic inertia, becoming openings for migrants. There, in this space of tension and contradiction, a radical transformative politics becomes possible in spite of desires to the contrary. Migrants, too, can bring borders down. Their politics activate "insurrectional" politics: the time and place of a new political.

Insurrectional politics is not a politics of the third, of the in-between, or even of the hybrid. Its politics is fugitive of the old and productive of the new where, as Antonio Gramsci would have put it, the old is not yet dying, the new cannot yet be born, but the transformation is underway. Its syncretic trajectories pressure the current divisions of the world of states, nations, borders, and strangers; it encourages a reformulation of traditional securities and insecurities that call people to action.

In an article, "Outlines of a Topography of Cruelty: Citizenship and Civility in the Era of Global Violence," Etienne Balibar calls for what he terms "insurrectional democracy," where politics is defined by *isonomia,* an equality of rights (or equality in rights) of human persons, and not by exclusion based on citizenship alone (Balibar 2001, 18). Balibar claims that citizenship is the predominant norm of authorization to have rights constitutive of a democratic state. Those who are "radically excluded, being denied citizenship, are also automatically denied the material conditions of life and the recognition of their human dignity" (16). What makes such a non-democratic condition of democracy acceptable to many is the claim that a community's security is only possible through the introduction of exclusionary borders not only physically in erecting concrete boundaries, but also, more importantly, through institutions of limit that proscribe access to rights.

That a dialectic resolution of antagonisms among various constitutive parts of a community, including illegal migrants, refugees, asylum seekers, and exiles, can be cultivated as radical communitarian politics never figures as practical politics. Instead, open horizon is made into striated topography where the cruelty of exclusion, exception, and denial, not civility informed by *isonomia,* emerges as the predominant mode of life. This spectral politics dominates the canvases of migrant lives in Western spaces, particularly in Europe.

Balibar maintains that such spectral politics at the institutional level introduces quasi-apartheid in the guise of an integrated European citizenship. Yet, paradoxically, he argues, such a territorializing exclusionary politics informs and even justifies its counter-politics in terms of the resilient politics by migrants and refugees to have "access to the means of existence." The recognition of the uneven "institutional distribution of survival and death" across the world's peoples shifts the perception of borders (ibid.). The recognition "grips the minds" and the imaginations of people, not always politically consciously, and "thereby become a material force" through the movements of people (Hall 1999, 27). What materializes as a result is a shift in thinking among those whom borders intend to keep at bay. Borders of separation, recast in light of the recognition of the systematic "distribution of survival and death," emerge as borders of connections, as sites of relationality. They become subject to renegotiation, or even result in their own radical contravention in struggles against the dominant ideology of the striated space.

I want to submit that two somewhat divergent politico-philosophical orientations help to highlight the insurrectional element in Balibar's work (which curiously remains unelaborated). The first to resonate in Balibar's work is a sense of ethics akin to Jacques Rancière's hermeneutic of an ethics of equality based on *"le politique* as a demand for justice" (Deranty 2003, para. 58). The second orientation lies in the Marxist notion of history as an active, if dialectical, transformative struggle. For Balibar, the first compels recognition of the "equality" of humans and the struggles this recognition empowers. The second has the potential to organize this recognition into a movement.

Writing on Rancière's ethics, Jean-Philippe Deranty sheds some light on such a recognition of equality as a tense moment: "The masters demand to be recognized as masters by those they dominate, but for this recognition of inequality to be possible, the masters must recognize the ability of the dominated to recognize at all. Underneath the existence of social hierarchy, there lies the more fundamental recognition of pure ontological equality" (ibid., para. 57). Deranty suggests that this tension is productive of a certain moral logic, "a logic of the wrong" (ibid.). The "logic of the wrong" points to the fact that the societal inequalities and hierarchies are testimonies to their own historical and political contingency, since they could only be made possible by power relations acting on the a priori ontological equality. They reflect the "wrongs" that have "wrung" the society and thus erode and curtail equality (Deranty 2003, para. 58). They point to the architects of the "wrong" and to their victims. Also, they accentuate the political grounds where the demands for justice can be articulated.

Rancière's hermeneutic of ethics resonates in Balibar's work in precisely this sense, where Balibar is most sensitive to the fundamental right of humans to have access to the means of existence in a way based not on their politically mediated locations but on their being human. Denied that right, humans have a right to seek it, even in radical, insurrectional movements. This, in Balibar's work, emerges as the new political.

It is here that radical historical materialism lends tactical support to Balibar's insurrectional politics. For the hierarchies and the inequalities to be redressed in any meaningful way, "multilateral" efforts must be organized and activated into "struggles" in the cooperation of the subordinated and dominated majority. "Populations themselves," writes Balibar, referring to those who are to author and shoulder struggles, "ought to be the agents of change, of even new representative institutions." But their struggles should "not be merely territorial, and certainly not purely national." They should be energized by a sense of the "cosmopolitical" in which *isomonia,* equality in rights, should undergird the relations of participation in communities.

Of course, Balibar is writing primarily about European experiences in relating to and accommodating non-European migrants, refugees, exiles, and foreigners of other kinds in Europe. It is not coincidental that he is not writing about accommodating Europeans in Africa, or Asians or North Americans in South America. After all, the flow of inequalities, too, has its directions, privileging Europeans over Africans, or Asians and North Americans over South Americans. The flow of people simply reflects the uneven distribution of "means of existence." Those who lack survival means depart for places where they imagine these to be plenty. The prevailing, even hegemonic, logic also shifts.

For Balibar, cosmopolitics should be the ideological site of such a negotiation. However, its deriving logic should be transversal, not captive to statist territoriality and nationalism. It is no wonder that when Balibar talks about the multilateral negations and openings of borders, even political and economic borders of inequalities, by "populations themselves," he adds the migrant populations into the mix. In a reference to Rancière, Balibar argues that it is only by making possible the active presence (participation) in the commonwealth of those who have no place in it that a radical democratization of those borders informing inequalities can be possible (Balibar 2001, 17). "This would mean an active transformation of exclusion processes into processes of inclusion of the discriminated . . . into the 'city' or the 'polity.'" Here, politics acquires its cosmopolitical dimension; it becomes *isonomic* (ibid.).

Balibar's investment in cosmopolitics is not rhetorical at all. It is deeply political in the sense of addressing the systemic limitations of statist, national, and sovereign political institutions in an increasingly "cruel" and "uncivil" world. In Balibar's thinking, cosmopolitics takes form as the new "historical bloc" of "populations" (in a Gramscian sense) to make possible *isonomic* ethics.

At the same time, Balibar seems well aware of the challenges faced by cosmopolitical orientations. This awareness brings him to the edge of a radical politics, a horizon of insurrectional politics. Yet he only gestures in this direction, never articulating its outlines as a political practice. What is therefore interesting is what Balibar does not expressly talk about: the counterpolitical practices that are already at work and that persistently force open and shift the borders and boundaries preserving the hierarchies in inequalities. In short,

Balibar only adumbrates but never identifies the already ongoing insurrectional struggles that are found in various fields and arenas of life within and across national borders.

Clearly, of all the insurrectionary struggles, migrants' struggles, especially the movements of illegal migrants, asylum seekers, and refugees, exemplify insurrectional politics most instructively. They do so, for migrants, particularly the clandestine, unexpected migrants, are antithetical to the sanctity of territorial and national relations and institutions. Refugees, asylum seekers, and illegal migrants epitomize the challenge to territorial and national orders, for as fugitive movements they often exceed the efforts to control and contain their ramifications.

Increasingly, these fugitive insurrectional populations embody the logic of ambiguity, which takes hold of the certainties and translates them into resources in confronting and challenging the very power relations that fuel certainties in the first place. It is not that these populations are all powerful, able to bring massive capabilities to bear on material realities of the world. Nor are they transformative in extraordinary ways. It is simply that they are increasingly more elusive in ontopolitical ways, which in turn makes their ontological and conceptual capture difficult. The more elusive they become ontologically, the more insurrectional they grow in their slow but steady impact on political knowledge and praxis.

Ironically, what fuels their shift into the underground is a relentless policing rooted in the politics of inequality. For example, one can only wonder how fugitive populations will respond to the UK government's plan soon to electronically "tag" refugees, "employing satellite technology to pinpoint the wearer's location" (Press Association 2003). Similarly, one wonders if refugees and asylum seekers will be able to develop enough "foreign" language skills for crossing the "legal" borders, since the UK government is also expanding language and accent tests to determine one's "authentic" country of origin (Travis and Smithers 2003).

Ultimately, refugees, illegal immigrants, and asylum seekers ontopolitically collude and collapse the borders of identity that surround them. In all their vulnerabilities, as I exemplify below, they expose the forever unstable realities of borders and boundaries in personal and collective experiences. De Certeau's claim that borders' stability is only an illusion surges to the front once more.

"AS FROM NOW, YOU ARE NOT YOU"

"As from now, you are no longer who you are. Forget who you are and assume your new identity"(Al-Jazairi 2002). This is how Iraqi dissident writer Zuhair Al-Jazairi's contribution to insurrectional politics began in Iraq, during Saddam Hussein's regime. Ultimately, Al-Jazairi's "run for his life" required stepping into the secretive underground of migratory lives, a shadowy, sometimes illegal dimension of insurrectional politics, having been disempowered and haunted by the system above the ground. What is important to note in Al-Jazairi's story is that his run to refuge mirrors a stream of others' under similar conditions.

The realities of moving refugees, asylum seekers, or clandestine migrant bodies are increasingly blurred in terms of laws, rules, legality. In many ways, the ontological borders across these categories have collapsed under new realities. The persistent and draconian efforts restricting mobility in forced and voluntary migration fields are folding these fields into one another and further blurring the boundaries of displacement categories. For example, draconian restrictions on asylum are channeling refugees like Al-Jazairi into more and more smuggling networks, while blockages in flows of "voluntary" migrations are forcing migrants to seek appropriate asylum as one of many possible instruments of mobility. People's survival imperatives are trumping conventional boundaries in ways and forms that exceed the simple conceptual formulations. In spite of the heavy costs, but also because of the promises, migrancy in all forms, from illegal migration to refugee-ness, is, for many, not perceived as a strange and ephemeral mode of life. For some, it is even a welcome mode.

If this is difficult for some of us to understand, it is because, simply, we are not those refugees or asylum seekers. This state of experiential gaps that makes understanding difficult was manifested quite brutally but effectively in the ontological chasm between an asylum seeker from Congo and a British immigration officer. "How would you feel if you were in my situation?" the Congolese asylum seeker Phillip reportedly asked the immigration officer. The response was simple: "I wouldn't be in your situation" (Baird 2002, 11).

Of course, the officer's response does not reflect the universe of feelings toward refugees, asylum seekers, and other migrants. Yet, it demonstrates the realness of the "power geometry" (Massey 1994) that engenders and preserves the harsh political-economic hierarchies in the world. As well, it intimates the complex imbrications of statist and nationalist governance with the dominant political-economic relations. Paradoxically, the officer's confident retort also aptly highlights the reasons for the shift in people's imagination of territorial and national belonging away from the sanctity of the national borders. Further, the response highlights the reasons behind such a shift that make people, if not embrace, capitalize on the broad migrancy.

Alarmingly, refugees, asylum seekers, illegal immigrants, and diasporic peoples readily resort to secrecy and illegality when that facilitates their movements across territorial borders.

Karim Haidari's escape from Afghanistan to Germany hints at the reasons for people's reluctance to play by the rules of such borders when they perceive them to be unjust, distributing "means of existence" unequally across the world in visible and invisible border acts. Haidari writes:

> The plane descended. I was the authority, giving myself the right to come here. But soon the power shifted to the voice of [the] man behind the immigration desk: "Passports please!" . . . I pretended not to speak [English]. Which airline did you travel with? . . . "No Anglish," I hesitated. Oh, my first conversation started with a lie. How many lies should I say before I could prove the truth? Why do reasons fail against the system? I was on the brink of saying in English: "Listen to me, I am screwed up by the system of my own country. I need shelter and food for now. I am capable of putting my bread on the table. So please let me get in. I wouldn't have left my home if I didn't have to. I understand your concerns but my reasons are strong. Can't we sit and talk as human beings?" But I remained silent. Humanity is not the superpower in this real world. . . . I was helpless and exhausted. After a long body and luggage search, I was led to a waiting room . . . where . . . I found an intimacy with the other people from various cultures. . . . "What is going to happen?" [his traveling companion Suson asks]. . . . I sighed: I don't know. [They get food wrapped in plastic sheets. Turning to Suson, Haidari remarks;] "They give rations to refugees all over the world," I said. She glanced at me, pausing while biting the plastic with her teeth. But I was delighted with a discovery. I had found my new identity: I'M A REFUGEE. (Haidari 2002, 20)

In this way, Haidari, like countless others, comes to recast the refugee status as a living space—partly out of desperation, partly through necessary silences and lies, and partly in the recognition of the insurrectional capacity of the refugee space vis-à-vis the borders that work to capture lives in the service of hierarchical visions. He, together with others, takes the word *refugee* and begins to tweak it, tinker with it, and inject it with new historical meanings, with new lies, silences, images, and borders. He makes the word reveal its history and, through its history, also its insurrectional potential. Let me explain.

Once examined critically, the word *refugee* reveals a tortured history. Its stories are not about simple refugee figures merely to be found and registered. In reality, the word is *politics,* a politics of capturing the human under duress and attempting to cast her into bare life in displacement. In this sense, the refugee history is a politics of an attempt at double denial: first, the denial of the political agency of the displaced human as a precondition for recognizing him/her as a refugee, and second, the denial that such a demand of conditionality is imposed upon the displaced human in the first place. The word as history is therefore also politics in the first order: it is a register of selective memories on human displacement. Ultimately, it works to incarcerate its own history by erasing the traces of its making. Yet the ironies and paradoxes that constantly haunt it expose it as a history of politics, and its politics as a movement now being challenged in insurrectional migrancy. Given all this, what is needed is an insurrectional refugee and asylum history narrated through the paradoxes and ironies that display the politics in displacement histories, instead of evacuating history from displacement stories.

Perhaps the most foundational paradox, and the ironic results that this paradox produces, are found in the very act of defining the refugee as a specific subject of displacement out of various forms of human displacement. The paradox is born in the very necessity to identify the refugee in a sea of displaced humans. If there is any commonality to people's lives in movement, it *is* movement, a sort of local and global nomadism that at once operates in the territorial universe of states yet simultaneously gnaws at its edges in a thousand plateaus of displacement (Deleuze and Guattari 1987, 384). As de Certeau points out, inertia is always illusory. Exactly who is a refugee and who is not, under international laws, is more and more difficult to determine in ontological terms, as a multiplicity of complex and integrated reasons—which collapse political-internecine violence with the violence of poverty and destitution, environmental disasters, and gendered dominations—are at work in producing displaced populations.

What is the paradox here? The paradox lies in the imperative of defining and delimiting the boundaries of displacement events that can be recognized as refugee or illegal immigration events. On the one hand, the proliferation of various kinds of human displacement "necessitates" the articulation of a strict definition and administration of refugee events from other migrant events. On the other hand, simultaneously, the very same complexity and proliferation of human displacement, which blurs the categorical boundaries separating refugees from other displaced humans, is employed as a pretext to curtail refugee protection. Against the background of the argument that further vigilance is imperative to ferret out and help the "genuine" refugees and to deny relief to the "bogus" refugees, fewer and fewer displaced humans are recognized as worthy of protection. At some point, the whole concept loses its meaning in relation to the material realities. It stops having any referents; it exists only in name.

Productively, even in name, it works to legitimize a protection regime that offers promises more than affords relief. The name becomes instrumental in the forging of a climate, both popular and legal, where the displaced peoples emerge as "bare bodies" without proper political agency, simultaneously excluded from the rule of protection but captured as hostages to that very denial by the rule itself. In such a climate, if refugees and asylum seekers did not resist, lie, remain silent, pretend, or turn invisible (and thus illegal) in furtive passages, they would be condemned to live in what Agamben calls a "zone of indistinction" where they are neither recognized as people in need nor let go as free political agents. They would forever be captured.

Given this, refugees' plight alone unmistakably spells out the demise of the hold of legal categories over people, and the advent of popular identities whose borders are shaped by refugees and asylum seekers through the logic of ambiguity. Informed by the ambiguous logic of contemporary borderizations and the ethical imperatives of insurrectional (democratic) politics, these popular categories unexpectedly and even strangely empower many lives. Unfamiliar or even illegal tactics and strategies make survival possible while simultaneously highlighting the limits and the pitfalls of the names *refugee, asylum-seeker, illegal immigrant.*

In many ways, insurrectional migrants expose a profound failure of territorially bound state-centricity, manifested in the catastrophe of displacements of those whom the state is presumably empowered to protect. The very existence of displaced human beings is, then, a challenge to the state. As Agamben argues regarding the refugee, it is possible to argue that insurrectional migrants together constitute a "border concept" that calls into question the ontological and epistemological borders of the state: a "limit concept" that exposes the limits of the state (1995, 117). This concept unhinges the seeming perfect alignment in the old trinity of state/nation/territory by highlighting its fractures, its failures, under the rubric of the refugee. Increasingly, it emerges as an insurrectional concept.

In *Homo Sacer,* Agamben asks, "In what ways does bare life dwell in the polis?" (1998, 8). I start the task of recasting from this question, and inquire: in what ways does the bare body of the refugee, the asylum seeker, or even the clandestine migrant dwell in the order of states? In response to his question, Agamben argues that bare life has its own voice even as it dwells in the polis as "excluded," as a site of intervention by sovereign power in which modernity's borders (inclusions and exclusions) are shaped (8–10).

Similarly, the insurrectional migrants as bare lives, as sacred but expandable bodies in the order of the state, have their own voices. As Agamben, too, suggests, they take away and conserve their own voices shaped in their experiences with the sovereign power, voices that have the sensation of the just and the unjust. The bare migrant bodies develop their own active agency or subjecthood to counteract the logic of the sovereign power and to articulate and enact alternative borders and roots in life. In such an articulation, which lays bare the limits of protective efforts within the dominant state-system, refugees, asylum seekers, and clandestine migrants point to radical democratic possibilities. Incautious stories injected into citizen tales, they rewrite their history from the standpoint not of the state and the citizen, but of the migrant.

BORDER'S CAPTURE AND THE NEW POLITICAL

In insurrectional borderizations, refugees, asylum seekers, and clandestine migrants clearly refuse the political and economic hierarchies as the given of their destinies. They rescue their lives from the statist stranglehold and democratize them. Democratization in this sense means an approach to the issue of broad migrancy that affords all migrants a real, even if vulnerable, agency in shaping the conditions of their displacement. It inserts the insurrectional element into democracy. In insurrectional democracy, "the right of the *demos* to participate directly"—dialectical participation—becomes definitive of the democratic process rather than exceptional to it (Balibar 2003).

In fundamental ways, insurrectional movements of refugees, asylum seekers, legal and illegal immigrants, and diasporic and exilic peoples diffuse states' total(izing) powers over landscapes of human displacement in ways that sufficiently alter the structure of power matrix. These movements deepen the spaces of a radical democracy that resist the given political, economic, cultural, and racial hierarchies. Contributing to these movements are many transversal grassroots organizations working to check the excesses of a politics that erroneously associates the possibilities of community, culture, welfare, and identity only with the exclusionary territorial statism and nationalism.

Undoubtedly, to many state-hegemonic ears, such an insurrectional democratization sounds dangerous. That it may sound so is a testimony to the intrinsic conservatism of much discourse on refugees and migrants that "operates within a statist paradigm. It also demonstrates a fundamental incapacity to imagine anything but state-centered solutions" (Pieterse 1997, 86). Yet, Zygmunt Bauman suggests, it is in such unfamiliar insurrection that new experiences of community for refugees, asylum seekers, and clandestine migrants (and their citizen others) can be imagined and given practical meaning even before they can be experienced (Bauman 1997, 11–12). Regardless, it is difficult to ignore the migrant dynamism amid contemporary reconfigurations of the social, political, and territorial conditions in much of the world. Migrants are already practicing the new political as they negotiate their displacements and re-placements in syncretic influxes in local and global landscapes of militant migrancy.

Insurrectional migrants, whether they are called refugees, asylum seekers, or illegal migrants, suffer the suffering, do the dying, and tolerate what must be tolerated to make living possible, both their own living and the living of others dependent on them. Ironically,

it is precisely their subalternity that inspires the insurrectional politics, for their structural and discursive political and economic vulnerabilities demonstrate the very hierarchies that capture them and
commodify their bodies in the first place. Much as de Certeau suggested, the borders that capture migrants thus simultaneously expose them as borders of unequal exchanges. Of course, this recognition does not necessarily dissolve the borders or even automatically
recalibrate them. But, increasingly, it energizes a counterpolitical
force, the insurrectional politics, which in multiple and syncretic
manifestations cuts through the foundations of those borders. A
new view of "dwelling in the polis" is conceived and practiced in the
day-to-day realities of migrant human beings. The statist promises,
long undelivered, have given way to new political horizons.

WORKS CITED

Agamben, Giorgio. 1995. "We Refugees." *Symposium* 49, no. 2: 114–19.
Agamben, Giorgio. 1998. *Homo Sacer: Sovereign Power and Bare Life.*
Stanford, Calif.: Stanford University Press.
Alibhai-Brown, Yasmin. 2002. "No Room at the Inn." *New Internationalist,*
no. 350 (October).
Al-Jazairi, Zuhair. 2002. "You Are Not You." *New Internationalist,* no.
350 (October).
Baird, Victoria. 2003. "Fear Eats the Soul." *New Internationalist,* no. 350
(October).
Balibar, Etienne. 2001. "Outlines of a Topography of Cruelty: Citizenship
and Civility in the Era of Global Violence." *Constellations* 8, no. 1:
15–29.
Bauman, Zygmunt. 1997. "What Prospects of Morality in Times of Un-
certainty?" *Theory, Culture, and Society* 15, no. 1: 11–22.
Bearup, Greg. 2004. "Gunned Down to Impress America." *Guardian,*
May 8.
Chorlton, David. 1994. "The Border." In *Outposts.* Exeter, UK: Taxus
Press.
de Certeau, Michel. 1988. *The Practice of Everyday Life.* Berkeley and Los
Angeles: University of California Press.
Deleuze, Gilles, and Felix Guattari. 1987. *A Thousand Plateaus: Capital-
ism and Schizophrenia.* Minneapolis: University of Minnesota Press.
Deranty, Jean Philippe. 2003. "Rancière and Contemporary Political On-
tology." *Theory and Event* 6, no. 4.

Evans, C. 2003. "Asylum Seekers and Border Panic in Australia." *Peace Review* 15, no. 2.

Foucault, Michel. 1984. *The Foucault Reader,* ed. Paul Rabinow. New York: Pantheon.

Haidari, Karim. 2002. "08:59." *New Internationalist,* no. 350 (October).

Hall, Stuart. 1999. "The Problem of Ideology: Marxism without Guarantees." In *Stuart Hall: Critical Dialogues in Cultural Studies,* ed. David Morley and Kuan-Hsing Chen. London and New York: Routledge, 25–46.

Migration News. 2003. "Sanctions, Borders, Refugees." October. http://migration.ecdavis.edu/mn/comments.php?id=78020 (accessed November 24, 2003).

Pieterse, Jan Nederveen. 1997. "Sociology of Humanitarian Intervention: Bosnia, Rwanda, and Somalia Compared." *International Political Science Review* 18, no. 1: 71–93.

Press Association. 2003. "Tagging Plan for Asylum Seekers." *Guardian,* November 27.

Rancière, Jacques. 1998. *Disagreement: Politics and Philosophy,* trans. Julie Rose. Minneapolis: University of Minnesota Press.

Senocak, Zafer. 1989. "Dopplemann." Quoted in Heidrun Suhr, "Ausländerliteratur: Minority Literature in the Federal Republic of Germany," *New German Critique* 46: 71–103.

Sullivan, T., and R. Casert. 2000. "2 Boys, a Letter, a Continent: African Stowaways Touch the World." Associated Press, March 19.

Thacker, A. 2003. *Moving through Modernity: Space and Geography in Modernism.* Manchester: Manchester University Press.

Travis, A., and R. Smithers. 2003. "Accents on Trial in Asylum Seeker Tests." *Guardian,* November 27.

UNITED for Intercultural Action. 2004. "List of Documented Refugee Deaths through Fortress Europe." December. http://www.unitedagainstracism.org.

United Nations High Commissioner for Refugees (UNHCR). 1995. *The State of the World's Refugees: In Search of Solutions.* Oxford: Oxford University Press.

Contributors

DIDIER BIGO is professor of international relations at Sciences-Po (Institut d'Études Politiques) in Paris, associated researcher at the Center for International Studies and Research, scientific coordinator of Challenge (http://www.libertysecurity.org/), and editor of *International Political Sociology* and *Cultures et Conflits*. He is the editor of *Controlling Frontiers: Free Movement into and within Europe*.

KARIN DEAN has a PhD in geography from the National University of Singapore. Her areas of interest in political geography are borderland studies and state and sovereignty, focusing on Burma; the Sino–Burmese, Thai–Burmese, and Indo–Burmese borders; and the Mekong region.

CARL GRUNDY-WARR is senior lecturer in geography at the National University of Singapore. He has written on border studies and political geography.

ELSPETH GUILD is senior research fellow at the Centre for European Policy Studies, professor of European immigration law at the University of Nijmegen, and a partner in the London law firm

Kingsley Napley. She is the author of *The Developing Immigration and Asylum Policies of the European Union* and *Legal Elements of European Identity: Citizenship and Migration*.

EMMA HADDAD obtained her PhD from the European Institute at the London School of Economics and Political Science. She is the author of *Between Sovereigns: The Refugee in International Society* and contributed to *Global Society, International Journal of Human Rights*, and the *Encyclopaedia of International Relations and Global Politics*.

ALEXANDER HORSTMANN is a senior research fellow and teaches the social anthropology of Southeast Asia at the Westphalian Wilhelms University of Münster. He is also an activist on the board of Asia House, Essen. He is a coeditor of *Centering the Margin: Agency and Narrative in Southeast Asian Borderlands*.

ALICE M. NAH is a postgraduate student of sociology at the National University of Singapore and the refugee affairs coordinator of the National Human Rights Society (HAKAM) in Malaysia. She has published in *Social Identities, Geografiska Annalar: Series B, Urban Studies*, and the *Singapore Journal of Tropical Geography*.

SUVENDRINI PERERA is a senior research fellow at Curtin University of Technology, Australia. She has contributed to several major anthologies and journals in race, ethnicity, and cultural studies, and her publications include *Reaches of Empire, Asian and Pacific Inscriptions: Identities/Ethnicities/Nationalities*, and *Our Patch: Australian Sovereignty Post-2001*.

PREM KUMAR RAJARAM is associate professor of sociology and social anthropology at Central European University. He has published on migration studies, social theory, and international relations in several journals, including *Journal of Refugee Studies, Alternatives: Global, Local, Political*, and *Review of International Studies*.

JAMES D. SIDAWAY is professor of geography at the University of Plymouth in England. He is the associate editor of *Political Geography* and coeditor of the *Singapore Journal of Tropical Geography*.

NEVZAT SOGUK is associate professor of political science at the University of Hawai'i at Manoa. He is the author of *States and Strangers: Refugees and Displacements of Statecraft* (Minnesota, 1999). He has contributed to *Alternatives: Global, Local, Political* and *New Political Science.*

DECHA TANGSEEFA teaches political science at Thammasat University, Bangkok. His research interests are political theory and philosophy, critical international studies, and cultural studies, especially relating to displaced peoples.

MIKA TOYOTA is a research fellow at the Asia Research Institute, National University of Singapore. She has published more than twenty academic articles in English and Japanese.

BORDERLINES

Index

Aboriginal/Indigenous and native peoples: claims and rights/title, xiii, 107–9, 210–12; colonial/official identification of *(chaokao)*, 98–101, 109n1; denial of citizenship, 91–109; essentialized, 100, 107–9, 254–55n20; flexible and fluid identities, 96–97, 99–100; forced displacement/resettlement of hill tribes, 103–4, 107; Mabo & Wik judgement, 210, 213–14; positioning as alien/national/racialized other, 98, 99, 100, 104–5, 106–8, 109, 212–13; rendered stateless in Thailand, 92, 98; Salish story, 285–86; self-representation, 107–8, 109; sovereignty, xxxv, 210–13, 224; transborder practices, regional identity, itineraries, and legal agreements of, xxv–xxvi, xxxiii, xxxv, 104, 208–10; welcome/farewelling of asylum seekers, 210, 218–19; as wild/primitives *(khon pa)*, 95–96. *See also* Akha; (il)legality; Kachin; Karens; noncitizens; nonbelonging; Rohingyas; Shan

Acehnese refugees/asylum seekers: in Malaysia, 40, 41, 43, 45–50, 52, 54, 55, 56

Afghanistan: refugees/asylum seekers from, 27, 203, 302; renditions to Guantanamo, 18, 207; war and bombing of (2001), 67, 142, 218

Agamben, Giorgo, xviii, xix–xxii, xxviii, 11–16, 19–20, 203, 216, 232–35, 241–42, 289, 304–5

Akha, xxv, 96–97

alien: abnormal, 5–6, 29; detention/repatriation, 4–5; illegal, 24–25; integrated, and legal expulsion of, 67, 88n5; legal but, 92. *See also* detention; migrants

Al-Jazairi, Zuhair, 301

Anggraeni, Dewi, 220–21

315

naries in discussion/forms of, 10, 108, 184, 185–86, 191, 192; and/or boundarying practices, 9–10, 108; camouflaged performance and reproduction of, x, 284, 285, 290, 291, 292; cartographic representations, 164–65, 207–8; concealed senses, xxix; as configured by itineraries, 205, 206, 222; constructions of, controls, discourses, local terminologies, myths, perceptions and practices in, xix, xxiv, xxxii, xxxiii, 6, 37–39, 41–56, 92, 119–121, 127, 128–130, 134–35, 184, 186, 190, 192, 196–98, 207, 215; controls, regimes of, 205; counting and, xxviii, xxx; debates on porosity of, 185; deterritorialization of, 213; between different worlds/means of existence/life or survival and death, 92, 185, 192, 194–95, 196, 297; as dynamic representations, 165; episodic (scales of exclusion), xxxii; ethnic identity, xxxiii, 97, 99, 109, 140; of foreignness, 5; geographical violence of, 214; hegemonic control and ideology as form of, xxvi; identity and, 52–53, 99–100, 120, 167; indeterminacy and undecidability of/at, 19–20, 36, 39, 53, 55, 56, 170–72, 197, 205; information highway and, 196; insurrectional concept, 305; international law and, 204, 235–37; intersubjective, xxxii, 30; knowledge and, xv; of law and order and protection,

ix–x, xii, xiii, xv, xvii, xix–xx, xxxiii, 4, 15–16, 17–18, 69–70, 73–74, 80–82, 98, 100, 121, 218, 241, 247–48; legitimating spatial practices and, 191–92; as liminal and regulated, xxxv, 172, 173, 174; as Möbius strip, xxxii, 5, 16, 30; of national security, 74; nation-making and regulation, 173; new modalities of connection at, 219; of norm, xxi, 4, 122, 126, 129, 296–97; as performative sites, 166; as periphery, 128; of politicized subjectivity, xvii; as politics, xxiii–xxiv, xxviii; racialized, 203; realpolitik in management of, 47, 52; rebordering and, 154; as relational, xvi, xxi; relict, 168, 176n7; representations of, xi, xxxi, xxxiii–xxxiv, xxxv, 162–63, 165–66, 167, 169, 170, 171, 175, 177, 184, 206–7, 208; as resource, 174; after September 11, as simulacrum, 10; as sites of interacting spaces, xxiv, 183, 184–85, 191, 192, 197–98; as spaces of becoming, xxxvi, 198; of state and terrorism, 9; topographic measurement of, 169–70; as transformative, xii, xvii, xxiii–xxiv; transnational sites of control and, 10. *See also* Australia; China; citizenship; de Certeau, Michel; detention; EU; globalization; landscape; Portuguese–Spanish border; sovereignty; spaces; territorialization; Thai–Burmese border; Thai–Malaysian border; zones

borderscapes, x, xvii, xxiv–xxv,